Abstract Recursion and Intrinsic Complexity

This book presents and applies a framework for studying the complexity of algorithms. It is aimed at logicians, computer scientists, mathematicians and philosophers interested in the theory of computation and its foundations, and it is written at a level suitable for non-specialists.

Part I provides an accessible introduction to abstract recursion theory and its connection with computability and complexity. This part is suitable for use as a textbook for an advanced undergraduate or graduate course: all the necessary elementary facts from logic, recursion theory, arithmetic and algebra are included. Part II develops and applies an extension of the homomorphism method due jointly to the author and Lou van den Dries for deriving lower complexity bounds for problems in number theory and algebra which (provably or plausibly) restrict all elementary algorithms from specified primitives.

The book includes over 250 problems, from simple checks of the reader's understanding, to current open problems.

YIANNIS N. MOSCHOVAKIS is Professor Emeritus and Distinguished Research Professor of Mathematics at the University of California, Los Angeles, and Professor Emeritus at the University of Athens. Among his many professional commendations he is a Fellow of the AMS and a corresponding member of the Academy of Athens. He has received Guggenheim and Sloan Fellowships and has been an invited speaker at the International Congress of Mathematicians. Professor Moschovakis has worked primarily in the theory of recursion, descriptive set theory and the foundations of the theory of algorithms and computation.

LECTURE NOTES IN LOGIC

A Publication of The Association for Symbolic Logic

This series serves researchers, teachers, and students in the field of symbolic logic, broadly interpreted. The aim of the series is to bring publications to the logic community with the least possible delay and to provide rapid dissemination of the latest research. Scientific quality is the overriding criterion by which submissions are evaluated.

Editorial Board

Zoe Chatzidakis
DMA, Ecole Normale Supérieure, Paris

Peter Cholak, Managing Editor
Department of Mathematics, University of Notre Dame, Indiana

Leon Horsten
School of Arts, University of Bristol

Paul Larson
Department of Mathematics, Miami University

Paulo Oliva
School of Electronic Engineering and Computer Science, Queen Mary University of London

Martin Otto
Department of Mathematics, Technische Universität Darmstadt, Germany

Slawomir Solecki
Department of Mathematics, Cornell University, New York

More information, including a list of the books in the series, can be found at http://aslonline.org/books/lecture-notes-in-logic/

LECTURE NOTES IN LOGIC 48

Abstract Recursion and Intrinsic Complexity

YIANNIS N. MOSCHOVAKIS

University of California, Los Angeles

and

University of Athens

ASSOCIATION FOR SYMBOLIC LOGIC

CAMBRIDGE
UNIVERSITY PRESS

CAMBRIDGE
UNIVERSITY PRESS

University Printing House, Cambridge CB2 8BS, United Kingdom

One Liberty Plaza, 20th Floor, New York, NY 10006, USA

477 Williamstown Road, Port Melbourne, VIC 3207, Australia

314–321, 3rd Floor, Plot 3, Splendor Forum, Jasola District Centre, New Delhi – 110025, India

79 Anson Road, #06–04/06, Singapore 079906

Cambridge University Press is part of the University of Cambridge.

It furthers the University's mission by disseminating knowledge in the pursuit of education, learning, and research at the highest international levels of excellence.

www.cambridge.org
Information on this title: www.cambridge.org/9781108415583
DOI: 10.1017/9781108234238

Association for Symbolic Logic
Richard A. Shore, Publisher
Department of Mathematics, Cornell University, Ithaca, NY 14853
http://aslonline.org

First published 2019

Printed and bound in Great Britain by Clays Ltd, Elcograf S.p.A.

A catalogue record for this publication is available from the British Library.

ISBN 978-1-108-41558-3 Hardback

CONTENTS

Part II. Intrinsic complexity

INTRODUCTION

This is the (somewhat polished) present state of an evolving set of lecture notes that I have used in several courses, seminars and workshops, mostly at UCLA and in the *Graduate Program in Logic and Algorithms* (MPLA) at the University of Athens. The general subject is the theory of *abstract* (first-order) *recursion* and its relevance to the *foundations of the theory of algorithms and computational complexity*, but the work on this broad project is very incomplete and so the choice of topics that are covered is somewhat eclectic.

The preliminary Chapter 1 gives a brief, elementary exposition of some basic facts and examples and helps make the material which follows accessible to students and researchers with varying backgrounds. After that, the book naturally splits into two, roughly equal parts according to the title: Part I (Chapters 2 – 3) on *abstract recursion* and Part II (Chapters 4 – 9) on *intrinsic complexity*.

– Chapter 2 introduces *recursive* (McCarthy) *programs* on abstract structures and develops their elementary theory. There is little that is new here, other than Vaughan Pratt's very interesting nondeterministic algorithm for coprimeness in Section 2E, but I do not know of another easily accessible, self-contained and reasonably complete source for this material.

– Chapter 3 introduces the natural *complexity measures* for recursive programs and establishes their basic properties. There is some novelty in approach, especially as the complexity measures are defined directly for the programs and so are independent of any particular "implementation of recursion"; and there are also some new results, most notably Theorems 3B.9 and 3B.12 which are due to Anush Tserunyan and have (I think) substantial foundational significance.

Part II is about the derivation of robust and widely applicable lower bounds for problems (especially) in arithmetic and algebra, and perhaps the simplest way to introduce my take on this is to give a fairly precise formulation of a fundamental conjecture about an ancient object of mathematical study.

The *Euclidean algorithm* (on the natural numbers, using division) can be specified succinctly by the *recursive equation*

$$\varepsilon : \quad \gcd(x, y) = \begin{cases} x, & \text{if } y = 0, \\ \gcd(y, \operatorname{rem}(x, y)), & \text{otherwise,} \end{cases}$$

where $\text{rem}(x, y)$ is the remainder in the division of x by y. It computes *the greatest common divisor* of x and y when $x, y \geq 1$ and it is an algorithm *from* (*relative to*) the remainder function rem and the relation eq_0 of equality with 0: meaning that in its execution, ε has access to "oracles" which provide on demand the value $\text{rem}(s, t)$ for any s and $t \neq 0$ and the truth value of $\text{eq}_0(s)$. It is not hard to prove that

$$(*) \qquad c_\varepsilon(x, y) \leq 2\log y \leq 2\log x \quad (x \geq y \geq 2),$$

where $c_\varepsilon(x, y)$ is the number of divisions (calls to the rem-oracle) required for the computation of $\gcd(x, y)$ by the Euclidean and logarithms are to the base 2. Much more is known about $c_\varepsilon(x, y)$, but this upper bound suggests one plausible formulation of the Euclidean's (worst-case) *weak optimality*:

Main Conjecture. *For every algorithm α from* rem *and* eq_0 *which computes* $\gcd(x, y)$ *when* $x, y \geq 1$, *there is a number* $r > 0$, *such that for infinitely many pairs* (x, y) *with* $x > y \geq 1$,

$$c_\alpha(x, y) > r \log x,$$

where $c_\alpha(x, y)$ is the number of calls to the rem-oracle that α makes in the computation of $\gcd(x, y)$.

This is a classical fact about the Euclidean algorithm, taking for example the pairs (F_{n+3}, F_{n+2}) of successive Fibonacci numbers, cf. Problems x1C.8, x1C.9. The general case is open, probably not easy and certainly not precise as it stands, without specifying *what algorithms it is about* and what it means for an algorithm *to call an oracle* in the course of a *computation*.

Now, there are Turing machines which compute $\gcd(x, y)$ making no oracle calls at all, simply because $\gcd(x, y)$ is Turing computable—so that's not it.

In fact, there is no generally accepted, rigorous definition of *what algorithms are*. This is not a problem when we study particular algorithms, which are typically specified precisely in some form or other without any need to investigate whether *all relevant algorithms* can be similarly specified. In *Complexity Theory*—and especially when we want to establish *lower bounds* for some measure of computational complexity—the standard methodology is to ground proofs on rigorously defined *models of computation*, such as Turing machines, register or random access machines, decision trees, straight line programs, etc., and sometimes also on specific *representations of the input*, e.g., *unary* or *binary* notation for natural numbers, *adjacency matrices* for graphs, etc. There is a problem with this practice, when we try to compare lower bound results obtained for different models, typically attacked by establishing *simulations* of one model by another, cf. van Emde Boas [1990]; and this problem becomes acute when we want to prove *absolute* (or at least widely applicable) lower bounds which are small, polynomial or even linear (in the length of the input) as in the Main Conjecture, generally less complex than the standard simulations.

So there are two equally important aims of research in this area:

One is to derive *lower bounds for mathematical problems*; the other is to develop a foundational framework in which one may be able to prove (or at least argue convincingly) that these bounds are *absolute*, that they restrict *all relevant algorithms*. The first of these naturally requires mathematical tools from the area in which the problems arise; and the second inevitably involves logic.

Recursion gets in the picture because there are both foundational arguments and mathematical results which support the view that *all elementary*[1] *algorithms can be faithfully expressed by recursive programs*, so that lower bounds established for them should be absolute, cf. Moschovakis [1984], [1989a], [1998], [2001]. This connection has motivated much of the work reported here, but it is not our topic.

Here I will take a different approach to the derivation and justification of robust lower bounds, which is more widely applicable and does not tie us to any specific foundational view of what algorithms are.

In Chapter 4, which is the heart of this book, we formulate three simple axioms about algorithms in the style of *abstract model theory*. These are bundled into the notion of a *uniform process* of an arbitrary (first order) structure: all *concrete* algorithms specified by computation models induce uniform processes, as do their usual *nondeterministic* versions. Uniform processes can "compute" functions that are not computable, they are not about that; but they carry a rich complexity theory which, when applied to concrete algorithms yields non-trivial lower bounds, in some cases optimal, absolutely or up to a multiplicative constant.

For a sample result, suppose

$$\mathbf{A} = (A, R_1^{\mathbf{A}}, \dots, R_k^{\mathbf{A}}, \phi_1^{\mathbf{A}}, \dots, \phi_l^{\mathbf{A}}) = (A, \mathbf{\Phi})$$

is a first order structure on the vocabulary $\Phi = \{R_1, \dots, R_k, \phi_1, \dots, \phi_l\}$, suppose $P \subseteq A^n$ is an n-ary relation on A and let $\Phi_0 \subseteq \Phi$. From these data, we will define a function

$$c = \text{calls}(\Phi_0)(\mathbf{A}, P) : A^n \to \mathbb{N} \cup \{\infty\} = \{0, 1, \dots, \infty\},$$

[1]There are algorithms whose implementations print output (or drop bombs), ask "the user" if she prefers business or coach class and may never terminate. In this book we confine ourselves to pure, finitary algorithms which compute partial functions or decide relations from given partial functions and relations, for which complexity theory is most fully developed. The extension of most of what we say to algorithms with *side effects* or *interaction* requires combining the methods we will use with classical *domain theory*, introduced by Scott and Strachey [1971] and richly developed by Scott and many others since then, especially the early Plotkin [1977], [1983]. It is not as different from what we will be doing as one might think, but we will not go into it here.

the *intrinsic* calls(Φ_0)-*complexity function* of P, such that *if* α *is any* (deterministic or nondeterministic) *algorithm from* $\mathbf{\Phi}$ *which decides* P, *then for all* $\vec{x} \in A^n$,

$$(*) \quad c(\vec{x}) \leq \text{ the number of calls to primitives in } \mathbf{\Phi}_0 \\ \text{that } \alpha \text{ must execute to decide } P(\vec{x}) \text{ from } \mathbf{\Phi}.$$

This is a theorem if α is expressed by a concrete algorithm from $\mathbf{\Phi}$ so that, in particular, the complexity measure on the right is precisely defined; it will be made plausible for all algorithms, by a brief conceptual analysis of what it means (minimally) to *compute from primitives*; and it is not trivial, e.g., we will show that *if* $x \perp\!\!\!\perp y$ *is the coprimeness relation on* $\mathbb{N}$ *and*

$$c = \text{calls}(\text{rem})((\mathbb{N}, \text{rem}, \text{eq}_0), \perp\!\!\!\perp) : \mathbb{N}^2 \to \mathbb{N},$$

then for infinitely many pairs (a, b) *with* $a > b$,

$$(**) \qquad c(a, b) > \frac{1}{10} \log \log a.$$

This follows from the (much stronger) Theorem 6C.5, an abstract version of one of the main results in van den Dries and Moschovakis [2004]. It gives a (very) partial result towards the Main Conjecture, one log below what we would like to prove—but Vaughn Pratt's nondeterministic algorithm for coprimeness in Theorem 2E.2 suggests that the conjecture may hold only for deterministic algorithms.

The main tool for defining the intrinsic complexities and deriving lower bounds for them is the *homomorphism method*, an abstract and mildly extended version of the *embedding method* developed in van den Dries and Moschovakis [2004], [2009]. We will use it in Chapters 5 – 8 to get somewhat strengthened versions of some of the lower bound results about arithmetic in these two papers and then again in Chapter 9 to get similar results in algebra. Few of these applications in the last two Parts are new: my main aim here is to explain the homomorphism method, illustrate its applicability in two different areas and (primarily) to identify some basic notions of the theory of computational complexity which (perhaps) have not been noticed.

Notice that **this is not a textbook on computability and complexity**, a core part of Computer Science which is covered in many excellent books including the classical Papadimitriou [1994]; **it is not a textbook on Turing computability and recursion on** $\mathbb{N}$, a huge subject amply covered in classical texts like Kleene [1952], Davis [1958] and Rogers [1967] and many more recent ones; and **it is definitely not a textbook on arithmetic and algebraic complexity**, not even a good introduction to these vast research areas about which I really know very little. It is natural to assume for some of the discussion that the reader knows something about these subjects, but the rigorous development of the material is substantially self-contained and limited to a few results which (I hope) throw some light on the central problem of *deriving and justifying absolute lower complexity bounds*.

The exposition is elementary, aimed at advanced undergraduates, graduate students and researchers in mathematics and computer science with some knowledge of logic, a good understanding of the basic facts about algorithms and computability and an interest in foundational questions. Many of the (roughly) 250 problems are very easy, to test understanding, but there are also more challenging ones, sometimes marked with an asterisk * and a few that I cannot do, marked "Open Problem".

I have tried hard to assign results to those who proved them and to give correct and useful references to the literature, but this is not the place to look for a history of recursion and its interaction with computability—another vast and complex topic which is way out of my expertise and certainly not of the moment.

Yiannis N. Moschovakis
Santa Monica, CA and Paleo Faliro, Greece

Acknowledgments.

My greatest debt is to Lou van den Dries, whose insights in van den Dries [2003] led to a fruitful (and very pleasant) collaboration that produced van den Dries and Moschovakis [2004], [2009] and ultimately led to Part II of this book.

I am grateful to Vaughan Pratt and Anush Tserunyan for letting me include in this book unpublished results of theirs; to Vasilis Paschalis and Tyler Arant for chasing errors and typos in Parts I and II respectively—and I know they must have missed some, it's OK; and to my wife, always, and for many things other than her help with this book.

To go farther than this, I would need to put down the many logicians, philosophers and computer scientists who have informed my understanding of logic, recursion, algorithms and the connections among these subjects, including Stephen Kleene, John McCarthy, Dana Scott and many, many others—too long a list to put down here and certainly not unique to me.

Finally, I want to thank the hundreds of students who have taken courses or wrote M.Sc. or Ph.D. Theses with me on these topics, mostly at UCLA and the University of Athens, including the *Graduate Program in Logic, Algorithms and Computation* (MPLA). It is sometimes said that we learn more from our students than they learn from us and perhaps this is true of me; in any case, there is no doubt that I have enjoyed the process, very much.

CHAPTER 1

PRELIMINARIES

We collect here some basic facts we need from set theory, recursion theory, logic and arithmetic, primarily to fix terminology and notation, and we also describe some simple examples of algorithms to which we can refer later. Most readers of this book will know most of these facts and should peruse this chapter quickly, coming back to it later, as needed. One exception is the "where" notation in Section 1B for definitions by mutual recursion which we will use extensively, and another might be the treatment of *equational logic of partial terms* in Section 1E: there are several ways to approach this topic, but only one works well with recursive equations and it is important to get it right early on.

1A. Standard notations

As usual, $\mathbb{N} = \{0, 1, \dots\}$ is the set of *natural numbers*,

$$\mathbb{Z} = \{\dots, -2, -1, 0, 1, 2, \dots\}$$

is the set of *integers*, $\mathbb{Q}$ is the set of fractions and $\mathbb{R}, \mathbb{C}$ are the sets of *real* and *complex numbers* respectively. We will use the same symbols $0, 1, +, -, \cdot, \div$ for the corresponding objects and functions in all these sets—and in all rings and fields, in fact. We also set

$$S(x) = x + 1, \quad x \dot{-} y = \text{if } (x < y) \text{ then } 0 \text{ else } x - y, \quad \text{Pd}(x) = x \dot{-} 1$$

for the *successor*, *arithmetic subtraction* and *predecessor* functions on $\mathbb{N}$, and

$$\log(x) = \text{the unique real number } y \text{ such that } 2^y = x \quad (x \in \mathbb{R}, x > 0).$$

This is the "true", binary logarithm function. We will sometimes compose it with one of the functions

$$\lfloor x \rfloor = \text{the largest integer } \leq x \qquad \text{(the } \textit{floor} \text{ of } x\text{)},$$
$$\lceil x \rceil = \text{the least integer } \geq x \qquad \text{(the } \textit{ceiling} \text{ of } x\text{)}$$

to get an integer value.

By the *Division Theorem* (for arithmetic), if $x, y \in \mathbb{N}$ and $y > 0$, then there exist unique numbers q and r such that

$$x = yq + r \text{ and } 0 \le r < y; \tag{1A-1}$$

if $x < y$, then $q = 0$ and $r = x$, while if $x \ge y$, then $q \ge 1$. We refer to (1A-1) as the *correct division equation* (**cde**) for x, y, and we set

$$\mathrm{iq}(x, y) = q, \quad \mathrm{rem}(x, y) = r \quad (y > 0), \tag{1A-2}$$

with the unique q and the r for which it holds. We also put

$$\mathrm{iq}_m(x) = \mathrm{iq}(x, m),\ \mathrm{rem}_m(x) = \mathrm{rem}(x, m)\ (m \ge 2),\ \mathrm{parity}(x) = \mathrm{rem}_2(x).$$

For the *divisibility relation*, we write

$$y \mid x \iff (\exists q)[x = yq], \text{ so if } y \ne 0, y \mid x \iff \mathrm{rem}(x, y) = 0.$$

Two positive numbers are *relatively prime* or *coprime* if their only common divisor is 1,

$$x \perp\!\!\!\perp y \iff x, y \ge 1 \;\&\; (\forall d > 1)[d \nmid x \vee d \nmid y].$$

We also call n numbers $x_1, \ldots, x_n$ *relatively prime* if they are all positive and no number other than 1 divides all of them.

The *greatest common divisor* of two natural numbers is what its name means,[2]

$$\gcd(x, y) =_{\mathrm{df}} \text{the largest } d \le \max(x, y) \text{ such that } d \mid x \text{ and } d \mid y. \tag{1A-3}$$

Thus,

$$x \perp\!\!\!\perp y \iff x, y \ge 1 \;\&\; \gcd(x, y) = 1.$$

The most commonly used notations for comparing the growth rate of unary functions on $\mathbb{N}$ are the *Landau symbols*:

$$\begin{aligned}
f(n) = o(g(n)) &\iff \lim_{n\to\infty} \frac{f(n)}{g(n)} = 0 \\
f(n) = O(g(n)) &\iff (\exists K, C)(\forall n \ge K)[f(n) \le Cg(n)] \\
f(n) = \Theta((g(n)) &\iff f(n) = O(g(n)) \;\&\; g(n) = O(f(n)) \\
f(n) = \Omega(g(n)) &\iff (\exists K, r)(\forall n \ge K)[f(n) \ge rg(n))
\end{aligned}$$

where the constants $K, C \in \mathbb{N}$ while r is a positive fraction. We will also use the notation

$$f(n) \sim_\infty g(n) \iff \lim_{n\to\infty} \frac{f(n)}{g(n)} = 1,$$

[2]This definition gives $\gcd(x, 0) = 0$ (even when $x = 0$), silly values that we do not care about; but it is useful to have $\gcd(x, y)$ defined for all x, y, and these values simplify some equations.

most notably by appealing in a few critical places to *Stirling's formula*,

$$\log(n!) \sim_\infty n \log n. \tag{1A-4}$$

Products and relations. The (Cartesian) *product* of n sets is the set

$$X_1 \times \cdots \times X_n = \{(x_1, \dots, x_n) : x_1 \in X_1, \dots, x_n \in X_n\}$$

of all sequences from them, set by convention to X_1 if $n = 1$ and to $\boldsymbol{I} = \{\emptyset\}$ when $n = 0$, for which we assume (by convention, if you wish) that

$$X_1 \times \cdots \times X_k \times \boldsymbol{I} \times Y_1 \times \cdots \times Y_l = X_1 \times \cdots \times X_k \times Y_1 \times \cdots \times Y_l.$$

As usual, $X^n = \underbrace{X \times \cdots \times X}_{n \text{ times}}$ with $X^0 = \boldsymbol{I}$.

We fix a two-element set

$$\mathbb{B} = \{\mathrm{tt}, \mathrm{ff}\} \tag{1A-5}$$

for *the truth values*, and with each relation $R \subseteq X = X_1 \times \cdots \times X_n$ on the sets $X_1, \dots, X_n$ we associate its *characteristic function*

$$\chi_R(\vec{x}) = \begin{cases} \mathrm{tt}, & \text{if } R(\vec{x}), \\ \mathrm{ff}, & \text{otherwise.} \end{cases} \qquad (\vec{x} \in X = X_1 \times \cdots \times X_n); \tag{1A-6}$$

we identify $R \subseteq X$ with $\chi_R : X \to \mathbb{B}$, so

$$R(\vec{x}) \iff R(\vec{x}) = \mathrm{tt}, \qquad \neg R(\vec{x}) \iff R(\vec{x}) = \mathrm{ff}. \tag{1A-7}$$

Most often we will be considering n-ary relations $R \subseteq A^n$ on a single set, subsets of A if $n = 1$.

The *cardinal number* of a set X is denoted by $|X|$, so that for distinct $x_1, \dots, x_n$, $|\{x_1, \dots, x_n\}| = n$.

Partial functions. For any two sets X, W, a *partial function*[3] $f : X \rightharpoonup W$ is a function $f : D_f \to W$, where $D_f \subseteq X$ is the *domain of convergence* of f. We call X and W the *input* and *output* sets of f and for $f, g : X \rightharpoonup W$ and $x \in X$, we set

$$\begin{aligned} f(x)\downarrow &\iff x \in D_f \quad (f(x) \text{ converges}), \\ f(x)\uparrow &\iff x \notin D_f \quad (f(x) \text{ diverges}), \\ \mathrm{Graph}_f(x, w) &\iff f(x) = w, \\ f \sqsubseteq g &\iff (\forall x \in X, w \in W)[f(x) = w \implies g(x) = w] \\ &\iff \mathrm{Graph}_f \subseteq \mathrm{Graph}_g, \end{aligned}$$

[3]A partial function $f : X \rightharpoonup W$ comes "tagged" with its input and output sets, which means that "given f" implies that we are also given X and W—and we will often use this, especially in definitions. On the other hand, if $D_f \subseteq X'$ and $W \subset W'$, then there is exactly one $f' : X' \rightharpoonup W'$ which agrees with f on their common domain of convergence, and it is natural to use the same name for it—which we will often do.

read "f is a subfunction of g". If $f : X \rightharpoonup W$ and $h : Y \rightharpoonup W$, then

$$\text{(1A-8)} \quad f(x) = h(y) \iff [f(x)\uparrow \ \& \ h(y)\uparrow] \text{ or } [f(x)\downarrow \ \& \ h(y)\downarrow \ \& \ f(x) = h(y)],$$

i.e., two possibly undefined expressions are equal if they are both undefined or they are both defined and equal. On occasion (especially in definitions) we also use the ungrammatical "$f(x) = \uparrow$" which is synonymous with "$f(x)\uparrow$".[4]

Partial functions compose *strictly*, they "call their arguments" *by value* in computer science terminology: if $g_i : X \rightharpoonup W_i$ for $i = 1, \dots, m$ and $f : W_1 \times \cdots \times W_m \rightharpoonup W$, then

$$\text{(1A-9)} \quad f(g_1(x), \dots, g_m(x)) = w \iff (\exists w_1, \dots, w_m)[g_1(x) = w_1 \ \& \ \cdots \ \& \ g_m(x) = w_m \ \& \ f(w_1, \dots, w_m) = w],$$

so that in particular,

$$f(g_1(x), \dots, g_m(x))\downarrow \implies g_1(x)\downarrow, \dots, g_m(x)\downarrow .$$

Monotone and continuous functionals. For any two sets Y and W, $(Y \rightharpoonup W)$ is the space of all partial functions $p : Y \rightharpoonup W$. A *functional* is a partial function

$$\text{(1A-10)} \qquad f : X \times (Y_1 \rightharpoonup W_1) \times \cdots \times (Y_k \rightharpoonup W_k) \rightharpoonup W$$

which takes members of X and partial functions (on various sets) as arguments and gives a value in some set W when it converges; f is *monotone* if

$$f(x, p_1, \dots, p_k)\downarrow \ \& \ p_1 \sqsubseteq q_1 \ \& \ \cdots \ \& \ p_k \sqsubseteq q_k \implies f(x, p_1, \dots, p_k) = f(x, q_1, \dots, q_k);$$

and it is *continuous* if it is monotone and *compact*, i.e.,

$$f(x, p_1, \dots, p_k)\downarrow \implies (\exists \text{ finite } p_1^0 \sqsubseteq p_1, \dots, p_k^0 \sqsubseteq p_k)[f(x, p_1^0, \dots, p_k^0) = f(x, p_1, \dots, p_k)],$$

where a partial function is *finite* if it has finite domain of convergence.

Partial functions are (degenerate) continuous functionals, with the conventional understanding of (1A-10) and these notions when $k = 0$.

Typical (and most basic) continuous functionals are the *applications*,

$$\text{(1A-11)} \qquad \mathrm{ap}_n(x_1, \dots, x_n, p) = p(x_1, \dots, x_n) \quad (n \geq 1),$$

usually with $x_1, \dots, x_n$ ranging over some set A and $p : A^n \rightharpoonup W$.

---

[4]This is Kleene's *strong equality* between "partial values" often denoted by "$\simeq$".

Notation gets messy when we work with functionals and we certainly do not want to have to repeat or refer to (1A-10) to specify their input and output sets every time we use them. Sometimes

we will write $f(x, \vec{p})$ when we mean f,

which is convenient if logically inappropriate, as it confuses "function value" with "function".[5]

Sections and λ-abstractions. Putting this convention to work, we define the *x-section* and the *λx-abstraction* of a functional $f(x, y, \vec{p})$ by

$$f_x(y, \vec{p}) = f(x, y, \vec{p}),$$
$$(\lambda x)f(x, y, \vec{p})(x) = f(x, y, \vec{p}).$$

More precisely (for once), if $f : X \times Y \times \mathcal{P} \rightharpoonup W$ with $\mathcal{P}$ a (possibly empty) product of partial function spaces, then the first of these defines for each $x \in X$ a functional $f_x : Y \times \mathcal{P} \rightharpoonup W$ and the second defines an operation $f_X : Y \times \mathcal{P} \to (X \rightharpoonup W)$ which takes partial functions as values. Notice that by our convention on the empty product above, every partial function $f : X \times Y \rightharpoonup W$ is a functional, the definitions apply and $f_x : Y \rightharpoonup W$, $f_X : Y \to (X \rightharpoonup W)$.

Operations on functionals. From the many operations on functionals we will use, we list here just four, for easy reference.[6]

Substitution: from $g(t_1, \dots, t_k, \vec{r})$ and $h_i(\vec{x}, \vec{r})$ $(i = 1, \dots, k)$, define

(1A-12) $$f(\vec{x}, \vec{r}) = g(h_1(\vec{x}, \vec{r}), \dots, h_k(\vec{x}, \vec{r}), \vec{r}).$$

λ-substitution: from $g(\vec{u}, \vec{x}, \vec{p})$ and $h(\vec{y}, q, \vec{r})$, define

(1A-13) $$f(\vec{x}, \vec{y}, \vec{p}, \vec{r}) = h(\vec{y}, (\lambda \vec{u})g(\vec{u}, \vec{x}, \vec{p}), \vec{r}).$$

Branching: from $g_0(\vec{x}, \vec{r})$ with output set $\mathbb{B}$ and $g_1(\vec{x}, \vec{r}), g_2(\vec{x}, \vec{r})$ with the same output set, define

(1A-14) $$f(\vec{x}, \vec{r}) = \text{if } g_0(\vec{x}, \vec{r}) \text{ then } g_1(\vec{x}, \vec{r}) \text{ else } g_2(\vec{x}, \vec{r})$$
$$= \begin{cases} g_1(\vec{x}, \vec{r}), & \text{if } g_0(\vec{x}, \vec{r}) = \mathrm{tt}, \\ g_2(\vec{x}, \vec{r}), & \text{if } g_0(\vec{x}, \vec{r}) = \mathrm{ff}, \\ \uparrow & \text{otherwise.} \end{cases}$$

Mangling: from $g(u_1, \dots, u_k, r_1, \dots, r_l)$, define

(1A-15) $$f(x_1, \dots, x_n, r_1, \dots r_m) = g(x_{\pi(1)}, \dots, x_{\pi(k)}, r_{\sigma(1)}, \dots, r_{\sigma(l)}),$$

where $\pi : \{1, \dots, k\} \to \{1, \dots, n\}$, $\sigma : \{1, \dots, l\} \to \{1, \dots, m\}$.

---

[5] When the error was once pointed out to Kleene by a student, he responded that the distinction between "function" and "function value" is surely important, but if he could not handle it he was probably in the wrong field.

[6] It is assumed that the input and output sets of the given functionals and the ranges of the variables fit, so that the definitions make sense.

A more dignified name for mangling might be *adding, permuting and identifying variables*, but that is way too long; whatever you call it, it is a very useful operation, it provides for definitions of the form

$$f(x, y, p, r) = h(x, x, r, p, p),$$

and combined with the three operations preceding it justifies complex explicit definitions, e.g.,

$$f(x, y, p, r) = h(x, g_1(x, y, p), (\lambda\vec{u})g_2(u, x, u), p, r).$$

1A.1. **Proposition.** *If a class of functionals $\mathcal{F}$ is closed under mangling* (1A-15), *then $\mathcal{F}$ is closed under substitution* (1A-12) *if and only if it is closed under composition*

(1A-16) $$f(\vec{x}, \vec{y}, \vec{p}, \vec{r}) = g(h(\vec{x}, \vec{p}), \vec{y}, \vec{r}).$$

PROOF is left for Problem x1A.2. ⊣

Strings. For any set L, $L^* = L^{<\omega}$ is the set of *strings* (words, finite sequences) from L and we will use mostly standard notations for them:

(1A-17)
$$\begin{gathered}
\text{nil} = (\,) \quad \text{(the empty string)},\\
|(u_0, \dots, u_{m-1})| = m, \quad (u_0, \dots, u_{m-1})_i = u_i \quad (i < m),\\
\text{head}((u_0, \dots, u_{m-1})) = (u_0) \quad (\text{with head(nil)} = \text{nil}),\\
\text{tail}((u_0, \dots, u_{m-1})) = (u_1, \dots, u_{m-1}) \quad (= \text{nil if } m \le 1),\\
\text{cons}(u, (v_0, \dots, v_{m-1})) = ((u)_0, v_0, \dots, v_{m-1})\\
(u_0, \dots, u_{m-1}) * (v_0, \dots, v_{n-1}) = (u_0, \dots, u_{m-1}, v_0, \dots, v_{n-1}),\\
u \sqsubseteq v \iff (\exists w)[u * w = v], \quad \text{(the initial segment relation)},\\
u \sqsubsetneq v \iff u \sqsubseteq v \;\&\; u \neq v.
\end{gathered}$$

These definitions of head(u) and cons(u, v) in effect identify a member t of L with the string $(t) \in L^*$, which simplifies in some ways dealing with strings.

Sometimes we denote strings by simply listing their elements

$$u_0u_1 \cdots v_{m-1} \equiv (u_0, u_1, \dots, u_{m-1}),$$

especially when we think of them as *words* from some *alphabet* L of *symbols*; and in such cases, we typically use "$\equiv$" to denote the equality relation on words, since "$=$" is often one of the symbols in the alphabet.

Trees. For our purposes, a (finite, non-empty, rooted, $\mathbb{N}$-labelled) *tree* on a set X is any finite set $\mathcal{T} \subset (\mathbb{N} \times X)^{<\omega}$ of non-empty finite sequences (*nodes*) from $\mathbb{N} \times X$ which has a unique node of length 1, its *root* and is closed under initial segments,

$$\emptyset \neq u \sqsubseteq v \in \mathcal{T} \Longrightarrow u \in \mathcal{T}.$$

The *children* of a node $u \in \mathcal{T}$ are all one-point extensions $u * (y) \in \mathcal{T}$ and u is *splitting* if it has at least two children.

The (out-) *degree* of $\mathcal{T}$ is the maximal number of children that any node has—unless $\mathcal{T} = \{\text{root}\}$, in which case (by convention) $\text{degree}(\mathcal{T}) = 1$.

A node v is *below* a node u if $v = u$ or there is a (necessarily unique) non-empty w such that $v = u * w$.

The *depth* of $\mathcal{T}$ is the largest m such that there is a sequence of nodes $u_1, \dots, u_m$ with each u_{i+1} strictly below u_i, so that $\text{depth}(\mathcal{T}) = 0$ exactly when $\mathcal{T} = \{\text{root}\}$; and the *splitting depth* of $\mathcal{T}$ is the largest m such that there is a sequence of splitting nodes $u_1, \dots, u_m$ with each u_{i+1} below u_i, so that $\text{spdepth}(\mathcal{T}) = 0$ exactly when $\text{degree}(\mathcal{T}) = 1$.

The *size* of a tree is the number of its nodes, $\text{size}(\mathcal{T}) = |\mathcal{T}|$.

A node is a *leaf* if it has no children, $\text{leaves}(\mathcal{T})$ is the set of all leaves of $\mathcal{T}$ and it is easy to check by induction on $\text{spdepth}(\mathcal{T})$ that

$$(1A\text{-}18) \qquad |\text{leaves}(\mathcal{T})| \le \text{degree}(\mathcal{T})^{\text{spdepth}(\mathcal{T})} \le \text{degree}(\mathcal{T})^{\text{depth}(\mathcal{T})},$$

cf. Problem x1A.4.

For each $u = (u_0, \dots, u_n) \in \mathcal{T}$, let

$$\mathcal{T}_u = \{(u_n) * v \in (\mathbb{N} \times X)^{<\omega} : u * v \in \mathcal{T}\}.$$

This is the *subtree* of $\mathcal{T}$ below u, with root u_n, so that in particular,

$$(1A\text{-}19) \qquad \mathcal{T}_{\text{root}} = \mathcal{T}, \quad |\mathcal{T}| = 1 + \sum \Big\{ |\mathcal{T}_u| : u \text{ is a child of root} \Big\}.$$

For each tree $\mathcal{T}$, let

$$\mathcal{T}' = \{u \in \mathcal{T} : u \text{ is not a leaf}\}.$$

This is empty if $\mathcal{T}$ has only one node, but it is a tree if $|\mathcal{T}| > 0$, the *derived* (pruned) subtree of $\mathcal{T}$ and

$$\text{depth}(\mathcal{T}') = \text{depth}(\mathcal{T}) - 1.$$

In dealing with these trees, we will think of them as sets of sequences from X, mostly disregarding the labels: their only purpose is to allow a node to have several "identical" children which are counted separately. For example, we will want to draw the tree

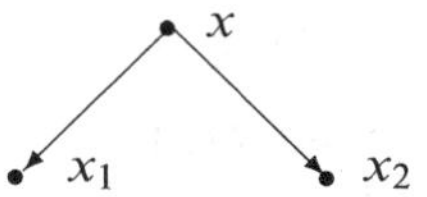

and assume it has this structure (with a root which has two children) even if it happens that $x_1 = x_2$; so formally, we need to set

$$T = \{((0, x)), ((0, x), (0, x_1)), ((0, x), (1, x_2))\},$$

but we will indicate this by the simpler

$$\mathcal{T} = \{(x), (x, x_1), (x, x_2)\}.$$

For example, if $z \in X$ and $\mathcal{T}, \mathcal{T}_1, \dots, \mathcal{T}_k$ are trees on X, then $\mathrm{Top}(z, \mathcal{T}_1, \dots, \mathcal{T}_k)$ is the tree with root z and $\mathcal{T}_1, \dots, \mathcal{T}_k$ immediately below it. We will draw the result of this operation as if

$$\text{(1A-20)}\quad \mathrm{Top}(z, \mathcal{T}_1, \dots, \mathcal{T}_k) = \{(z, x_1, \dots, x_n) \mid \text{for some } i = 1, \dots, k, (x_1, \dots, x_n) \in \mathcal{T}_i\};$$

the formal definition, with the labels, is the more formidable

$$\mathrm{Top}(z, \mathcal{T}_1, \dots, \mathcal{T}_k) = \{((0, z), (\langle i, j_1\rangle, x_1), \dots, (\langle i, j_n\rangle, x_n)) \mid i = 1, \dots, k, ((j_1, x_1), \dots, (j_n, x_n)) \in \mathcal{T}_i\},$$

where $\langle i, j\rangle = 2^{i+1}3^{j+1}$ is an injection of $\mathbb{N} \times \mathbb{N}$ into $\mathbb{N} \setminus \{0\}$.

Problems for Section 1A

x1A.1. **Problem.** Verify that $(X \rightharpoonup W)$ with the subfunction relation $\sqsubseteq$ is a *Scott domain* whose maximal points are the total functions, i.e.,

(1) For all $f, g, h : X \rightharpoonup W$,

$$f \sqsubseteq f,\ [f \sqsubseteq g \sqsubseteq h] \Longrightarrow f \sqsubseteq h,\ [f \sqsubseteq g \sqsubseteq f] \Longrightarrow f = g.$$

(2) Every $f : X \rightharpoonup W$ is the least upper bound of the set of its finite subfunctions.

(3) If $I \subseteq (X \rightharpoonup W)$ and every pair $\{f, g\} \subseteq_p I$ has an upper bound, then I has a least upper bound.

(4) If $f : X \to W$ is total and $f \sqsubseteq g$, then $f = g$.

x1A.2. **Problem.** Prove Proposition 1A.1, that if $\mathcal{F}$ is a class of functionals closed under mangling (1A-15), then $\mathcal{F}$ is closed under substitution, (1A-12) if and only if it is closed under composition, (1A-16).

x1A.3. **Problem.** Prove that if $\mathcal{F}$ is a class of functionals closed under all four schemes of explicit definition on page 11, then it is also closed under the scheme

$$f(x, y, p, r) = h(x, g_1(x, y, p), (\lambda\vec{u})g_2(u, x, u), p, r).$$

x1A.4. **Problem.** Suppose $\mathcal{T}$ is a finite tree.

(1) Prove (1A-18) and (1A-19).

(2) Prove that if $\mathrm{degree}(\mathcal{T}) = 1$, then $|\mathcal{T}| = \mathrm{depth}(\mathcal{T}) + 1$.

(3) Prove that if $\mathrm{degree}(\mathcal{T}) \geq 2$, then

$$|\mathcal{T}| \leq \mathrm{degree}(\mathcal{T})^{\mathrm{depth}(\mathcal{T})+1} - 1 < \mathrm{degree}(\mathcal{T})^{\mathrm{depth}(\mathcal{T})+1}.$$

1B. Continuous, call-by-value recursion

From recursion theory, we will need the following, fundamental result which justifies recursive definitions:

1B.1. **Theorem** (The Fixed Point Lemma). (1) *For every continuous functional $f : X \times (X \rightharpoonup W) \rightharpoonup W$, the recursive equation*

$$p(x) = f(x, p) \tag{1B-1}$$

has a $\sqsubseteq$-least solution $\overline{p} : X \rightharpoonup W$, characterized by the conditions

(FP) $\quad \overline{p}(x) = f(x, \overline{p}) \quad (x \in X),$

(MIN) $\quad$ if $(\forall x)[f(x, q)\downarrow \Rightarrow f(x, q) = q(x)]$, then $\overline{p} \sqsubseteq q \quad (q : X \rightharpoonup W).$

We say that $\overline{p}$ is the *canonical solution* of (1B-1) and we write

$$p(x) = f(x, p) \gg \overline{p}. \tag{1B-2}$$

(2) *Similarly, every system of mutual continuous recursive equations*

$$p_1(x_1) = f_1(x_1, p_1, \dots, p_K), \ \dots, \ p_K(x_K) = f_K(x_K, p_1, \dots, p_K) \tag{1B-3}$$

(with input and output sets for $f_1, \dots, f_K$ matching so that the equations make sense) *has a $\sqsubseteq$-least* (canonical) *solution tuple $\overline{p}_1, \dots, \overline{p}_K$ characterized by the conditions*

(FP) $\quad \overline{p}_i(x_i) = f(x_i, \overline{p}_1, \dots \overline{p}_K) \quad (i = 1, \dots, K, x_i \in X_i),$

(MIN) $\quad$ if for $(i = 1, \dots, K,$ and all $x_i \in X_i)$

$$\Big(f(x_i, q_1, \dots, q_K)\downarrow \implies f_i(x_i, q_1, \dots, q_K) = q_i(x_i)\Big),$$

$$\text{then } \overline{p}_1 \sqsubseteq q_1, \dots, \overline{p}_K \sqsubseteq q_K,$$

and we write

$$\left\{ \begin{array}{rcl} p_1(x_1) & = & f_1(x_1, p_1, \dots, p_K) \\ & \vdots & \\ p_K(x_K) & = & f_K(x_K, p_1, \dots, p_K) \end{array} \right. \gg \begin{array}{c} \overline{p}_1 \\ \vdots \\ \overline{p}_K. \end{array} \tag{1B-4}$$

(3) *Moreover, if the given functionals have additional partial function arguments $\vec{r} = (r_1, \dots, r_m)$ and $\overline{p}_1(\vec{r}), \dots, \overline{p}_K(\vec{r})$ are defined for each $\vec{r}$ by* (2) *so that*

$$\left\{ \begin{array}{rcl} p_1(x_1) & = & f_1(x_1, p_1, \dots, p_K, \vec{r}) \\ & \vdots & \\ p_K(x_K) & = & f_K(x_K, p_1, \dots, p_K, \vec{r}) \end{array} \right. \gg \begin{array}{c} \overline{p}_1(\vec{r}) \\ \vdots \\ \overline{p}_K(\vec{r}), \end{array} \tag{1B-5}$$

then the functionals

$$g_i(x_i, \vec{r}) = \overline{p}_i(\vec{r})(x_i) \quad (i = 1, \dots, K)$$

are continuous.

OUTLINE OF PROOF. (1) For the one-equation case, define by recursion on $\mathbb{N}$ the *iterates*

$$\overline{p}^0(x) = \uparrow \quad \text{(i.e., } \overline{p}^0 \text{ is the totally undefined partial function on } X \text{ to } W\text{)},$$
$$\overline{p}^{k+1}(x) = f(x, \overline{p}^k);$$

prove by induction, using monotonicity, that $\overline{p}^k \sqsubseteq \overline{p}^{k+1}$, so that

$$\overline{p}^0 \sqsubseteq \overline{p}^1 \sqsubseteq \overline{p}^2 \sqsubseteq \cdots;$$

and set $\overline{p} = \bigcup\{\overline{p}^k : k \in \mathbb{N}\}$, i.e.,

$$\overline{p}(x) = w \iff (\exists k)[\overline{p}^k(x) = w].$$

If $\overline{p}(x) = w$, then, for some k,

$$\overline{p}^{k+1}(x) = f(x, \overline{p}^k) = w$$

by the definition, and hence $f(x, \overline{p}) = w$, by monotonicity, since $\overline{p}^k \sqsubseteq \overline{p}$. On the other hand, if $f(x, \overline{p}) = w$, then there is a finite $q \sqsubseteq \overline{p}$ such that $f(x, q) = w$, by continuity, and then there is some k such that $q \sqsubseteq \overline{p}^k$; thus, by monotonicity, $f(x, \overline{p}^k) = \overline{p}^{k+1}(x) = w$, and so $\overline{p}(x) = w$, which completes the proof that, for all x,

$$f(x, \overline{p}) = \overline{p}(x).$$

To verify the minimality of $\overline{p}$, suppose that

$$(\forall x)[f(x, q)\downarrow \implies f(x, q) = q(x)],$$

and show (by an easy induction on k, using monotonicity) that $\overline{p}^k \sqsubseteq q$, so that in the end $\overline{p} \sqsubseteq q$.

(3) For the one-equation case

$$p(x) = f(x, p, \vec{r}) \gg \overline{p}(\vec{r}),$$

check by an easy induction on k that if

$$g^k(x, \vec{r}) = \overline{p}^k(\vec{r})(x) \quad (k \in \mathbb{N})$$

is defined as in the proof of (1) for each fixed $\vec{r}$, then g^k is continuous.

(2) and the full statement (for systems) of (3) are proved by the same argument with some added (messy) notation. ⊣

We refer to (FP) and (MIN) in Part (1) of the Fixed Point Lemma as the *fixed point* and the *minimality* properties of the recursive definition—and similarly with the corresponding conditions in Part (2).

The $\overline{\text{where}}$-notation for mutual recursion. To express and prove properties of recursive definitions, we now introduce the following two notations for mutual recursion: given a continuous system of recursive equations and their solutions

$$\begin{cases} p_1(x_1) &= f_1(x_1, \vec{p}, \vec{r}) \\ &\vdots \\ p_k(x_k) &= f_k(x_k, \vec{p}, \vec{r}) \end{cases} \gg \begin{matrix} \overline{p}_1(\vec{r}) \\ \vdots \\ \overline{p}_k(\vec{r}) \end{matrix}$$

as in (1B-5) (with $\vec{p} = p_1, \dots, p_k, \vec{r} = r_1, \dots, r_m$) and a continuous functional $f_0(y, \vec{p}, \vec{r})$ (with y varying over some set Y), we set

$$\text{(1B-6)}\quad f_0(y, \vec{p}, \vec{r}) \,\overline{\text{where}}\, \Big\{ p_1(x_1) = f_1(x_1, \vec{p}, \vec{r}), \dots, p_k(x_k) = f_k(x_k, \vec{p}, \vec{r}) \Big\}$$
$$= f_0(y, \vec{p}, \vec{r}) \,\overline{\text{where}}\, \Big\{ p_i(x_i) = f_i(x_i, \vec{p}, \vec{r}) : 1 \leq i \leq k \Big\}$$
$$=_{\text{df}} f_0(y, \overline{p}_1(\vec{r}), \dots, \overline{p}_k(\vec{r}), \vec{r}).$$

A functional $f(y, \vec{r})$ is *defined by recursion* from given continuous functionals $f_0, f_1, \dots, f_k$ (with suitable input and output sets) if

$$\text{(1B-7)}\quad f(y, \vec{r})$$
$$= f_0(y, \vec{p}, \vec{r}) \,\overline{\text{where}}\, \Big\{ p_1(x_1) = f_1(x_1, \vec{p}, \vec{r}), \dots, p_k(x_k) = f_k(x_k, \vec{p}, \vec{r}) \Big\}.$$

The first of the two notations in (1B-6) is useful for giving succinct recursive definitions of specific functionals while the second simplifies the statement of general rules for mutual recursion, as in the next result:

1B.2. **Theorem** (Recursion rules). *If $f, f_0, \dots, g, g_0, \dots$ are all continuous and $\vec{p} = p_1, \dots, p_k, \vec{q} = q_1, \dots, q_t, \vec{r} = r_1, \dots, r_m$, then:*

$$\text{(subst)}\qquad f(g(x, \vec{r}), y, \vec{r}) = f(q(x), y, \vec{r}) \,\overline{\text{where}}\, \Big\{ q(x) = g(x, \vec{r}) \Big\}$$

$$(\lambda\text{-subst})\quad f(x, (\lambda\vec{u})g(\vec{u}, y, \vec{r}), \vec{r})$$
$$= f(x, (\lambda\vec{u})q(\vec{u}, y), \vec{r}) \,\overline{\text{where}}\, \Big\{ q(\vec{u}, y) = g(\vec{u}, y, \vec{r}) \Big\}$$

$$\text{(head)}\quad f_0(x, \vec{p}, \vec{q}, \vec{r}) \,\overline{\text{where}}\, \Big\{ p_i(x_i) = f_i(x_i, \vec{p}, \vec{q}, \vec{r}) : 1 \leq i \leq k \Big\}$$
$$\overline{\text{where}}\, \Big\{ q_j(y_j) = g_j(y_j, \vec{q}, \vec{r}) : 1 \leq j \leq t \Big\}$$
$$= f_0(x, \vec{p}, \vec{q}, \vec{r}) \,\overline{\text{where}}\, \Big\{ p_i(x_i) = f_i(x_i, \vec{p}, \vec{q}, \vec{r}) : 1 \leq i \leq k,$$
$$q_j(y_j) = g_j(y_j, \vec{q}, \vec{r}) : 1 \leq j \leq t \Big\}$$

(Bekič-Scott) $f_0(x, q, \vec{p}, \vec{r})$ $\overline{\text{where}}$ $\Big\{ p_i(x_i) = f_i(x_i, q, \vec{p}, \vec{r}) : 1 \leq i \leq k,$

$$q(y) = g_0(y, q, \vec{q}, \vec{p}, \vec{r})\ \overline{\text{where}}\ \Big\{ q_j(y_j) = g_j(y_j, q, \vec{q}, \vec{p}, \vec{r}) : 1 \leq j \leq t \Big\}\Big\}$$

$$= f_0(x, q, \vec{p}, \vec{r})\ \overline{\text{where}}\ \Big\{ p_i(x_i) = f_i(x_i, q, \vec{p}, \vec{r}) : 1 \leq i \leq k,$$

$$q(y) = g_0(y, q, \vec{q}, \vec{p}, \vec{r}), q_j(y_j) = g_j(y_j, q, \vec{q}, \vec{p}, \vec{r}) : 1 \leq j \leq t \Big\}$$

PROOF. The (head) and (Bekič-Scott) rules are stated for the record in very general forms which are hard to understand, and it helps to read the proof for a special case of (head) below and to work out a special case for (Bekič-Scott) in Problem x1B.9. Basically, these rules allow us to "flatten" nested recursions, simply remove all the "$\overline{\text{where}}$s" and all the braces except the outermost ones.

(λ-subst). There is only one recursive definition in this identity, which is in fact trivial,

$$q(\vec{u}, y) = g(\vec{u}, y, \vec{r}) \gg \overline{q}(\vec{r}).$$

The FP property for it gives

$$\overline{q}(\vec{r})(\vec{u}, y) = g(\vec{u}, y, \vec{r}),$$

so that the right-hand side of (λ-subst) has the value

$$f(x, (\lambda\vec{u})\overline{q}(\vec{r})(\vec{u}, y), \vec{r}) = f(x, (\lambda\vec{u})g(u, y, \vec{r}), \vec{r});$$

which is the same as the value of its left-hand side.

(head). For the simple case with $k = t = 1$ and suppressing the dependence on $\vec{r}$ which does not enter in the argument, we need to show

$$\text{(1B-8)} \quad \Big(f_0(x, p, q)\ \overline{\text{where}}\ \Big\{p(u) = f_1(u, p, q)\Big\}\Big)\ \overline{\text{where}}\ \Big\{q(v) = g_1(v, q)\Big\}$$
$$= f_0(x, p, q)\ \overline{\text{where}}\ \Big\{p(u) = f_1(u, p, q), q(v) = g_1(v, q)\Big\}.$$

There are three recursive definitions involved in this equation:

$$\text{(A)} \qquad q(v) = g_1(v, q) \gg \overline{q}$$

$$\text{(B)} \qquad p(u) = f_1(u, p, q) \gg \overline{p}(q)$$

$$\text{(C)} \qquad \begin{cases} p(u) = f_1(u, p, q) \\ q(v) = g_1(v, q) \end{cases} \gg \begin{matrix} \widetilde{p} \\ \widetilde{q} \end{matrix}$$

From the definitions, what we need to show is that

$$f_0(x, \overline{p}(\overline{q}), \overline{q}) = f_0(x, \widetilde{p}, \widetilde{q}),$$

and for this it suffices to prove that

$$\text{(1B-9)} \qquad \overline{q} = \widetilde{q}, \quad \overline{p}(\overline{q}) = \widetilde{p}.$$

(1) By the FP property on the second equation of (C), we have (for every v), $\widetilde{q}(v) = g_1(v, \widetilde{q})$, and then the MIN property on (A) gives $\boxed{\overline{q} \sqsubseteq \widetilde{q}}$.

(2) The FP property on the first equation of (C) gives $\widetilde{p}(u) = f_1(u, \widetilde{p}, \widetilde{q})$, and then the MIN property on (B) gives $\boxed{\overline{p}(\widetilde{q}) \sqsubseteq \widetilde{p}}$.

(3) Consider now the two partial functions $\overline{p}(\overline{q})$ and $\overline{q}$ which by the FP properties on (B) and (A) satisfy both equations of (C); the MIN property on (C) then implies that $\boxed{\widetilde{p} \sqsubseteq \overline{p}(\overline{q}),\ \widetilde{q} \sqsubseteq \overline{q}}$.

The boxed inequalities imply first that $\overline{q} = \widetilde{q}$, and then

$$\widetilde{p} \sqsubseteq \overline{p}(\overline{q}) = \overline{p}(\widetilde{q}) \sqsubseteq \widetilde{p},$$

so that all these partial functions are equal and we have $\overline{p}(\overline{q}) = \widetilde{p}$.

Proof of the general case of (head) is only a notational variant of this and the *Substitution* (subst) and *Bekič-Scott* rules (Bekič-Scott) in the theorem are proved similarly and we leave them for the problems. ⊣

Monotone recursion. It is well known that Theorems 1B.1 and 1B.2 hold for monotone functionals which need not be continuous, cf. Theorem 7.36 and the problems for Chapter 7 in Moschovakis [2006]. These are classical results of elementary set theory whose proofs require definition by transfinite recursion and we will not need them here.

Problems for Section 1B

To solve a recursive equation (1B-1) or a system (1B-3) means to identify the canonical solution(s) in explicit terms. For example, the solution of

(1B-10) $$f(x, y) = \text{if } (y = 0) \text{ then } x \text{ else } S(f(x, \text{Pd}(y)))$$

in $\mathbb{N}$ is the sum function $g(x, y) = x + y$: because

(1) every solution of (1B-10) is total by an easy induction on y, and

(2) addition satisfies (1B-10),

so $\overline{f} \sqsubseteq g$ and hence $\overline{f} = g$ since $\overline{f}$ is total.

In the problems which follow, individual variables vary over $\mathbb{N}$ and function variables vary over partial functions on $\mathbb{N}$ (of various arities).

x1B.1. **Problem.** Solve in $\mathbb{N}$ the recursive equation

$$f(x, y) = \text{if } (y = 0) \text{ then } 0 \text{ else } f(x, \text{Pd}(y)) + x.$$

x1B.2. **Problem.** Solve in $\mathbb{N}$ the recursive equation

$$\begin{aligned} f(x, y) = \text{if } (y = 0) &\text{ then } 0 \\ &\text{else if } (y = 1) \text{ then } x \\ &\text{else } 2 \cdot f(x, \text{iq}_2(y)) + f(x, \text{parity}(y)). \end{aligned}$$

x1B.3. **Problem.** Consider the following recursive equation in $\mathbb{N}$:

$$f(x,y,r)=\begin{cases} r, & \text{if } x=y=0,\\ 2f(\mathrm{iq}_2(x),\mathrm{iq}_2(y),0), & \text{ow., if parity}(x)+\text{parity}(y)+r=0,\\ 2f(\mathrm{iq}_2(x),\mathrm{iq}_2(y),0)+1, & \text{ow., if parity}(x)+\text{parity}(y)+r=1,\\ 2f(\mathrm{iq}_2(x),\mathrm{iq}_2(y),1), & \text{ow., if parity}(x)+\text{parity}(y)+r=2,\\ 2f(\mathrm{iq}_2(x),\mathrm{iq}_2(y),1)+1, & \text{ow,} \end{cases}$$

and let $\overline{f}$ be its least solution. Prove that $\overline{f}(x,y,r)=x+y+r$ if $r\leq 1$ so that that $\overline{f}(x,y,0)=x+y$.

x1B.4. **Problem.** Solve in $\mathbb{N}$ the recursive equation

$$f(x,y)=\text{if } (\phi(x,y)=0) \text{ then } y \text{ else } f(x,y+1),$$

where $\phi:\mathbb{N}^2\rightharpoonup\mathbb{N}$ is some fixed, given partial function. HINT: Compute $f(x,x)$ when $\phi(x,y)=0 \iff \text{Prime}(y) \ \&\ \text{Prime}(y+2)$.

x1B.5. **Problem** (Morris, per Manna [1974]). Solve in $\mathbb{N}$ the recursive equation

$$f(x,y)=\text{if } (x=0) \text{ then } 1 \text{ else } f(\mathrm{Pd}(x),f(x,y)).$$

x1B.6. **Problem.** Solve in L^* the recursive equation

$$f(u)=\text{if } (u=\text{nil}) \text{ then nil else } f(\text{tail}(u))*\text{head}(u).$$

x1B.7. **Problem.** If you are not already familiar with the Fixed Point Lemma 1B.1, work out the details of the proof.

x1B.8. **Problem.** Prove the *substitution rule* (subst) in Theorem 1B.2.

x1B.9. **Problem.** Prove the following special case of the *Bekič-Scott rule* in Theorem 1B.2:

$$\begin{aligned} &f_0(x,q,p_1,p_2)\ \overline{\text{where}}\Big\{p_1(x_1)=f_1(x_1,p_1,p_2),\, p_2(x_2)=f_2(x_2,p_1,p_2),\\ &\qquad q(y)=g_0(y,r,q,p_1,p_2)\ \overline{\text{where}}\Big\{r(u)=g_1(u,r,q,p_1,p_2)\Big\}\Big\}\\ &=f_0(x,q,p_1,p_2)\ \overline{\text{where}}\Big\{p_1(x_1)=f_1(x_1,p_1,p_2),\, p_2(x_2)=f_2(x_2,p_1,p_2),\\ &\qquad\qquad q(y)=g_0(y,r,q,p_1,p_2),\, r(u)=g_1(u,r,q,p_1,p_2)\Big\}. \end{aligned}$$

The next problem is another very special (but important) case of the Bekič-Scott rule; a version of this is known in classical recursion theory as the *First Recursion Theorem*:

x1B.10. **Problem.** Let $f:X\times(X\rightharpoonup W)\rightharpoonup W$ be a functional defined by a recursion

$$f(x,p)=f_0(x,p,q)\ \overline{\text{where}}\Big\{q(y)=f_1(y,p,q)\Big\}$$

where f_0, f_1 are continuous, so that f is also continuous, and suppose

$$p(x) = f(x, p) \gg \overline{p};$$

prove that

$$\overline{p}(x) = p(x) \ \overline{\text{where}} \left\{ p(x) = f_0(x, p, q), q(y) = f_1(y, p, q) \right\}.$$

1C. Some basic algorithms

We review here briefly some classical examples of *algorithms from specified primitives*, primarily to illustrate how recursive equations can be interpreted as instructions for the computation of their least fixed points. This process will be made precise in the next chapter.

The merge-sort algorithm. Suppose L is a set with a fixed total ordering $\leq$ on it. A string

$$v = v_0 v_1 \cdots v_{n-1} = (v_0, \ldots, v_{n-1}) \in L^*$$

is *sorted* (in non-decreasing order) if $v_0 \leq v_1 \leq \cdots \leq v_{n-1}$, and for each $u \in L^*$, sort(u) is the sorted "rearrangement" of u,

(1C-1) $$\text{sort}(u) =_{\text{df}} \text{the unique, sorted } v \in L^* \text{ such that for some permutation } \pi : \{0, \ldots, n-1\} \rightarrowtail\!\!\!\!\!\twoheadrightarrow \{0, \ldots, n-1\}, \quad v = (u_{\pi(0)}, u_{\pi(1)}, \ldots, u_{\pi(n-1)}).$$

The efficient computation of sort(u) is important in many computing applications and many sorting algorithms have been studied. We consider here just one of these algorithms, which is easily expressed by a system of two, simple, recursive equations.

The merge-sort uses as a "subroutine" an algorithm for *merging* two strings specified as follows:

1C.1. **Proposition.** *The equation*

(1C-2) $$\begin{aligned}\text{merge}(w, v) = {} & \text{if } (|w| = 0) \text{ then } v \\ & \text{else if } (|v| = 0) \text{ then } w \\ & \text{else if } (w_0 \leq v_0) \text{ then } (w_0) * \text{merge}(\text{tail}(w), v) \\ & \text{else } (v_0) * \text{merge}(w, \text{tail}(v))\end{aligned}$$

determines a value merge(w, v) *for all strings* $w, v \in L^*$, *and if* w *and* v *are both sorted, then*

(1C-3) $$\text{merge}(w, v) = \text{sort}(w * v).$$

Moreover, the value merge(w, v) *can be computed by successive applications of* (1C-2), *using no more than* $|w| + |v| \dot{-} 1$ *comparisons.*

PROOF. That (1C-2) determines a function and that (1C-3) holds are both trivial, by induction on $|w| + |v|$. For the comparison counting, notice first that (1C-2) computes merge(w, v) using no comparisons at all, if one of w or v is nil; if both $|w| > 0$ and $|v| > 0$, we make one initial comparison to decide whether $w_0 \leq v_0$, and no more than $|w| + |v| - 2$ additional comparisons after that (by the induction hypothesis, in either case), for a total of $|w| + |v| - 1$.⊣

In the next Chapter 2, we will make precise what it means *to compute* merge(w, v) *by repeated applications of* (1C-2), but it is really quite obvious: for example, when $L = \mathbb{N}$ with the natural ordering:

$$\begin{aligned}\text{merge}((3,1),(2,4)) &= (2) * \text{merge}((3,1),(4))\\ &= (2,3) * \text{merge}((1),(4))\\ &= (2,3,1) * \text{merge}((\,),(4))\\ &= (2,3,1,4).\end{aligned}$$

For each sequence u with $|u| = m > 1$ and $k = \lfloor \frac{m}{2} \rfloor$ the integer part of $\frac{1}{2}|u|$, let:

$$\text{(1C-4)} \qquad \text{half}_1(u) = (u_0, \ldots, u_{k-1}), \quad \text{half}_2(u) = (u_k, \ldots, u_{m-1}),$$

and for $|u| \leq 1$, set

$$\text{(1C-5)} \quad \text{half}_1(\text{nil}) = \text{half}_2(\text{nil}) = \text{nil}, \ \text{half}_1((x)) = \text{nil}, \ \text{half}_2((x)) = (x),$$

so that in any case

$$u = \text{half}_1(u) * \text{half}_2(u)$$

and each of the two halves of u has length within 1 of $\frac{1}{2}|u|$.

1C.2. **Proposition.** *The sort function satisfies the equation*

$$\text{(1C-6)} \qquad \begin{aligned}\text{sort}(u) = {} & \text{if } |u| \leq 1 \text{ then } u\\ & \text{else merge}(\text{sort}(\text{half}_1(u)), \text{sort}(\text{half}_2(u)))\end{aligned}$$

and it can be computed from (1C-2) *and* (1C-6) *using no more than* $|u| \log |u|$ *comparisons.*

PROOF. The validity of (1C-6) is immediate, by induction on $|u|$. To prove the bound on comparisons, also by induction, note that it is trivial when $|u| \leq 1$, and suppose that $\lceil \log |u| \rceil = k + 1$, so that (easily) both halves of u have length $\leq 2^k$. Thus, by the induction hypothesis and Proposition 1C.1, we can compute sort(u) using no more than

$$k2^k + k2^k + 2^k + 2^k - 1 < (k+1)2^{k+1}$$

comparisons. ⊣

By a basic, classical result, the merge-sort is optimal (in a very strong sense) for the number of comparisons required to sort a string, cf. Theorem 4G.1.

The Euclidean algorithm. Euclid actually defined the subtractive version of the algorithm which works by iterating the classical *anthyphairesis* operation on unordered pairs, $\{a, b\} \mapsto \{\min(a, b), \max(a, b) - \min(a, b)\}$, The recursive specification of the version with division fits better with what we will be doing—and we could argue that it is implicit in Euclid.

1C.3. **Lemma.** *The greatest common divisor function on* $\mathbb{N}$ (defined by (1A-3)) *satisfies the following recursive equation, which determines it*:

$$\text{(1C-7)} \qquad \gcd(x, y) = \text{if } (y = 0) \text{ then } x \text{ else } \gcd(y, \text{rem}(x, y)).$$

PROOF. If $y \neq 0$ and $y \nmid x$, then the pairs $\{x, y\}$ and $\{y, \text{rem}(x, y)\}$ have the same common divisors. ⊣

Equation (1C-7) yields a procedure for computing $\gcd(x, y)$ for all x, y using rem and eq_0:

if $y = 0$ *give output* x*, else set* $x := y, y := \text{rem}(x, y)$ *and repeat.*

For example:

$$\begin{aligned} \gcd(165, 231) &= \gcd(231, 165) && \textbf{cde}\text{: } 165 = 231 \cdot 0 + 165 \\ &= \gcd(165, 66) && \textbf{cde}\text{: } 231 = 165 \cdot 1 + 66 \\ &= \gcd(66, 33) && \textbf{cde}\text{: } 165 = 66 \cdot 2 + 33 \\ &= \gcd(33, 0) \\ &= 33. \end{aligned}$$

The computation required three divisions in this case. In general, we set

$$c_{\{\text{rem}\}}(\varepsilon, x, y) = \text{the number of divisions required to compute } \gcd(x, y) \text{ using (1C-7)},$$

so that, directly from (1C-7), for $x \geq y \geq 1$,

$$\text{(1C-8)} \quad c_{\{\text{rem}\}}(\varepsilon, x, y) = \text{ if } (y = 0) \text{ then } 0 \text{ else } 1 + c_{\{\text{rem}\}}(\varepsilon, y, \text{rem}(x, y)).$$

1C.4. **Proposition.** *For all* $x \geq y \geq 2$, $c_{\{\text{rem}\}}(\varepsilon, x, y) \leq 2 \log y$.

PROOF is by (complete) induction on y, and we must consider three cases (with $c(x, y) = c_{\{\text{rem}\}}(\varepsilon, x, y)$):

Case 1, $y \mid x$; now $\text{rem}(x, y) = 0$ and $c(x, y) = 1 + c(y, 0) = 1 \leq 2 \log y$, since $y \geq 2$ and so $\log y \geq 1$.

Case 2, $x = q_1 y + r_1$ with $0 < r_1 < y$ but $r_1 \mid y$. Now

$$c(x, y) = 1 + c(y, r_1) = 2 \leq 2 \log y \text{ since } y \geq 2.$$

Case 3, $y = q_2 r_1 + r_2$ with $0 < r_2 < r_1$. If $r_2 = 1$, then $c(x, y) = 2 + c(r_1, 1) = 3$, but $y > r_1 > r_2$, so $y \geq 3$ and $3 \leq 2 \log 3 \leq 2 \log y$. If $r_2 \geq 2$, then the induction hypothesis applies and it gives

$$c(x, y) = 2 + c(r_1, r_2) \leq 2 + 2 \log r_2 = 2 \log(2r_2);$$

on the other hand,

$$y = q_2 r_1 + r_2 \geq r_1 + r_2 > 2r_2, \text{ so } 2\log y > 2\log(2r_2) \geq c(x, y). \quad \dashv$$

The lower bounds for the complexity measure $c_{\{\text{rem}\}}(\varepsilon, x, y)$ are best expressed in terms of the classical *Fibonacci sequence*, defined by the recursion

$$F_0 = 0, \ F_1 = 1, \ F_{k+2} = F_k + F_{k+1}, \tag{1C-9}$$

so that $F_2 = 0 + 1 = 1$, $F_3 = 1 + 1 = 2$, $F_4 = 3$, $F_5 = 5$, etc. We leave them for the problems.

Coprimeness by the Euclidean. In the formal terminology that we will introduce in the next chapter, the Euclidean is a recursive algorithm of the *structure* $(\mathbb{N}, \text{rem}, \text{eq}_0)$. If we use it to check the coprimeness relation, we also need to test at the end whether $\gcd(x, y) = 1$, so that as a decision method for coprimeness, the Euclidean is a recursive algorithm of the structure

$$\mathbf{N}_\varepsilon = (\mathbb{N}, \text{rem}, \text{eq}_0, \text{eq}_1). \tag{1C-10}$$

As we mentioned in the Introduction, it is not known whether the Euclidean algorithm is optimal (in any natural sense) among algorithms from its natural primitives, either for computing the gcd or for deciding coprimeness. One of our main aims is to make these questions precise and establish the strongest, known partial results about them.

The binary (Stein) algorithm. This modern algorithm computes $\gcd(x, y)$ and decides $x \perp\!\!\!\perp y$ in $O(\log x + \log y)$ steps, from "linear" operations, which are much simpler than division.

1C.5. **Proposition** (Stein [1967], Knuth [1981], Sect. 4.5.2). *The gcd satisfies the following recursive equation for $x, y \geq 1$, by which it can be computed in $O(\log x + \log y)$ steps:*

$$\gcd(x, y) = \begin{cases} x & \text{if } x = y, \\ 2\gcd(x/2, y/2) & \text{otherwise, if parity}(x) = \text{parity}(y) = 0, \\ \gcd(x/2, y) & \text{otherwise, if parity}(x) = 0, \text{parity}(y) = 1, \\ \gcd(x, y/2) & \text{otherwise, if parity}(x) = 1, \text{parity}(y) = 0, \\ \gcd(x \mathbin{\dot{-}} y, y) & \text{otherwise, if } x > y, \\ \gcd(x, y \mathbin{\dot{-}} x) & \text{otherwise.} \end{cases}$$

PROOF. That the gcd satisfies these equations and is determined by them is trivial. To check the number of steps required, notice that (at worst) every other application of one of the clauses involves halving one of the arguments—the worst case being subtraction, which, however must then be immediately followed by a halving, since the difference of two odd numbers is even. $\dashv$

Anticipating again terminology from the next chapter, the Stein is a recursive algorithm of the structure

(1C-11) $$\mathbf{N}_{\mathrm{st}} = (\mathbb{N}, \mathrm{parity}, \mathrm{em}_2, \mathrm{iq}_2, \dot{-}, =, <)$$

whose primitives are *Presburger functions*, cf. page 44.

Horner's rule. For any field F, Horner's rule computes the value

$$V_F(a_0, \dots, a_n, x) = \chi(x) = a_0 + a_1 x + \cdots + a_n x^n \qquad (n \geq 1)$$

of a polynomial $\chi(x)$ of degree n using no more than n multiplications and n additions in F as follows:

$$\begin{aligned}
\chi_0(x) &= a_n, \\
\chi_1(x) &= a_{n-1} + x\chi_0(x) && = a_{n-1} + a_n x \\
&\vdots \\
\chi_j(x) &= a_{n-j} + x\chi_{j-1}(x) && = a_{n-j} + a_{n-j+1}x + \cdots + a_n x^j \\
&\vdots \\
\chi(x) = \chi_n(x) &= a_0 + x\chi_{n-1}(x) && = a_0 + a_1 x + \cdots + a_n x^n.
\end{aligned}$$

This is an example of a simple but important *straight line algorithm* from the field primitives of F. It can also be used to decide the (plausibly simpler) *nullity relation* of degree n on F,

(1C-12) $$N_F(a_0, \dots, a_n, x) \iff a_0 + a_1 x + \cdots + a_n x^n = 0,$$

from the primitives of (the expansion of F by the identity relation on) $\mathbf{F}$,

$$\mathbf{F} = (F, 0, 1, +, -, \cdot, \div, =)$$

by adding a last line[7]

$$N_F(x) = \mathrm{eq}_F(\chi_n(x), 0) = \mathrm{eq}_F(a_0 + a_1 x + \cdots + a_n x^n, 0).$$

It is known that Horner's rule is optimal for many fields and inputs, both for the number of multiplications and the number of additions that are needed to compute $V_F(\vec{a}, x)$ or to decide $N_F(\vec{a}, x)$, in fact the earliest results on this (from the 1960s) were the first significant lower bounds for natural problems in algebra. We will establish some of them in Chapter 9.

Problems for Section 1C

x1C.1. **Problem.** Prove that if $x > v_0 > v_1 > \cdots > v_{n-1}$, then the computation of merge$((x), v)$ by (1C-2) will require n comparisons.

In the next two problems we define and analyze a simple algorithm for sorting, which is much less efficient than the merge-sort.

[7] We sometimes use $\mathrm{eq}_A(x, y)$ rather than $x = y$, especially when the infix notation $x = y$ might cause confusion.

x1C.2. **Problem.** Prove that the equation

$$\text{insert}(x, u) = \text{if } (|u| = 0) \text{ then } (x) \quad \text{else if } x \le u_0 \text{ then } (x) * u \quad \text{else } (u_0) * \text{insert}(x, \text{tail}(u)) \tag{1C-13}$$

determines a value $\text{insert}(x, u) \in L^*$ for any $x \in L$ and $u \in L^*$, and if u is sorted, then

$$\text{insert}(x, u) = \text{sort}((x) * u). \tag{1C-14}$$

Moreover, $\text{insert}(x, u)$ can be computed from (1C-13) using no more than $|u|$ comparisons.

x1C.3. **Problem** (The insert-sort algorithm). Prove that the sort function satisfies the equation

$$\text{sort}(u) = \text{if } |u| \le 1 \text{ then } u \quad \text{else insert}(u_0, \text{sort}(\text{tail}(u))), \tag{1C-15}$$

and can be computed from (1C-15) and (1C-13) using no more than $\frac{1}{2}|u|(|u| - 1)$ comparisons. Illustrate the computation with some examples, and show also that if u is inversely ordered, then this computation of $\text{sort}(u)$ requires exactly $\frac{1}{2}|u|(|u| - 1)$ comparisons.

To see the difference between the merge-sort and the insert-sort, note that when $|u| = 64 = 2^6$, then the insert-sort may need as many as 2016 comparisons, while the merge-sort will need no more than 384. On the other hand, as the next two problems show, there is nothing wrong with the idea of sorting by repeated inserting—it is only that (1C-13) expresses a very inefficient algorithm for insertion.

x1C.4*. **Problem** (Binary insertion). Prove that the equation

$$\begin{aligned} \text{binsert}(x, u) = \; & \text{if } (|u| = 0) \text{ then } (x) \\ & \text{else if } (x \le \text{half}_2(u)_0) \\ & \qquad \text{then binsert}(x, \text{half}_1(u)) * \text{half}_2(u) \\ & \text{else half}_1(u) * (\text{half}_2(u)_0) * \text{binsert}(x, \text{tail}(\text{half}_2(u))) \end{aligned}$$

determines a value $\text{binsert}(x, u) \in L^*$ for every $x \in L$ and $u \in L^*$, and if u is sorted, then

$$\text{binsert}(x, u) = \text{insert}(x, u) = \text{sort}((x) * u).$$

Moreover, $\text{binsert}(x, u)$ can be computed using (for $|u| > 0$) no more than $b(|u|)$ comparisons, where

$$b(m) = \begin{cases} \log m + 1, & \text{if } m \text{ is a power of 2,} \\ \lceil \log m \rceil, & \text{otherwise.} \end{cases}$$

x1C.5*. **Problem** (Binary-insert-sort). Prove that the sort function satisfies the equation

$$\text{(1C-16)} \qquad \text{sort}(u) = \text{if } |u| \le 1 \text{ then } u \\ \text{else binsert}(u_0, \text{sort}(\text{tail}(u))),$$

and can be computed from (1C-16) and the equation in Problem x1C.4* using no more than $s(|u|)$ comparisons, where for $m > 0$,

$$\text{(1C-17)} \qquad s(m) = \lceil \log((m-1)!) \rceil + (m-1) \le \log((m-1)!) + m.$$

x1C.6*. **Problem.** For the function $s(m)$ defined in (1C-17), prove that

$$\lim_{m\to\infty} \tfrac{s(m)}{\log(m!)} = 1.$$

By Stirling's formula (1A-4), $m \log m \sim_\infty \log(m!)$, and so the merge-sort and the binary-insert-sort algorithms are asymptotically equally efficient for the required number of comparisons.

We next turn to some problems related to the Euclidean algorithm.

Recall the definition of the Fibonacci sequence $\{F_k\}_k$ in (1C-9).

x1C.7. **Problem.** Prove that if $\varphi = \frac{1}{2}(1+\sqrt{5})$ is the positive root of the quadratic equation $x^2 = x + 1$, then for all $k \ge 2$,

$$\varphi^{k-2} \le F_k \le \varphi^k.$$

x1C.8. **Problem.** Prove that if $\varphi = \frac{1+\sqrt{5}}{2}$ and $\hat{\varphi} = \frac{1-\sqrt{5}}{2}$ are the two roots of the quadratic equation $x^2 = x + 1$, then $1 < \varphi < 2$, $|\hat{\varphi}| < 1$. and for all k,

$$F_k = \frac{\varphi^k - \hat{\varphi}^k}{\sqrt{5}} \ge \frac{\varphi^k}{\sqrt{5}} - 1.$$

HINT: Use induction on k for the equation, and infer the inequality from the fact that $\left|\frac{\hat{\varphi}^k}{\sqrt{5}}\right| < 1$.

x1C.9. **Problem.** Prove that successive Fibonacci numbers F_k, F_{k+1} with $k \ge 2$ are relatively prime, and $c_{\{\text{rem}\}}(\varepsilon, F_{k+1}, F_k) = k - 1$.

x1C.10. **Problem** (Lamé's Lemma). Prove that if $y \le F_k$ with $k \ge 2$, then, for every $x \ge y$, $c(\varepsilon, x, y) \le k - 1$. HINT: Use induction on $k \ge 2$, checking separately (by hand) the two basis cases $k = 2, 3$ and imitating the argument in the proof of Proposition 1C.4.

Lamé's Lemma predicts the following upper bounds for $c(\varepsilon, x, y)$ for small values of y (and any $x \geq y$):

Values of y	$c(\varepsilon, x, y)$
1	1
2	2
3	3
4 - 5	4
6 - 8	5
9 - 13	6

These are a bit better than the simple $2\log y$ bound. The next two problems clarify the situation, but require some arithmetic (of the sort that we will often "leave for an exercise"):

x1C.11. **Problem.** Prove that if $x \geq y \geq 2$, then

$$c(\varepsilon, x, y) \leq \frac{\log(\sqrt{5}y)}{\log\varphi},$$

where φ is the positive root of $x + 1 = x^2$.

x1C.12*. **Problem.** Prove that for all real numbers $y \geq 16$,

$$\frac{\log(\sqrt{5}y)}{\log\varphi} < 2\log y.$$

HINT: Check the inequality by hand for $y = 16$, and then check (using the Mean value Theorem) that the function

$$f(y) = 2\log y - \frac{\log(\sqrt{5}y)}{\log\varphi}$$

on $\mathbb{R}$ is increasing for $y > 0$.

x1C.13. **Problem** (Bezout's Lemma). Prove that for all natural numbers $x, y \geq 1$, there exist integers $\alpha, \beta \in \mathbb{Z}$ such that

$$\gcd(x, y) = \alpha x + \beta y.$$

In fact, we can set $\alpha = \alpha(x, y)$, $\beta = \beta(x, y)$ where the functions

$$\alpha, \beta : \mathbb{N} \times \mathbb{N} \to \mathbb{Z} = \{\dots, -2, -1, 0, 1, 2, \dots\}$$

are the least solutions of the following system of recursive equations, for $x \geq y \geq 1$:

$$\begin{aligned}
\alpha(x, y) &= \text{ if } (y \mid x) \text{ then } 0 \text{ else } \beta(y, \text{rem}(x, y)),\\
\beta(x, y) &= \text{ if } (y \mid x) \text{ then } 1\\
&\qquad \text{else } \alpha(y, \text{rem}(x, y)) - \text{iq}(x, y)\beta(y, \text{rem}(x, y)).
\end{aligned}$$

Use this recursion to express $\gcd(231, 165)$ as an integer, linear combination of 231 and 165.

x1C.14. **Problem.** Prove that two numbers $x, y \geq 1$ are coprime if and only if there exist integers $\alpha, \beta \in \mathbb{Z}$ such that $1 = \alpha x + \beta y$.

x1C.15. **Problem.** For positive numbers, show: if $x \perp\!\!\!\perp a$ and $x \mid ab$, then $x \mid b$.

x1C.16. **Problem.** Prove that for all $x \geq y \geq 1$, there are infinitely many choices of integers α and β such that

$$\gcd(x, y) = \alpha x + \beta y,$$

but only one choice such that $0 \leq \alpha < \dfrac{y}{\gcd(x, y)}$.

x1C.17*. **Problem.** Define an algorithm from the primitives of a field F which decides the nullity relation (1C-12) of degree n using no more than $n - 1$ additions, along with multiplications and equality tests, and count how many multiplications and equality tests it uses. Hint: Consider separately the cases where the characteristic of F is 2 or $\neq 2$ and show first that you can test whether $a + bx = 0$ using no additions or subtractions, just multiplications and equality tests.

1D. Partial structures

To use the basic notions of equational logic in the study of recursion and computation, we must introduce two small wrinkles: allow the interpretations of function symbols by partial functions, since computations often diverge, and add *branching* (conditionals) to the term-formation rules. We deal with the first of these here and will move to the second in the next section.

(Many-sorted, partial) structures. A pair (S, Φ) is a *vocabulary* if the *set of sorts* S is not empty, containing in particular the *boolean sort* `boole` and at least one other sort, and Φ is a finite set of function symbols, each with an assigned *type* of the form

$$\text{type}(\phi) = (s_1, \dots, s_n, \text{sort}(\phi))$$

where $s_1, \dots, s_n \in S \setminus \{\texttt{boole}\}$ and $\text{sort}(\phi) \in S$.

A (partial) (S, Φ)-*structure* is a pair

$$\mathbf{A} = (\{A_s\}_{s \in S}, \boldsymbol{\Phi}) = (\{A_s\}_{s \in S}, \{\phi^{\mathbf{A}}\}_{\phi \in \Phi}), \tag{1D-1}$$

where each A_s is a set; and for each $\phi \in \Phi$,

$$\text{if type}(\phi) = (s_1, \dots, s_n, s), \text{ then } \phi^{\mathbf{A}} : A_{s_1} \times \cdots \times A_{s_n} \rightharpoonup A_s.$$

For $s \neq \texttt{boole}$, the convergent objects $\phi^{\mathbf{A}}$ with $\text{type}(\phi) = (s)$ are the *distinguished elements of sort* s of $\mathbf{A}$.

Φ-structures. Most often there is just one sort ind (other than boole): we describe these structures as in model theory, by identifying the *universe* (of individuals) $A = A_{\mathtt{ind}}$, listing $\mathbf{\Phi}$, and letting the notation suggest

$$\mathrm{type}(\phi) = (\underbrace{\mathtt{ind}, \dots , \mathtt{ind}}_{n_\phi}, s)$$

for every $\phi \in \Phi$—now called the *vocabulary*. The number n_ϕ is the *arity* of ϕ, and we allow $\mathrm{arity}(\phi) = 0$, in which case ϕ is a (boolean or individual, perhaps diverging) constant.

A structure is *relational* if all its primitives are of boolean sort, e.g., if $\mathbf{A} = (A, \leq)$ is an ordering.

Typical are the basic structures of unary and binary arithmetic

(1D-2) $\quad \mathbf{N}_u = (\mathbb{N}, 0, S, \mathrm{Pd}, \mathrm{eq}_0), \quad \mathbf{N}_b = (\mathbb{N}, 0, \mathrm{parity}, \mathrm{iq}_2, \mathrm{em}_2, \mathrm{om}_2, \mathrm{eq}_0),$

where

$$\mathrm{em}_2(x) = 2x, \ \mathrm{om}_2(x) = 2x + 1$$

are the operations of *even* and *odd multiplication by* 2. More generally, for any $k \geq 3$, the structure of *k-ary arithmetic* is

(1D-3) $$\mathbf{N}_{k\text{-}ary} = (\mathbb{N}, 0, \mathrm{m}_{k,0}, \dots , \mathrm{m}_{k,k-1}, \mathrm{iq}_k, \mathrm{rem}_k, \mathrm{eq}_0),$$

where $\mathrm{m}_{k,i}(x) = kx + i$, $\mathrm{iq}_k(x) = \mathrm{iq}(x, k)$ and $\mathrm{rem}_k(x) = \mathrm{rem}(x, k)$. These are *total structures*, as is the standard structure of Peano arithmetic

(1D-4) $$\mathbf{N} = (\mathbb{N}, 0, 1, +, \cdot, =),$$

as is the *Lisp structure* of strings (or lists) from a set L,

(1D-5) $$\mathbf{L}^* = (L^*, \mathrm{nil}, \mathrm{eq}_{\mathrm{nil}}, \mathrm{head}, \mathrm{tail}, \mathrm{cons}),$$

defined on page 12 (with each $x \in L$ identified with the string (x)).

The Euclidean structure

(1D-6) $$\mathbf{N}_\varepsilon = (\mathbb{N}, \mathrm{rem}, \mathrm{eq}_0, \mathrm{eq}_1)$$

is partial, because $\mathrm{rem}(x, y)$ converges only when $y \neq 0$, and so is any *field*

$$\mathbf{F} = (F, 0, 1, +, -, \cdot, \div, =)$$

(including the reals $\mathbf{R}$ and the complexes $\mathbf{C}$) because the quotient $x \div y$ converges only when $y \neq 0$.

There are many interesting examples of many-sorted structures, e.g., a *vector space* V over a field F

$$\mathbf{V} = (V, F, 0_F, 1_F, +_F, -_F, \cdot_F, \div_F, 0_V, +_V, -_V, \cdot)$$

where the last primitive $\cdot : F \times V \to V$ is scalar-vector multiplication and the other symbols have their natural meanings. On the other hand, dealing directly with many sorts is tedious, and we will work with one-sorted Φ-structures. The more general versions follow by "identifying" a many-sorted **A** as in (1D-1) with the single-sorted

$$(1D\text{-}7) \qquad (\uplus_{s\in S'} A_s, \{A'_s : s \in S'\}, \mathbf{\Phi}) \quad (S' = S \setminus \{\texttt{boole}\}),$$

where $\uplus_{s\in S'} A_s = \bigcup\{(s,x) : s \in S' \,\&\, x \in A_s\}$ is the *disjoint union* of the basic universes of **A**, $A'_s(t,x) \Leftrightarrow [t = s \,\&\, x \in A_s]$ for $s \neq \texttt{boole}$, and the primitives in $\mathbf{\Phi}$ are defined in the obvious way and diverge on arguments not of the appropriate kind. We will often assume this representation of many-sorted structures without explicit mention.

Caution! Contrary to the usual assumption in logic, we have allowed the *empty Φ-structure* $\mathbf{A} = (\emptyset, \{\phi^{\mathbf{A}}\}_{\phi\in\Phi})$, in which every $\phi^{\mathbf{A}}$ is an "empty function". The empty structure is of no interest, of course, but it comes up in constructions of families of structures, where $A_t = \{s \in B : P(s,t)\}$ with some $P(s,t)$ for which it is difficult (perhaps effectively impossible) to check whetherA_t is empty. The convention is unusual but does not cause any problems.

Unified notation for functions and relations. There are still two sorts in Φ-structures, ind and boole, and we will need to deal with both partial functions and relations of all arities on their universe. Typically we will just call all these objects (partial) *functions* and write

$$(1D\text{-}8) \qquad f : A^n \rightharpoonup A_s \quad (s \in \{\texttt{ind}, \texttt{boole}\}, A_{\texttt{ind}} = A, A_{\texttt{boole}} = \mathbb{B}),$$

most often skipping the tiresome side notation which explains what this "s" stands for.

Restrictions. If $\mathbf{A} = (A, \mathbf{\Phi})$ is a Φ-structure and $U \subseteq A = A_{\texttt{ind}}$, we set

$$\mathbf{A} \restriction U = (U, \{\phi^{\mathbf{A}} \restriction U\}_{\phi\in\Phi}),$$

where, for any $f : A^n \rightharpoonup A_s$,

$$f \restriction U(x_1, \dots, x_n) = w \iff x_1, \dots, x_n \in U, w \in U_s \,\&\, f(x_1, \dots, x_n) = w.$$

Expansions and reducts. An *expansion* of a Φ-structure **A** is obtained by adding new primitives to **A**,

$$(\mathbf{A}, \mathbf{\Psi}) = (A, \mathbf{\Phi} \cup \mathbf{\Psi}).$$

Conversely, the *reduct* $\mathbf{A} \restriction \Phi_0$ of a structure[8] $\mathbf{A} = (A, \mathbf{\Phi})$ to a subset $\Phi_0 \subseteq \Phi$ of its vocabulary is defined by removing all the operations in $\mathbf{\Phi} \setminus \mathbf{\Phi}_0$. For example, the reduct of the field of real numbers to $\{0, +, -\}$ is the additive group on $\mathbb{R}$,

$$\mathbf{R} \restriction \{0, +, -\} = (\mathbb{R}, 0, +, -).$$

[8] It will be convenient to allow on occasion an infinite vocabulary Φ; and in these (clearly identified) cases, claims about Φ-structures can be interpreted, as usual, as claims about their reducts $\mathbf{A} \restriction \Phi_0$ to finite subsets of Φ.

Notice that we use the same symbol $\restriction$ for restriction to a subset and to a sub-vocabulary, but there is no danger of confusion, even though we will sometimes combine these two operations:

$$\mathbf{A} \restriction \Phi_0 \restriction U = (\mathbf{A} \restriction \Phi_0) \restriction U, \tag{1D-9}$$

so that, for example, $\mathbf{R} \restriction \{0, 1, +, -, \cdot\} \restriction \mathbb{Z} = (\mathbb{Z}, 0, 1, +, -, \cdot)$, the ring of integers.

Substructures. A (partial) *substructure* $\mathbf{U} \subseteq_p \mathbf{A}$ of a Φ-structure $\mathbf{A}$ is a structure of the same vocabulary Φ, such that $U \subseteq A$ and for every $\phi \in \Phi$, $\phi^{\mathbf{U}} \sqsubseteq \phi^{\mathbf{A}}$, i.e.,

$$\Big(\vec{x} \in U^n \;\&\; w \in U_s \;\&\; \phi^{\mathbf{U}}(\vec{x}) = w\Big) \Longrightarrow \phi^{\mathbf{A}}(\vec{x}) = w.$$

A substructure $\mathbf{U}$ is *strong* (or *induced*) if in addition, for all $\phi \in \Phi$,

$$\Big(\vec{x} \in U^n \;\&\; w \in U_s \;\&\; \phi^{\mathbf{A}}(\vec{x}) = w\Big) \Longrightarrow \phi^{\mathbf{U}}(\vec{x}) = w,$$

in which case $\mathbf{U} = \mathbf{A} \restriction U$.

Notice that *we do not require of a substructure* $\mathbf{U} \subseteq_p \mathbf{A}$ *that it be closed under the primitives of* $\mathbf{A}$—in particular, it need not contain all the distinguished elements of $\mathbf{A}$. This is contrary to the usual terminology in mathematics and logic, where, for example, a *subfield* of a field F must (by definition) contain 0, 1 and be closed under $+. -, \cdot$ and $\div$.

Diagrams. The (equational) *diagram* of a Φ-structure $\mathbf{A}$ is the set

$$\mathrm{eqdiag}(\mathbf{A}) = \{(\phi, \vec{x}, w) : \phi \in \Phi, \vec{x} \in A^n, w \in A_{\mathrm{sort}(\phi)} \text{ and } \phi^{\mathbf{A}}(\vec{x}) = w\},$$

where for nullary ϕ, the entry in $\mathrm{eqdiag}(\mathbf{A})$ is $(\phi, \phi^{\mathbf{A}})$.

For example, we might have

$$\mathrm{eqdiag}(\mathbf{U}) = \{2 + 1 = 3,\ 2 + 3 = 5,\ 2 \leq 5,\ 5 \not\leq 1\}, \tag{1D-10}$$

where $\mathbf{U} \subseteq_p (\mathbb{N}, 0, 1, +, \cdot, \leq)$ with $2, 1, 3 \in U$; and we have used here the obvious conventions, to write

$$\phi(\vec{x}) = w,\ R(\vec{x}),\ \neg R(\vec{x})$$

rather than the more pedantic

$$(\phi, \vec{x}, w),\ (R, \vec{x}, \mathrm{tt}),\ (R, \vec{x}, \mathrm{ff})$$

and to use "infix notation", i.e., write $x + y$ rather than $+(x, y)$.

Structures are most easily specified by giving their universe and their equational diagram—especially if they are finite; and clearly,

$$\mathbf{U} \subseteq_p \mathbf{A} \iff U \subseteq A \;\&\; \mathrm{eqdiag}(\mathbf{U}) \subseteq \mathrm{eqdiag}(\mathbf{A}).$$

Sometimes we will specify a structure $\mathbf{A}$ by just giving its diagram, the implication being that its universe A comprises those members of A which occur in $\mathrm{eqdiag}(\mathbf{A})$.

Homomorphisms and embeddings. A *homomorphism* $\pi : \mathbf{U} \to \mathbf{V}$ of one Φ-structure into another is any mapping $\pi : U \to V$ such that

$$\phi^{\mathbf{U}}(\vec{x}) = w \Longrightarrow \phi^{\mathbf{V}}(\pi(\vec{x})) = \pi(w) \quad (\phi \in \Phi), \tag{1D-11}$$

where $\pi(\vec{x}) = (\pi(x_1), \dots, \pi(x_n))$ and, by convention $\pi(\mathrm{tt}) = \mathrm{tt}, \pi(\mathrm{ff}) = \mathrm{ff}$, so that for partial relations it insures

$$R^{\mathbf{U}}(\vec{x}) \Longrightarrow R^{\mathbf{V}}(\pi(\vec{x})), \neg R^{\mathbf{U}}(\vec{x}) \Longrightarrow \neg R^{\mathbf{V}}(\pi(\vec{x})).$$

A homomorphism is an *embedding* $\pi : \mathbf{U} \rightarrowtail \mathbf{V}$ if it is injective (one-to-one), and it is an *isomorphism* $\pi : \mathbf{U} \rightarrowtail\!\!\!\!\rightarrow \mathbf{V}$ if it is a surjective embedding and, in addition, the inverse map $\pi^{-1} : U \rightarrowtail\!\!\!\!\rightarrow V$ is also an embedding. Clearly

$$\mathbf{U} \subseteq_p \mathbf{V} \iff U \subseteq V \text{ and the identity } \mathrm{id}_U : U \rightarrowtail V \text{ is an embedding.}$$

If $\pi : \mathbf{U} \to \mathbf{A}$ is a homomorphism, then $\pi[\mathbf{U}]$ is the substructure of $\mathbf{A}$ with universe $\pi[U]$ and

$$\mathrm{eqdiag}(\pi[\mathbf{U}]) = \{(\phi, \pi(\vec{x}), \pi(w)) : (\phi, \vec{x}, w) \in \mathrm{eqdiag}(\mathbf{U})\}.$$

This construction is especially useful when $\pi : \mathbf{U} \rightarrowtail \mathbf{A}$ is an embedding, in which case $\pi : \mathbf{U} \rightarrowtail\!\!\!\!\rightarrow \pi[\mathbf{U}]$ is an isomorphism.

These notions of structures and homomorphisms are natural in our context, but they have unfamiliar-looking consequences; e.g., there are (obviously) no homomorphisms $\pi : \mathbf{N} \to \mathbf{N}$ of the Peano structure other than the identity, but there are many (peculiar but ultimately useful) embeddings $\pi : \mathbf{U} \to \mathbf{N}$ of finite substructures of $\mathbf{N}$ into $\mathbf{N}$, cf. Problem x1D.1.

Substructure generation. For a fixed Φ-structure $\mathbf{A}$ and any $X \subseteq A$, we set

$$\begin{aligned} G_0[X] &= X, \\ G_{m+1}[X] &= G_m[X] \bigcup \{w \in A : \text{ for some } \vec{u} \in G_m[X]^n \text{ and } \phi \in \Phi, w = \phi^{\mathbf{A}}(\vec{u})\}, \\ G_\infty[X] &= \bigcup\nolimits_m G_m[X]. \end{aligned}$$

$G_m[X]$ is *the subset of A generated in $\mathbf{A}$ by X in m steps*, and by a simple induction on m,

$$G_{m+k}[X] = G_m[G_k[X]]. \tag{1D-12}$$

The *substructure of* $\mathbf{A}$ *generated by X in m steps* is

$$\mathbf{G}_m[X] = \mathbf{A} \restriction G_m[X], \tag{1D-13}$$

so that

$$\begin{aligned} &\mathrm{eqdiag}(\mathbf{G}_m[X])) \\ &\quad = \{(\phi, u_1, \dots, u_{n_i}, w) \in \mathrm{eqdiag}(\mathbf{A}) : u_1, \dots, u_{n_i}, w \in G_m[X] \cup \mathbb{B}\}. \end{aligned}$$

For a tuple $\vec{x} = (x_1, \dots, x_n) \in A^n$, we write

$$G_m(\vec{x}) = G_m[\{x_1, \dots, x_n\}], \quad G_\infty(\vec{x}) = \bigcup_m G_m(\vec{x}),$$
$$\mathbf{G}_m(\vec{x}) = \mathbf{A} \restriction G_m[\{x_1, \dots, x_n\}], \quad \mathbf{G}_\infty(\vec{x}) = \mathbf{A} \restriction G_\infty[X],$$

and if the structure in which these sets and substructures are computed is not obvious from the context, we write $G_m(\mathbf{A}, \vec{x})$, $\mathbf{G}_m(\mathbf{A}, \vec{x})$, etc.

A structure $\mathbf{A}$ is *generated by* $\vec{x}$ if $\mathbf{A} = \mathbf{G}_\infty(\mathbf{A}, \vec{x})$, so that if it is also finite, then $\mathbf{A} = \mathbf{G}_m(\mathbf{A}, \vec{x})$ for some m.

Certificates. A (Φ-)*certificate* is a pair $(\mathbf{U}, \vec{x})$, such that

(1D-14) $\quad$ $\mathbf{U}$ is a finite Φ-structure, $\vec{x} \in U^n$, and $\mathbf{U} = \mathbf{G}_\infty(\mathbf{U}, \vec{x})$;

and it is a certificate *of* or *in* a Φ-structure $\mathbf{A}$ if, in addition, $\mathbf{U} \subseteq_p \mathbf{A}$.

Notice that *if* $(\mathbf{U}, \vec{x})$ *is a certificate, then* $\mathbf{U} = \mathbf{G}_m(\mathbf{U}, \vec{x})$ for some finite m. We will motivate the terminology in Section 4D but, basically, we will use certificates $(\mathbf{U}, \vec{x})$ to model *computations on the input* $\vec{x}$.

Certificates carry some natural measures of "size" or *complexity*, e.g.,

$$\text{depth}(\mathbf{U}, \vec{x}) = \min\{m : \mathbf{U} = \mathbf{G}_m(\mathbf{U}, \vec{x})\},$$
$$\text{values}(\mathbf{U}, \vec{x}) = |\{w \in U : (\phi, \vec{u}, w) \in \text{eqdiag}(\mathbf{U}) \text{ for some } \phi, \vec{u}\}|,$$
$$\text{calls}(\mathbf{U}, \vec{x}) = |\text{eqdiag}(\mathbf{U})|.$$

Notice that depth$(\mathbf{U}, \vec{x})$ depends on both $\mathbf{U}$ and $\vec{x}$, while values$(\mathbf{U}, \vec{x})$ and calls$(\mathbf{U}, \vec{x})$ depend only on $\mathbf{U}$.

The meanings of the first and last of these measures are obvious, and the intermediate values$(\mathbf{U}, \vec{x})$ counts the number of *distinct values* of entries in eqdiag$(\mathbf{U})$; it may be smaller than the universe U or the number of entries in eqdiag$(\mathbf{U})$, for example if $(\mathbf{U}, (2, 3))$ is the certificate of the Peano structure $\mathbf{N}$ with

$$U = \{2, 3, 5, 6\}, \ \text{eqdiag}(\mathbf{U}) = \{2 + 3 = 5, 3 + 2 = 5, 3 + 3 = 6\},$$

cf. Problem x1D.6.

We also need the *depth of an element below a tuple*,

(1D-15) $\quad$ $\text{depth}(w; \mathbf{A}, \vec{x}) = \min\{m : w \in G_m(\mathbf{A}, \vec{x})\}, \quad (w \in G_\infty(\mathbf{A}, \vec{x})),$

and by convention,

$$\text{depth}(\mathrm{tt}; \mathbf{A}, \vec{x}) = \text{depth}(\mathrm{ff}; \mathbf{A}, \vec{x}) = 0.$$

Clearly, depth$(x_i; \mathbf{A}, \vec{x}) = 0$, and if $\phi^{\mathbf{A}}(u_1, \dots, u_{n_\phi})\downarrow$, then

$$\text{depth}(\phi^{\mathbf{A}}(u_1, \dots, u_{n_\phi}); \mathbf{A}, \vec{x}) \leq \max\{\text{depth}(u_i; \mathbf{A}, \vec{x}) : i = 1, \dots, n_\phi\} + 1.$$

The inequality is strict if sort$(\phi) = \texttt{boole}$ or

$$m = \max\{\text{depth}(u_i; \mathbf{A}, \vec{x}) : i = 1, \dots, n_\phi\} \Longrightarrow \phi(\vec{u}) \in G_m(\mathbf{A}, \vec{x}).$$

1D.1. **Proposition.** *If* $(\mathbf{U}, \vec{x})$ *is a certificate of* $\mathbf{A}$ *and* $w \in U$, *then*

$$\begin{aligned}(1\text{D-16}) \quad \text{depth}(w; \mathbf{A}, \vec{x}) &\leq \text{depth}(w; \mathbf{U}, \vec{x}) \\ &\leq \text{depth}(\mathbf{U}, \vec{x}) \leq \text{values}(\mathbf{U}, \vec{x}) \leq \text{calls}(\mathbf{U}, \vec{x}).\end{aligned}$$

PROOF is easy and we leave it for Problem x1D.7. ⊣

Problems for Section 1D

x1D.1. **Problem.** (1) Give an example of an embedding $\phi : \mathbf{U} \rightarrowtail \mathbf{V}$ of one Φ-structure into another which is bijective but not an isomorphism.

(2) Let $\mathbf{U}$ be the finite substructure of the Peano structure $\mathbf{N}$ with universe $U = \{1, 2, 4\}$ and $\text{eqdiag}(\mathbf{U}) = \{1 + 1 = 2,\ 2 \cdot 2 = 4\}$, define $\pi : U \to \mathbb{N}$ by $\pi(1) = 2, \pi(2) = 4, \pi(4) = 16$, and prove that $\pi : \mathbf{U} \to \mathbf{N}$ is an embedding.

x1D.2. **Problem.** Let $\mathbf{A} = (A, \phi, \psi)$ where ϕ is total, unary of sort `ind` and ψ is total, unary of sort `boole`, and suppose x, y, u are distinct elements of A such that

$$\phi(x) = u, \phi(y) = x, \phi(u) = x, \psi(x) = \psi(y) = \psi(u) = \mathrm{tt}.$$

Compute $\mathbf{G}_0(x, y)$, $\mathbf{G}_1(x, y)$, $\mathbf{G}_2(x, y)$ and $\mathbf{G}_\infty(x, y)$ by giving their universes and equational diagrams.

x1D.3. **Problem.** Prove that for every Φ-structure $\mathbf{A}$, $\vec{x} \in A^n$ and m,

$$\begin{aligned}\mathbf{U} = \mathbf{G}_m(\mathbf{A}, \vec{x}) \iff & U = G_m(\mathbf{A}, \vec{x}) \\ & \& \ (\forall \vec{u} \in U^k, \phi \in \Phi)\Big(\phi^{\mathbf{A}}(\vec{u})\downarrow \implies \phi^{\mathbf{A}}(\vec{u}) \in U \cup \mathbb{B}\Big).\end{aligned}$$

Infer that for all m, $\mathbf{G}_m(\mathbf{A}, \vec{x}) \subseteq_p \mathbf{G}_{m+1}(\mathbf{A}, \vec{x})$.

x1D.4. **Problem.** Let $\mathbf{U}$ be the finite substructure of the Peano structure $\mathbf{N}$ specified by

$$U = \{1, 2, 4, 8, 24\},\ \text{eqdiag}(\mathbf{U}) = \{1 + 1 = 2, 4 + 4 = 8, 2 \cdot 4 = 8, 24 \neq 8\}.$$

(1) Prove that $\mathbf{G}_\infty(\mathbf{U}, \{1\}) \subsetneq_p \mathbf{U}$, so that $\mathbf{U}$ is not generated by $\{1\}$.

(2) Prove that $\mathbf{G}_\infty(\mathbf{U}, \{1, 4\}) \subsetneq_p \mathbf{U}$, so that $\mathbf{U}$ is not generated by $\{1, 4\}$.

(3) Prove that $\mathbf{G}_1(\mathbf{U}, \{1, 4, 24\}) = \mathbf{G}_\infty(\mathbf{U}, \{1, 4, 24\}) = \mathbf{U}$, so $\{1, 4, 24\}$ generates $\mathbf{U}$.

(4) Set $\pi(1) = a, \pi(4) = b, \pi(24) = c$, where a, b, c are any numbers. What conditions must be satisfied by a, b, c so that π can be extended to a homomorphism $\pi : \mathbf{U} \to \mathbf{N}$? In particular, can $a \neq 1$?

x1D.5. **Problem.** Prove that for every certificate $(\mathbf{U}, \vec{x})$,

$$\text{values}(\mathbf{U}, \vec{x}) \leq |U| \leq \text{values}(\mathbf{U}, \vec{x}) + |\{x_1, \dots, x_n\}|,$$

(where $|\{x_1, \dots, x_n\}|$ may be smaller than n, because the x_i's need not be distinct).

x1D.6. **Problem.** Let $\mathbf{U}$ be the substructure of the Peano structure $\mathbf{N}$ specified by

$$U = \{2, 3, 5, 6\}, \ \text{eqdiag}(\mathbf{U}) = \{2 + 3 = 5, 3 + 2 = 5, 3 + 3 = 6\}.$$

Prove that $(\mathbf{U}, (2, 3))$ is a certificate in $\mathbf{N}$ and compute $|U|$, $\text{depth}(\mathbf{U}, (2, 3))$, $\text{values}(\mathbf{U}, (2, 3))$ and $\text{calls}(\mathbf{U}, (2, 3))$.

x1D.7. **Problem.** Prove Proposition 1D.1.

For some constructions with equational diagrams, it is useful to think of them as sequences (rather than sets), as in the following

x1D.8. **Problem.** Prove that if $(\mathbf{U}, \vec{x})$ is a certificate, then we can enumerate its diagram

$$\text{eqdiag}(\mathbf{U}) = \Big((\phi_0, \vec{u}_0, w_0), \dots, (\phi_m, \vec{u}_m, w_m)\Big),$$

so that each t in $\vec{u}_0$ is x_j for some j, and for each $s < m$, each t in the tuple $\vec{u}_{s+1}$ occurs in the sequence $\vec{x}, \vec{u}_0, w_0, \vec{u}_1, w_1, \dots, \vec{u}_s, w_s$; in particular, each structure $\mathbf{U}^s$ with

$$\text{eqdiag}(\mathbf{U}^s) = \Big((\phi_0, \vec{u}, w_0), \dots, (\phi_s, \vec{u}_s, w_s)\Big)$$

is generated by $\vec{x}$ and so $(\mathbf{U}^s, \vec{x})$ is a certificate.

x1D.9. **Problem.** Prove that for every (finite) vocabulary Φ, there is a number a such that for every Φ-structure $\mathbf{A}$, every $\vec{x} \in A^n$ and every m,

$$|G_m(\vec{x})| \le C^{2^{am}} \qquad (C = n + |\Phi|).$$

Give an example of a structure $\mathbf{A}$ where $|G_m(x)|$ cannot be bounded by a single exponential in m.

1E. Equational logic of partial terms with conditionals

With each Φ-structure $\mathbf{A}$, we associate a formal language of terms, an extension of the classical language of equational logic by a *branching* (*conditional*) construct.

Syntax. The Φ-*terms with parameters from* $\mathbf{A}$ (or $\mathbf{A}$-terms) are defined by the *structural recursion*

$$(\mathbf{A}\text{-terms}) \quad E :\equiv \mathrm{tt} \mid \mathrm{ff} \mid x \ (x \in A)$$
$$\mid \mathsf{v}_i \mid \phi(E_1, \dots, E_{n_\phi}) \mid \text{if } E_0 \text{ then } E_1 \text{ else } E_2,$$

where $\mathsf{v}_0, \mathsf{v}_1, \dots$ is a fixed sequence of formal individual variables of sort `ind` and $\phi(E_1, \dots, E_{n_\phi}) \equiv \phi$ when $n_\phi = 0$; in other words, the terms comprise the smallest set of strings which contains tt, ff, every $x \in A$ and every variable v_i (as strings of length 1), and is closed under the formation rules for application and branching.[9] There are obvious type restrictions: for branching, for example, it is required that $\mathrm{sort}(E_0) \equiv$ `boole`, $\mathrm{sort}(E_1) \equiv \mathrm{sort}(E_2)$, and then $\mathrm{sort}(\text{if } E_0 \text{ then } E_1 \text{ else } E_2) \equiv \mathrm{sort}(E_1) \equiv \mathrm{sort}(E_2)$.

The *parameters* of an **A**-term term E are the members of A which occur in it. A term E is *pure*—a Φ*-term*—if it has no parameters, and *closed* if no individual variables occur in it.

The *length* of a term E (as a string of symbols) and its *subterms* and *proper subterms* are defined as usual.

We will also need the terms without conditionals, which we will now call *algebraic* **A***-terms*. They are defined by the simpler recursion

(Algebraic **A**-terms) $\qquad E :\equiv \mathrm{tt} \mid \mathrm{ff} \mid \mathsf{v}_i \mid x \mid \phi(E_1, \dots, E_{n_\phi}).$

Notice that these include terms of sort `boole`, e.g., tt, ff and $\mathsf{R}(E_1, \dots, E_n)$ if $\mathsf{R} \in \Phi$ is of boolean sort, so they are more general than the usual terms of logic which are all of sort `ind`.

The *depth* of an algebraic term is defined by the recursion

$$\mathrm{depth}(\mathrm{tt}) = \mathrm{depth}(\mathrm{ff}) = \mathrm{depth}(\mathsf{v}_i) = \mathrm{depth}(x) = 0,$$
$$\mathrm{depth}(\phi(E_1, \dots, E_n)) = \max\{\mathrm{depth}(E_1), \dots, \mathrm{depth}(E_n)\} + 1.$$

Misspellings and abbreviations. Following common practice, we will simplify notation by "misspelling" (or "abbreviating") terms to help understanding: we will omit (or add) parentheses, use $x, y, x_1, \dots \mathsf{x}, \mathsf{y}, \mathsf{y}_1, \dots$ rather than the formal variables $\mathsf{v}_0, \mathsf{v}_1, \dots$, sometimes use infix notation

$$E + M :\equiv +(E, M)$$

for binary function symbols, etc. We also define the propositional connectives on terms of boolean sort using the conditional:

$$\text{(1E-1)} \quad \neg E :\equiv \text{if } E \text{ then ff else tt},$$
$$E_1 \,\&\, E_2 :\equiv \text{if } E_1 \text{ then } E_2 \text{ else ff},$$
$$E_1 \vee E_2 :\equiv \text{if } E_1 \text{ then tt else } E_2,$$
$$E_1 \to E_2 :\equiv \neg E_1 \vee E_2, \quad E_1 \leftrightarrow E_2 :\equiv (E_1 \to E_2) \,\&\, (E_2 \to E_1).$$

---

[9]We do not allow variables of boolean sort, a convenient choice which does not affect in any serious way the breadth of applicability of the results we will prove.

Semantics. For a fixed Φ-structure $\mathbf{A}$, we define

$$\mathrm{den} : \{\text{closed } \mathbf{A}\text{-terms}\} \rightharpoonup A \cup \mathbb{B}$$

by the obvious recursive clauses:

$$\mathrm{den}(\mathrm{tt}) = \mathrm{tt}, \quad \mathrm{den}(\mathrm{ff}) = \mathrm{ff}, \quad \mathrm{den}(x) = x \ (x \in A),$$

$$\mathrm{den}(\phi(M_1, \dots, M_{n_\phi})) = \phi^{\mathbf{A}}(\mathrm{den}(M_1), \dots, \mathrm{den}(M_{n_\phi}))$$

$$\mathrm{den}(\text{if } M_0 \text{ then } M_1 \text{ else } M_2) = \begin{cases} \mathrm{den}(M_1), & \text{if } \mathrm{den}(M_0) = \mathrm{tt}, \\ \mathrm{den}(M_2), & \text{if } \mathrm{den}(M_0) = \mathrm{ff} \\ \uparrow, & \text{otherwise.} \end{cases}$$

We call $\mathrm{den}(M)$ the *denotation* of the closed $\mathbf{A}$-term M (if $\mathrm{den}(M)\downarrow$), and in that case, clearly

$$\mathrm{sort}(M) = \texttt{boole} \Longrightarrow \mathrm{den}(M) \in \mathbb{B}, \quad \mathrm{sort}(M) = \texttt{ind} \Longrightarrow \mathrm{den}(M) \in A.$$

When we need to exhibit the structure in which the denotation is computed, we write $\mathrm{den}(\mathbf{A}, M)$ or we use model-theoretic notation,

(1E-2) $\quad \mathbf{A} \models E = M \iff_{\mathrm{df}} \mathrm{den}(\mathbf{A}, E) = \mathrm{den}(\mathbf{A}, M) \quad (E, M \text{ closed}).$

Partiality introduces some complications which deserve notice. For example, if we view subtraction as a partial function on $\mathbb{N}$, then for all $x, y, z \in \mathbb{N}$,

$$(\mathbb{N}, 0, 1, +, -) \models (x + y) - y = x;$$

but if $x < y$, then

$$(\mathbb{N}, 0, 1, +, -) \not\models (x - y) + y = x$$

because $(x - y)\uparrow$—and then, by the strictness of composition, $(x - y) + y \uparrow$ also. On the other hand,

$$\mathrm{den}(M_0) = \mathrm{tt} \Longrightarrow \mathrm{den}(\text{if } M_0 \text{ then } M_1 \text{ else } M_2) = \mathrm{den}(M_1),$$

whether $\mathrm{den}(M_2)$ converges or not.

Extended terms. An *extended term* of arity n is a pair

$$E(\mathsf{x}_1, \dots, \mathsf{x}_n) :\equiv (E, (\mathsf{x}_1, \dots, \mathsf{x}_n))$$

of a term E and a sequence of distinct variables which includes all the variables that occur in E. The notion provides a useful notation for substitution: if $M_1, \dots, M_n$ are terms of sort $\texttt{ind}$, then

$$\begin{aligned} E(M_1, \dots, M_n) &:\equiv E\{\mathsf{x}_1 :\equiv M_1, \dots, \mathsf{x}_n :\equiv M_n\} \\ &= \text{the result of replacing every occurrence of each } \mathsf{x}_i \text{ in } E \text{ by } M_i \end{aligned}$$

which is also (easily) a term. In particular, if $x_1, \dots, x_n \in A$, then

$$E(x_1, \dots, x_n) \equiv E\{\mathsf{x}_1 :\equiv x_1, \dots, \mathsf{x}_n :\equiv x_n\}$$

is the closed $\mathbf{A}$-term constructed by replacing each x_i by x_i.

Extended terms provide a notational convention which facilitates dealing with substitutions and the pedantic distinction between "terms" and "extended terms" is not always explicitly noted: we will sometimes refer to "a term $E(\vec{x})$", letting the notation indicate that we are specifying both a term E and a list $\vec{x} = (x_1, \dots, x_n)$ of distinct variables, or refer to "a term E" when we mean an extended term $E(\vec{x})$, assuming that the variable list $\vec{x}$ is supplied by the context. The default is that by "term" we mean "extended term".

We extend the notation in (1E-2) to terms which need not be closed by

(1E-3) $$\mathbf{A} \models E = F \iff (\forall \vec{x})[\mathbf{A} \models E(\vec{x}) = M(\vec{x})],$$

where $\vec{x}$ is any list of (distinct) variables which includes all the variables that occur in E or M (where the choice of $\vec{x}$ does not matter, easily), and

(1E-4) $$\models E = M \iff (\forall \mathbf{A})[\mathbf{A} \models E = M].$$

Explicit definability. A partial function $f : A^n \rightharpoonup A_s$ is *explicitly defined* or just *explicit* in $\mathbf{A}$ if for some extended, pure Φ-term $E(\vec{x})$,

(1E-5) $$f(\vec{x}) = \mathrm{den}(\mathbf{A}, E(\vec{x})) \quad (\vec{x} \in A^n);$$

and sometimes we will save some dots, by calling $f : A^n \rightharpoonup A^m$ of arity n and *co-arity* m explicit in $\mathbf{A}$, if $f(\vec{x}) = (f_1(\vec{x}), \dots, f_m(\vec{x}))$ with $\mathbf{A}$-explicit $f_1, \dots, f_m$.

More generally, a functional

(1E-6) $$f : A^n \times (A^{n_1} \rightharpoonup A_{s_1}) \times \cdots \times (A^{n_k} \rightharpoonup A_{s_k}) \rightharpoonup A_s$$

on the set A is explicit in $\mathbf{A}$ if

(1E-7) $$f(\vec{x}, r_1, \dots, r_k) = \mathrm{den}((\mathbf{A}, r_1, \dots, r_k), E(\vec{x})),$$

where $E(\vec{x})$ is now a (pure, extended) $\Phi \cup \{r_1, \dots, r_k\}$-term in the expansion of the vocabulary Φ by fresh[10] function symbols $\vec{r}$ of the appropriate arities and sorts, so that it can be evaluated in arbitrary expansions $(\mathbf{A}, r_1, \dots, r_k)$ of $\mathbf{A}$. We set

$$\mathbf{Expl}(\mathbf{A}) = \text{ the class of all } \mathbf{A}\text{-explicit partial functions and functionals.}$$

Notice that not all the variables in the list $\vec{x}$ need occur in $E(\vec{x})$ in (1E-5), (1E-7) so that, for example, the *projection functions*

(1E-8) $$P_i^n(x_1, \dots, x_n) = x_i \quad (1 \le i \le n)$$

are explicit in every structure. Combined with the strict understanding of composition of partial functions in (1A-9), this implies that $\mathbf{Expl}(\mathbf{A})$ does not

[10] The notion of a *fresh* object (or sequence of objects) is deep and hard to make precise—but well-understood and routinely used in mathematical discourse: roughly, the objects $x_1, \dots, x_n$ (of any kind) are fresh if they are different from each other and from any other object which has been specified in the current context.

have nice properties when **A** is not sufficiently rich, e.g., it need not be closed under composition, cf. Problem x1E.9.

1E.1. **Proposition.** *For every* Φ*-structure* **A**, **Expl**(**A**) *is the smallest class* $\mathcal{E}$ *of functionals on* A (as in (1E-6)) *which contains all n-ary functions* $\vec{x} \mapsto \mathrm{tt}, \vec{x} \mapsto \mathrm{ff}$ *and all projections* P^n_i *and is closed under the following operations*:

(1), Mangling, *as in* (1A-15),

$$f(x_1, \dots, x_n, r_1, \dots r_m) = h(x_{\pi(1)}, \dots, x_{\pi(k)}, r_{\sigma(1)}, \dots, r_{\sigma(l)}).$$

(2), Substitution into the primitives, $f(\vec{x}, \vec{r}) = \phi^{\mathbf{A}}(h_1(\vec{x}, \vec{r}), \dots, h_m(\vec{x}, \vec{r}))$.
(3), Substitution into applications, $f(\vec{x}, p, \vec{r}) = p(h_1(\vec{x}, \vec{r}), \dots, h_m(\vec{x}, \vec{r}))$.
(4), Branching, $f(\vec{x}, \vec{r}) = \text{if } h_1(\vec{x}, \vec{r}) \text{ then } h_2(\vec{x}, \vec{r}) \text{ else } h_3(\vec{x}, \vec{r})$.

PROOF. To prove that **Expl(A)** is closed under (1) – (4), we need to verify some simple, natural closure properties of the set of terms; and the proof that if $\mathcal{E}$ has these properties, then it contains every explicit f is by induction on the term E which defines f. Both arguments are simple exercises in using the definition and properties of the denotation function (especially Problem x1E.4) and we skip them, except for one small part which we put in Problem x1E.8.⊣

1E.2. **Proposition.** *Every* **A**-*explicit functional is continuous.*

PROOF. It is enough to check that the collection of continuous functionals on A is closed under definitions (1) – (4) in Proposition 1E.1. To take just one case which explains the (obvious) idea, suppose

$$f(x, r) = \phi^{\mathbf{A}}(h_1(x, r), h_2(x, r))$$

with continuous $h_1(x, r), h_2(x, r)$.

To verify first that $f(x, r)$ is monotone, assume that $r \sqsubseteq r'$ and compute:

$$\begin{aligned}
&\phi^{\mathbf{A}}(h_1(x, r), h_2(x, r)) = w \\
&\qquad \Longrightarrow h_1(x, r)\downarrow \ \& \ h_2(x, r)\downarrow \ \& \ \phi^{\mathbf{A}}(h_1(x, r), h_2(x, r)) = w, \\
&\qquad \text{but, by the hypothesis, } h_1(x, r') = h_1(x, r) \ \& \ h_2(x, r') = h_2(x, r), \\
&\qquad \text{and so } \phi^{\mathbf{A}}(h_1(x, r'), h_2(x.r')) = \phi^{\mathbf{A}}(h_1(x, r), h_2(x.r)) = w.
\end{aligned}$$

For the compactness of $f(x, r)$, similarly,

$$\begin{aligned}
&\phi^{\mathbf{A}}(h_1(x, r), h_2(x, r)) = w \\
&\qquad \Longrightarrow h_1(x, r)\downarrow \ \& \ h_2(x, r)\downarrow \ \& \ \phi^{\mathbf{A}}(h_1(x, r), h_2(x, r)) = w;
\end{aligned}$$

by the assumed compactness and monotonicity of h_1, h_2, there are finite $r_1 \sqsubseteq r$ and $r_2 \sqsubseteq r$ such that

$$h_1(x, r) = h_1(x, r_1) = h(x, r_1 \cup r_2), h_2(x, r) = h_2(x, r_2) = h_2(x, r_1 \cup r_2),$$

where $r_1 \cup r_2$ is the finite "union" of r_1 and r_2 and

$$\phi^{\mathbf{A}}(h_1(x, r), h_2(x, r)) = \phi^{\mathbf{A}}(h_1(x, r_1 \cup r_2), h_2(x, r_1 \cup r_2)). \qquad ⊣$$

By Problem x1E.9, **Expl**(**A**) is not, in general, closed under substitutions and λ-substitutions. The next Proposition is about the best result we can prove in this direction, and we will need it:

1E.3. **Proposition** (Immediate λ-substitution). *If $h(\vec{y}, \vec{r}_1, q, \vec{r}_2)$ is explicit in* **A**, *then so is the functional*

$$f(\vec{x}, \vec{y}, p, \vec{r}_1, \vec{r}_2) = h(\vec{y}, \vec{r}_1, (\lambda\vec{u})p(\vec{x}, \vec{u}, \vec{y}), \vec{r}_2). \tag{1E-9}$$

Proof is by checking that the class of functionals h for which the Proposition holds satisfies the hypothesis of Proposition 1E.1, and it is trivial in all cases except (3) when the variable q is involved: i.e., with $k = 2$ and skipping $\vec{r}_1, \vec{r}_2$ which do not enter the argument,

$$h(\vec{y}, q) = q(h_1(\vec{y}, q), h_2(\vec{y}, q)).$$

We compute:

$$\begin{aligned} &f(\vec{x}, \vec{y}, p) \\ &\quad = (\lambda u, v)p(\vec{x}, u, v)\Big(h_1(\vec{y}, (\lambda u, v)p(\vec{x}, u, v)), h_2(\vec{y}, (\lambda u, v)p(\vec{x}, u, v))\Big) \\ &\qquad = p\Big(\vec{x}, h_1(\vec{y}, (\lambda u, v)p(\vec{x}, u, v)), h_2(\vec{y}, (\lambda u, v)p(\vec{x}, u, v))\Big) \\ &\qquad\qquad = p(\vec{x}, f_1(\vec{x}, \vec{y}, p), f_2(\vec{x}, \vec{y}, p)) \end{aligned}$$

with explicit f_1, f_2 by the induction hypothesis, so $f(\vec{x}, \vec{y}, p)$ is explicit. ⊣

Such appeals to Proposition 1E.1 are most often the simplest way to verify properties of all **A**-explicit functionals, but sometimes it is just simpler to use directly their definition. A case in point is the following easy and useful result:

1E.4. **Proposition.** *If the vocabulary Φ has a relation symbol R of arity $k > 0$ and* **A** *is a Φ-structure such that $R^{\mathbf{A}} : A^k \to \mathbb{B}$ is total, then* **Expl**(**A**) *is closed under the substitution operation* (1A-12) *and λ-substitution* (1A-13):

$$f(\vec{x}, \vec{r}) = g(h_1(\vec{x}, \vec{r}), \ldots, h_k(\vec{x}, \vec{r}), \vec{r}), \ f(\vec{x}, \vec{y}, \vec{p}, \vec{r}) = h(\vec{y}, (\lambda\vec{u})g(\vec{u}, \vec{x}, \vec{p}), \vec{r}).$$

Proof. For substitution, with $k = 2$ and suppressing $\vec{r}$ which does not enter the argument, we need to prove that if

$$f(\vec{x}) = g(h(\vec{x}))$$

and $h(\vec{x}), g(u)$ are **A**-explicit, then so is $f(\vec{x})$.

Fix pure, extended terms $M(\vec{\mathrm{x}})$ and $E(\mathrm{u})$ such that

$$h(\vec{x}) = \mathrm{den}(M(\vec{x})), \quad g(u) = \mathrm{den}(E(u)).$$

Supposing that R is unary, let

$$F(\mathrm{x}) :\equiv \text{if } R(M(\mathrm{x})) \text{ then } E(M(\mathrm{x})) \text{ else } E(M(\mathrm{x}))$$

and appeal to Problem x1E.4 to check that

$$f(\vec{x}) = \mathrm{den}(F(\vec{x}));$$

and if R is not unary, replace $R(M(\mathsf{x}))$ by $R(M(\mathsf{x}), M(\mathsf{x}), \dots, M(\mathsf{x}))$ in the definition of $F(\vec{\mathsf{x}})$.

The argument for λ-substitutions is easier, by appealing to Proposition 1E.1 and we leave it for Problem x1E.11. ⊣

Problems for Section 1E

x1E.1. **Problem** (Parsing for terms). Prove that for any Φ-structure $\mathbf{A}$, every $\mathbf{A}$-term E satisfies exactly one of the following conditions.

1. $E \equiv \mathrm{tt}$, or $E \equiv \mathrm{ff}$, or $E \equiv x$ for some $x \in A$, or $E \equiv v$ for a variable v.
2. $E \equiv \phi(E_1, \dots, E_n)$ for a uniquely determined $\phi \in \Phi$ and uniquely determined terms $E_1, \dots, E_n$.
3. $E \equiv \text{if } E_0 \text{ then } E_1 \text{ else } E_2$ for uniquely determined E_0, E_1, E_2.

x1E.2. **Problem.** Prove that if $\pi : \mathbf{A} \rightarrowtail\!\!\!\!\twoheadrightarrow \mathbf{B}$ is an isomorphism between two Φ-structures, then for every extended Φ-term $E(\vec{\mathsf{x}})$,

$$\mathbf{A} \models E(\vec{x}) = w \iff \mathbf{B} \models E(\pi(\vec{x})) = \pi(w), \quad (x_1, \dots, x_n, w \in A).$$

x1E.3. **Problem.** Give an example of two (extended) terms $E_1(\mathsf{x})$ and $E_2(\mathsf{x})$ such that for every $x \in A$, $\mathrm{den}(E_1(x)) = \mathrm{den}(E_2(x))$, but if M is closed and $\mathrm{den}(M)\uparrow$, then $\mathrm{den}(E_1(M)) \neq \mathrm{den}(E_2(M))$.

x1E.4. **Problem.** Prove that for every term $E(\mathsf{x})$ and closed term M,

$$\mathrm{den}(M) = w \Longrightarrow \mathrm{den}(E(M)) = \mathrm{den}(E(w)).$$

x1E.5. **Problem.** Prove that for any two terms $E_1(\mathsf{x}), E_2(\mathsf{x})$, if M is closed, $\mathrm{den}(M)\downarrow$ and $\mathrm{den}(E_1(x)) = \mathrm{den}(E_2(x))$ for every $x \in A$, then

$$\mathrm{den}(E_1(M)) = \mathrm{den}(E_2(M)).$$

These results extend trivially to simultaneous substitutions.

x1E.6. **Problem** (Homomorphism property). Prove that if $\pi : \mathbf{A} \to \mathbf{B}$ is a homomorphism of one Φ-structure into another, then for every Φ-term $M(\vec{\mathsf{x}})$ and all $\vec{x} \in A^n$,

$$\text{if } \mathrm{den}(\mathbf{A}, M(\vec{x}))\downarrow, \text{ then } \pi(\mathrm{den}(\mathbf{A}, M(\vec{x})) = \mathrm{den}(\mathbf{B}, M(\pi(\vec{x})))$$

where, naturally, $\pi(x_1, \dots, x_n) = (\pi(x_1), \dots, \pi(x_n))$.

x1E.7. **Problem.** For each of the following (and with the definitions in (1E-1), (1E-4)) determine whether it is true or false for all terms of boolean sort:

(1) $\models \text{if } \phi \text{ then } \psi_1 \text{ else } \psi_2 = \text{if } \neg\phi \text{ then } \psi_2 \text{ else } \psi_1$.

(2) $\models \neg(\phi \,\&\, \psi) = (\neg\phi) \vee (\neg\psi)$.
(3) $\models \neg(\phi \,\&\, \psi) \leftrightarrow (\neg\phi) \vee (\neg\psi) = \mathrm{tt}$.
(4) $\models \phi \,\&\, \psi = \psi \,\&\, \phi$.

x1E.8. **Problem.** Prove that if $h_1(x, r)$ and $h_2(x, r)$ are continuous functionals on A and

$$f(x, p, r) = p(h_1(x, r), h_2(x, r)),$$

then $f(x, p, r)$ is also continuous. (This is one part of the proof of Proposition 1E.1.)

x1E.9. **Problem.** Let $\mathbf{A} = (A, \phi)$ where A is any non-empty set, $\phi : A \rightharpoonup A$ is not total, and for some ordering $\leq$ on A,

$$\phi(x)\downarrow \implies x < \phi(x).$$

(1) Let $\mathrm{tt}_1(x) = \mathrm{tt}$ be the explicit, total unary function with constant value tt and check that the composition

$$f(x) = \mathrm{tt}_1(\phi(x))$$

is not explicit.

(2) Prove that with $P^2_1(x, y) = x$, the (explicit) projection on the first variable, there are explicit partial functions $h_1(x), h_2(x)$ such that the composition

$$f(x) = P^2_1(h_1(x), h_2(x))$$

is not $\mathbf{A}$-explicit.

(3) Prove that there is an $\mathbf{A}$-explicit functional $g(x, p)$ (with p binary) and an $\mathbf{A}$-explicit partial function $h(u, v)$ such that the partial function $f(x) = g(x, (\lambda u, v)h(u, v))$ is not $\mathbf{A}$-explicit.

HINT: The only pure terms of sort `boole` are tt and ff, and the only terms of sort `ind` are $x, \phi(x), \phi(\phi(x)), \ldots$.

x1E.10. **Problem.** Prove that if $\mathbf{A}$ is total, then the composition

$$f(\vec{x}) = g(h_1(\vec{x}), \ldots, h_m(\vec{x}))$$

of $\mathbf{A}$-explicit functions is also $\mathbf{A}$-explicit. HINT: Apply Problem x1E.4 to the term that defines $g(\vec{w})$ by (1E-5).

x1E.11. **Problem.** Prove that if the vocabulary Φ has a relation symbol R of arity $k > 0$ and $\mathbf{A}$ is a Φ-structure such that $R^{\mathbf{A}} : A^k \to \mathbb{B}$ is total, then $\mathbf{Expl}(\mathbf{A})$ is closed under λ-substitutions (1A-13).

x1E.12*. **Problem** (Explicit expansions). Suppose the vocabulary Φ has a relation symbol R of arity $k > 0$ and $\mathbf{A}$ is a Φ-structure such that $R^{\mathbf{A}} : A^k \to \mathbb{B}$ is total. Suppose $f : A^n \rightharpoonup A$ is explicit in $\mathbf{A}$. Let f be a fresh function symbol and let $(\mathbf{A}, f)$ be the expansion of $\mathbf{A}$ in which f is interpreted by f. Define a mapping

$$M \mapsto M^*$$

which assigns to each term M in the vocabulary $\Phi \cup \{\mathrm{f}\}$ a Φ-term M^* with the same variables, so that

$$\mathrm{den}((\mathbf{A}, f), M(\vec{y})) = \mathrm{den}(\mathbf{A}, M^*(\vec{y})).$$

Infer that $\mathbf{Expl}(\mathbf{A}, f) = \mathbf{Expl}(\mathbf{A})$, which extends and implies the result in Problem 1E.4.

Prove also by a counterexample that the hypothesis about R cannot be removed.

x1E.13. **Problem.** Prove that for every m and $\vec{x} \in A^n$,

$$G_m(\mathbf{A}, \vec{x}) = \{\mathrm{den}(\mathbf{A}, E(\vec{x})) : E(\vec{x}) \text{ is pure, algebraic, } \mathrm{sort}(E) = \mathtt{ind} \text{ and } \mathrm{depth}(E) \leq m\},$$

by the definition of "depth" for algebraic terms on page 37.

x1E.14. **Problem.** Prove that a partial function $f : A^n \rightharpoonup A_s$ is $\mathbf{A}$-explicit if and only if there are pure, algebraic terms $C_i(\vec{x})$ of boolean sort and algebraic terms $V_i(\vec{x})$ such that

$$f(\vec{x}) = \begin{cases} \mathrm{den}(V_0(\vec{x})) & \text{if } \mathrm{den}(C_0(\vec{x})) = \mathrm{tt}, \\ \mathrm{den}(V_1(\vec{x})) & \text{ow., if } \mathrm{den}(C_0(\vec{x}))\downarrow \ \& \ \mathrm{den}(C_1(\vec{x})) = \mathrm{tt}, \\ \ \vdots & \\ \mathrm{den}(V_k(\vec{x})) & \text{ow., if } \mathrm{den}(C_{k-1}(\vec{x}))\downarrow \ \& \ \mathrm{den}(C_k(\vec{x})) = \mathrm{tt}, \\ \mathrm{den}(V_{k+1}(\vec{x})) & \text{ow., if } \mathrm{den}(C_k(\vec{x}))\downarrow . \end{cases}$$

Infer that for a total structure $\mathbf{A}$, a relation $R \subseteq A^n$ is $\mathbf{A}$-explicit if and only if it is definable by a quantifier-free formula, as these are defined in (1E-1).

This representation of explicit functions and relations is especially interesting (and has been much studied) for the *Presburger structure*[11]

(1E-10) $$\mathbf{N}_{\mathrm{Pres}} = (\mathbb{N}, 0, 1, +, \dot{-}, <, \mathrm{eq}_{\mathbb{N}}, \{\mathrm{rem}_m, \mathrm{iq}_m\}_{m \geq 2}),$$

where $\mathrm{rem}_m(x) = \mathrm{rem}(x, m), \mathrm{iq}_m(x) = \mathrm{iq}(x, m)$. This is because a total function $f : \mathbb{N}^n \to \mathbb{N}$ is explicit in $\mathbf{N}_{\mathrm{Pres}}$—a *Presburger function*—exactly when its graph is definable in additive arithmetic

$$\mathbf{N}_+ = (\mathbb{N}, 0, 1, +, \mathrm{eq}_{\mathbb{N}}),$$

a corollary of the classical *quantifier elimination result* for Presburger arithmetic, cf. Enderton [2001]. The Presburger functions are *piecewise linear* in the following, precise sense:

x1E.15*. **Problem.** Prove that if $f : \mathbb{N}^n \to \mathbb{N}$ is a Presburger function, then there is a partition of $\mathbb{N}^n$ into disjoint sets $D_1, \ldots, D_k \subseteq \mathbb{N}^n$ which are

[11]The Presburger structure has an infinite vocabulary, see Footnote 8 on page 31.

definable by quantifier free formulas of $\mathbf{N}_{\text{Pres}}$, such that for each $i = 1, \dots, k$, and suitable rational numbers $q_0, q_1, \dots, q_n$,

$$f(\vec{x}) = q_0 + q_1 x_1 + \cdots + q_n x_n \qquad (\vec{x} \in D_i).$$

HINT: Check that the result holds for the primitives of $\mathbf{N}_{\text{Pres}}$ and that the class of functions which satisfy it is closed under composition. Notice that $f(\vec{x}) \in \mathbb{N}$ in this expression, although some of the $q_i \in \mathbb{Q}$ may be proper, positive or negative fractions.

The claims in the next three problems will follow from results that we will prove later, but perhaps there are elementary proofs of them that can be given now—I do not know such easy proofs:

x1E.16*. **Problem.** Prove that the successor function $S : \mathbb{N} \to \mathbb{N}$ is not explicit in binary arithmetic $\mathbf{N}_b$.

x1E.17*. **Problem.** Prove that the parity relation

$$\text{parity}(x) \iff 2 \mid x$$

is not quantifier-free definable in unary arithmetic $\mathbf{N}_u$, and the successor relation

$$S(x, y) \iff x + 1 = y$$

is not quantifier-free definable in binary arithmetic $\mathbf{N}_b$.

x1E.18*. **Problem.** Prove that the divisibility relation

$$x \mid y \iff y \neq 0 \;\&\; \text{rem}(x, y) = 0$$

is not quantifier-free definable in the Presburger structure $\mathbf{N}_{\text{Pres}}$.

Part I. Abstract (first order) recursion

CHAPTER 2

RECURSIVE (McCARTHY) PROGRAMS

Recursive programs are deterministic versions of the classical *Herbrand-Gödel-Kleene systems of recursive equations* and they can be used to develop very elegantly the classical theory of *recursive* (computable) *functions and functionals* on the natural numbers. Here we will study them on arbitrary (partial) structures, and we will use them primarily to introduce in the next chapter some natural and robust notions of complexity for algorithms which compute functions from specified primitives.

We will also introduce and establish the basic properties of (finitely) *nondeterministic* recursive programs in Section 2E.

Recursive programs were introduced in McCarthy [1963], a fundamental contribution to the foundations of the theory of computation. It extended the earlier McCarthy [1960] and introduced many ideas and techniques which we take for granted today. Much of what we do in this chapter can be read as an exposition of the theory developed in McCarthy [1963], except that we take *least-fixed-point recursion* to be the fundamental semantic notion.[12]

2A. Syntax and semantics

Programs. A (deterministic) *recursive program* on the vocabulary Φ (or the Φ-structure $\mathbf{A}$) is a syntactic expression

$$E \equiv E_0 \text{ where } \left\{\mathsf{p}_1(\vec{\mathsf{x}}_1) = E_1, \dots, \mathsf{p}_K(\vec{\mathsf{x}}_K) = E_K\right\} \tag{2A-1}$$

which satisfies the following conditions:

[12]This work of McCarthy is not as well known or appreciated as it should be: there are almost no citations of it in early textbooks and articles on the theory of computation by computer scientists, with the notable exception of Manna [1974] whose last chapter on the *Fixpoint Theory of Programs* explains (and extends considerably) McCarthy's ideas; and the logicians who developed abstract (and higher type) recursion theory starting in the late 1950s worked in complete ignorance of McCarthy's work—including the author of Moschovakis [1989a] who will forever be embarrassed by this.

(1) $\mathsf{p}_1, \dots, \mathsf{p}_K$ are distinct function symbols not in Φ, the *recursive variables* of E.
(2) Each E_i $(0 \le i \le K)$ is a pure term in the *program vocabulary*

$$\mathrm{voc}(E) = \Phi \cup \{\mathsf{p}_1, \dots, \mathsf{p}_K\}.$$

(3) Each $\vec{\mathsf{x}}_i$ $(1 \le i \le K)$ is a list of distinct individual variables which includes all the individual variables that occur in E_i.
(4) The arities and sorts of the recursive variables $\mathsf{p}_1, \dots, \mathsf{p}_K$ and the sorts of the parts E_i of E are such that the equations within the braces make sense.

The term E_0 is the *head* of E, the system of equations

$$\text{(2A-2)} \qquad \mathsf{p}_1(\vec{\mathsf{x}}_1) = E_1, \dots, \mathsf{p}_K(\vec{\mathsf{x}}_K) = E_K$$

within the braces is its *body*, and we allow $K = 0$ in this definition, so that every pure Φ-term is identified with a program with empty body,

$$E \equiv E \text{ where } \{\ \}.$$

A program with non-empty body will be interpreted as a mutual recursive definition, a formal version of the $\overline{\text{where}}$-notation introduced in Section 1B.

We allow $\mathrm{arity}(\mathsf{p}_i) = 0$, in which case the equation for p_i in (2A-1) has the form "$\mathsf{p}_i = E_i$," the list of variable $\vec{\mathsf{x}}_i$ is empty and E_i is a pure, closed $\mathrm{voc}(E)$-term, i.e., a nullary constant ϕ or function variable p_j of either sort.

Free and bound occurrences of variables. All occurrences of the recursive variables $\mathsf{p}_1, \dots, \mathsf{p}_K$ and all occurrences of individual variables in the body of a program E are bound in E; the free variables of E (if any) are the individual variables which occur in the head term E_0.

In general, we think of recursive programs as generalized pure Φ-terms and we define *extended Φ-programs* as we did for terms: if the list of distinct variables $\vec{\mathsf{x}} \equiv \mathsf{x}_1, \dots, \mathsf{x}_n$ includes all the free individual variables of E (those which occur in the head E_0), then

$$\text{(2A-3)} \quad E(\vec{\mathsf{x}}) \equiv_{\mathrm{df}} (E, \vec{\mathsf{x}}) \equiv E_0(\vec{\mathsf{x}}) \text{ where } \Big\{\mathsf{p}_1(\vec{\mathsf{x}}_1) = E_1, \dots, \mathsf{p}_K(\vec{\mathsf{x}}_K) = E_K\Big\}.$$

As with terms, we will often refer ambiguously to "the program $E(\vec{\mathsf{x}})$" or "the program E", letting the notation and the context specify whether we are talking about a "plain" or an extended program, but the default is "extended". We set

$$\mathrm{sort}(E) = \mathrm{sort}(E_0), \quad \mathrm{arity}(E(\vec{\mathsf{x}})) = \mathrm{arity}(E(\mathsf{x}_1, \dots, \mathsf{x}_n)) = n.$$

All the recursive equations and systems of equations in the problems of Sections 1A and 1C are really extended recursive programs, just not sufficiently formalized. Problem x1C.13, for example, determines two programs on the structure $(\mathbb{Z}, 0, 1, +, -, \cdot, \text{rem}, \text{iq}, \text{eq}_0)$, one for each of the needed coefficients in Bezout's Lemma: the first is

$$\alpha(\mathsf{x},\mathsf{y}) \text{ where } \Big\{\alpha(\mathsf{x},\mathsf{y}) = \text{if } (\text{rem}(\mathsf{x},\mathsf{y}) = 0) \text{ then } 0 \text{ else } \beta(\mathsf{y}, \text{rem}(\mathsf{x},\mathsf{y})),\\ \beta(\mathsf{x},\mathsf{y}) = \text{if } (\text{rem}(\mathsf{x},\mathsf{y}) = 0) \text{ then } 1\\ \text{else } \alpha(\mathsf{y}, \text{rem}(\mathsf{x},\mathsf{y})) - \text{iq}(\mathsf{x},\mathsf{y}) \cdot \beta(\mathsf{y}, \text{rem}(\mathsf{x},\mathsf{y}))\Big\}$$

and the second is obtained from this by changing the head to $\beta(\mathsf{x},\mathsf{y})$. Both programs have the binary recursive variables α and β.

In some cases we express algorithms by a single recursive equation, e.g.,

$$\text{gcd}(x, y) = \text{if } (y = 0) \text{ then } x \text{ else } \text{gcd}(y, \text{rem}(x, y))$$

for the Euclidean, and then we need to add a trivial head term to accord with the "official" definition: so the formal recursive program which expresses the Euclidean is

(2A-4) $$E_\varepsilon(\mathsf{x},\mathsf{y}) \equiv \mathsf{p}(\mathsf{x},\mathsf{y}) \text{ where}\\ \Big\{\mathsf{p}(\mathsf{x},\mathsf{y}) = \text{if } \text{eq}_0(\mathsf{y}) \text{ then } \mathsf{x} \text{ else } \mathsf{p}(\mathsf{y}, \text{rem}(\mathsf{x},\mathsf{y}))\Big\}.$$

We will assume that this addition of a head term is done when needed.[13]

Semantics. Fix an extended recursive program $E(\vec{\mathsf{x}})$ on the vocabulary Φ as in (2A-3) and a Φ-structure $\mathbf{A}$, let $\vec{p} = (p_1, \dots, p_K)$ and set

(2A-5) $$f_0(\vec{x}, \vec{p}) = \text{den}((\mathbf{A}, \vec{p}), E_0(\vec{x})),\\ f_i(\vec{x}_i, \vec{p}) = \text{den}((\mathbf{A}, \vec{p}), E_i(\vec{x}_i)) \quad (i = 1, \dots, K).$$

These functionals are continuous by Proposition 1E.2, so we can set

(2A-6) $$\text{den}(\mathbf{A}, E(\vec{x})) = \text{den}^{\mathbf{A}}_{E(\vec{\mathsf{x}})}(\vec{x})\\ =_{\text{df}} f_0(\vec{x}, \vec{p}) \,\overline{\text{where}}\, \Big\{p_1(\vec{x}_1) = f_1(\vec{x}_1, \vec{p}), \dots, p_K(\vec{x}_K) = f_K(\vec{x}_K, \vec{p})\Big\}$$

using the $\overline{\text{where}}$ construct introduced in Section 1B.

We say that the partial function $\text{den}^{\mathbf{A}}_{E(\vec{\mathsf{x}})} : A^n \rightharpoonup A_s$ is *defined* (or *computed*) *in* $\mathbf{A}$ *by* $E(\vec{\mathsf{x}})$ and $\text{den}(\mathbf{A}, E(\vec{x}))$ is *the denotation of* $E(\vec{\mathsf{x}})$ *in* $\mathbf{A}$ *at* $\vec{x}$. We also use model–theoretic notation,

(2A-7) $$\mathbf{A} \models E(\vec{x}) = w \iff_{\text{df}} \text{den}(\mathbf{A}, E(\vec{x})) = w.$$

If $E \equiv E_0$ is a program with no body, then this agrees with the definition of $\text{den}(E_0)$ in Section 1E.

---

[13]Notice that the variables x, y occur both free (in the head) and bound (in the bodies) of these examples, a practice that is sometimes not allowed in formal languages but is very convenient in the formal study of recursive equations.

More generally, if $\vec{\mathsf{r}} = \mathsf{r}_1, \dots, \mathsf{r}_m$ is a list of distinct function symbols (other than the p_j) and every E_i is a pure term in

$$\mathrm{voc}(E) = \Phi \cup \{\mathsf{p}_1, \dots, \mathsf{p}_K, \mathsf{r}_1, \dots, \mathsf{r}_m\},$$

then (2A-3) is an extended recursive program on Φ with *parameters* (or *free function variables*) $\vec{\mathsf{r}}$ which is naturally interpreted in expansions

$$(\mathbf{A}, \vec{r}) = (A, \mathbf{\Phi}, r_1, \dots, r_k)$$

of a Φ-structure $\mathbf{A}$. The parts of $E(\vec{\mathsf{x}})$ define now functionals

$$\text{(2A-8)} \qquad \begin{aligned} f_0(\vec{x}, \vec{p}, \vec{r}) &= \mathrm{den}((\mathbf{A}, \vec{p}, \vec{r}), E_0(\vec{x})), \\ f_i(\vec{x}_i, \vec{p}, \vec{r}) &= \mathrm{den}((\mathbf{A}, \vec{p}, \vec{r}), E_i(\vec{x}_i)) \quad (i = 1, \dots, K) \end{aligned}$$

and $E(\vec{\mathsf{x}})$ *defines* (or *computes*) *in* $\mathbf{A}$ the functional

$$\text{(2A-9)} \quad \mathrm{den}^{\mathbf{A}}_{E(\vec{\mathsf{x}})}(\vec{x}, \vec{r})$$
$$=_{\mathrm{df}} f_0(\vec{x}, \vec{p}, \vec{r}) \; \overline{\text{where}} \; \Big\{ p_1(\vec{x}_1) = f_1(\vec{x}_1, \vec{p}, \vec{r}), \dots, p_K(\vec{x}_K) = f_K(\vec{x}_K, \vec{p}, \vec{r}) \Big\}.$$

A-recursive functions and functionals. A functional $f(\vec{x}, \vec{r})$ on A is *recursive in* $\mathbf{A}$ or *recursive from* (or *in*) the primitives $\mathbf{\Phi} = \{\phi^{\mathbf{A}} : \phi \in \Phi\}$ of $\mathbf{A}$, if it is defined by some $\mathbf{A}$-recursive program with function parameters, i.e., if

$$\text{(2A-10)} \quad f(\vec{x}, \vec{r}) = f_0(\vec{x}, \vec{p}, \vec{r})$$
$$\overline{\text{where}} \; \Big\{ p_1(\vec{x}_1) = f_1(\vec{x}_1, \vec{p}, \vec{r}), \dots, p_K(\vec{x}_K) = f_K(\vec{x}_K, \vec{p}, \vec{r}) \Big\}$$

with $\mathbf{A}$-explicit functionals $f_0, \dots, f_K$. We set

$$\text{(2A-11)} \qquad \mathbf{Rec}(\mathbf{A}) = \text{the set of all } \mathbf{A}\text{-recursive functionals},$$

and to state results for partial functions only,

$$\text{(2A-12)} \qquad \mathbf{Rec}^0(\mathbf{A}) = \text{the set of all } \mathbf{A}\text{-recursive partial functions}.$$

The classical example is the (total) structure $\mathbf{N}_u = (\mathbb{N}, 0, S, \mathrm{Pd}, \mathrm{eq}_0)$ of unary arithmetic whose recursive partial functions and functionals are exactly the Turing computable ones, an elegant characterization of Turing computability due to McCarthy [1963], cf. Proposition 2C.3.[14]

2A.1. **Theorem** (The First Recursion Theorem). *The class* $\mathbf{Rec}(\mathbf{A})$ *is closed under recursive definitions: i.e., if* (2A-10) *holds with* $\mathbf{A}$*-recursive functionals* $f_0, \dots f_k$, *then* f *is also* $\mathbf{A}$*-recursive.*

---

[14]More precisely, these are the classical *deterministic recursive functionals* on $\mathbb{N}$; we will introduce the *nondeterministic* ones in Section 2E.

PROOF is by induction on the number of functionals in (2A-10) which are not explicit.

If the head $f_0(\vec{x}, \vec{p}, \vec{r})$ in (2A-10) is not explicit, then

$$f_0(\vec{x}, \vec{p}, \vec{r}) = g_0(\vec{x}, \vec{p}, \vec{r}, \vec{q}) \,\overline{\text{where}}\, \Big\{q_1(\vec{y}_1) = g(\vec{x}, \vec{p}, \vec{r}, \vec{q}), \dots\Big\}$$

with suitable explicit functionals $g_0, g_1, \dots$; so

$$\begin{aligned} f(\vec{x}, \vec{r}) &= f_0(\vec{x}, \vec{p}, \vec{r}) \,\overline{\text{where}}\, \Big\{p_1(\vec{x}_1) = f_1(\vec{x}_1, \vec{p}, \vec{r}), \dots,\Big\} \\ &= \Big(g_0(\vec{x}, \vec{p}, \vec{r}, \vec{q}) \,\overline{\text{where}}\, \Big\{q_1(\vec{y}_1) = g(\vec{x}, \vec{p}, \vec{r}, \vec{q}), \dots\Big\}\Big) \\ &\qquad\qquad \overline{\text{where}}\, \Big\{p_1(\vec{x}_1) = f_1(\vec{x}_1, \vec{p}, \vec{r}), \dots,\Big\} \end{aligned}$$

$$= g_0(\vec{x}, \vec{p}, \vec{r}, \vec{q}) \,\overline{\text{where}}\, \Big\{q_1(\vec{y}_1) = g(\vec{x}, \vec{p}, \vec{r}, \vec{q}), \dots, p_1(\vec{x}_1) = f_1(\vec{x}_1, \vec{p}, \vec{r}), \dots,\Big\},$$

the last step by the (head) rule of Theorem 1B.2, and this recursive definition of f involves one fewer functional which is not explicit.

If $f_0(\vec{x}, \vec{p}, \vec{r})$ is explicit, we can simplify the definition of f by applying in the same way the (Bekič-Scott) rule of Theorem 1B.2 to some f_i, cf. Problem x2A.4. ⊣

In effect, the (head) and (Bekič-Scott) rules allow us to remove from any expression with nested occurrences of the $\overline{\text{where}}$-construct all the braces $\Big\{$ and $\Big\}$ other than the first and last without changing the value—messy to write up in full generality but basically trivial and easy to apply in specific recursive definitions.

The name of the theorem comes from the special case

$$f(\vec{x}) = p(\vec{x}) \,\overline{\text{where}}\, \Big\{p(\vec{x}) = f_1(\vec{x}, p)\Big\} \tag{2A-13}$$

by which *the least fixed point of every recursive functional is recursive* and which is often dubbed the *First Recursion Theorem*, at least for $\mathbf{N}_u$.

The First Recursion Theorem combines easily with the recursive rules in Theorem 1B.2 to yield the elementary closure properties of $\mathbf{Rec}(\mathbf{A})$:

2A.2. **Corollary.** *Every* $\mathbf{A}$*-explicit functional is recursive in* $\mathbf{A}$, *and* $\mathbf{Rec}(\mathbf{A})$ *is closed under the elementary operations on functionals in* (1A-12) – (1A-15):

(Substitution) $\quad f(\vec{x}, \vec{r}) = g(h_1(\vec{x}, \vec{r}), \dots, h_k(\vec{x}, \vec{r}), \vec{r})$

(λ-substitution) $\quad f(\vec{x}, \vec{y}, \vec{p}, \vec{r}) = h(\vec{x}, (\lambda\vec{u})g(\vec{u}, \vec{y}, \vec{p}), \vec{r})$

(Branching) $\quad f(\vec{x}, \vec{r}) = \text{if } h(\vec{x}, \vec{r}) \text{ then } g_1(\vec{x}, \vec{r}) \text{ else } g_2(\vec{x}, \vec{r})$

(Mangling) $\quad f(x_1, \dots, x_n, r_1, \dots, r_m)$
$$= h(x_{\pi(1)}, \dots, x_{\pi(k)}, r_{\sigma(1)}, \dots, r_{\sigma(l)}).$$

PROOF. For a simple case of *mangling* which illustrates the trivial argument that works in general, suppose

$$f(x, r_1, r_2) = h(x, x, r_1),$$
$$\text{with } h(u_1, u_2, q) = h_0(u_1, u_2, p, q) \text{ where } \Big\{ p(w) = h_1(w, p, q) \Big\}$$

and explicit h_0, h_1; it follows that

$$f(x, r_1, r_2) = h_0(x, x, p, r_1) \text{ where } \Big\{ p(w) = h_1(w, p, r_1) \Big\}$$

and so f is recursive because the class of explicit functionals is closed under mangling by Problem x1E.8.

The argument for *branching* is equally trivial.

Substitution. Suppressing the variables $\vec{r}$ which do not enter the argument, we start with the obvious

$$f(\vec{x}) = g(h_1(\vec{x}), \dots, h_k(\vec{x})) = p(q_1(\vec{x}), \dots, q_k(\vec{x}))$$
$$\text{where } \Big\{ q_1(\vec{x}) = h_1(\vec{x}), \dots, q_k(\vec{x}) = h_k(\vec{x}), p(\vec{y}) = g(\vec{y}) \Big\}$$

and then apply the First Recursion Theorem 2A.1.

λ-substitution. Suppressing again the variables $\vec{p}, \vec{r}$, we start with

(2A-14)
$$f(\vec{x}, \vec{y}) = h(\vec{x}, (\lambda\vec{u})g(\vec{u}, \vec{y}))$$
$$= h(\vec{x}, (\lambda\vec{u})p(\vec{u}, \vec{y})) \text{ where } \Big\{ p(\vec{u}, \vec{y}) = g(\vec{u}, \vec{y}) \Big\}.$$

To simplify the notation further, suppose that the given recursive definition of $h(\vec{x}, q)$ has only one equation in its body and rewrite it to insert a dummy dependence of the recursive variable r on $\vec{x}, \vec{y}$,[15]

$$h(\vec{x}, q) = h_0(\vec{x}, r, q) \text{ where } \Big\{ r(\vec{z}) = h_1(\vec{z}, r, q) \Big\}$$
$$= h_0(\vec{x}, (\lambda\vec{z})r(\vec{z}, \vec{x}, \vec{y}), q) \text{ where } \Big\{ r(\vec{z}, \vec{x}, \vec{y}) = h_1(\vec{z}, (\lambda\vec{z})r(\vec{z}, \vec{x}, \vec{y}), q) \Big\};$$

if we now set $q = (\lambda\vec{u})p(\vec{u}, \vec{x}, \vec{y})$ in this equation, we get

$$h(\vec{x}, (\lambda\vec{u})p(\vec{u}, \vec{x}, \vec{y})) = h_0(\vec{x}, (\lambda\vec{z})r(\vec{z}, \vec{x}, \vec{y}), (\lambda\vec{u})p(\vec{u}, \vec{x}, \vec{y}))$$
$$\text{where } \Big\{ r(\vec{z}, \vec{x}, \vec{y}) = h_1(\vec{z}, (\lambda\vec{z})r(\vec{z}, \vec{x}, \vec{y}), \lambda(u)p(\vec{u}, \vec{x}, \vec{y}) \Big\};$$

[15] This maneuver of adding dummy variables is needed because we did not allow the functionals $f_i(x_i, \vec{p}, \vec{r})$ in the body of a recursive definition (1B-7) with head $f_0(y, \vec{p}, \vec{r})$ to depend on the variable y—which complicates this proof, but simplifies considerably the formulation of the recursion rules in Theorem 1B.2.

and if we use this expression in (2A-14), we get

$$f(\vec{x},\vec{y}) = \Big(h_0(\vec{x},(\lambda\vec{z})r(\vec{z},\vec{x},\vec{y}),(\lambda\vec{u})p(\vec{u},\vec{x},\vec{y}))$$
$$\overline{\text{where}}\,\Big\{r(\vec{z},\vec{x},\vec{y}) = h_1(\vec{z},(\lambda\vec{z})r(\vec{z},\vec{x},\vec{y}),\lambda(u)p(\vec{u},\vec{x},\vec{y})\Big\}\Big)$$
$$\overline{\text{where}}\,\Big\{p(\vec{u},\vec{x},\vec{y}) = g(\vec{u},\vec{y})\Big\}.$$

The functional $h_0(\vec{x},(\lambda\vec{z})r(\vec{z},\vec{x},\vec{y}),(\lambda\vec{u})p(\vec{u},\vec{x},\vec{y}))$ is explicit (and hence recursive) by Proposition 1E.3 and all the other functionals on the right-hand side in this equation are recursive by the hypothesis; so the First Recursion Theorem 2A.1 applies and $f(\vec{x},\vec{y})$ is recursive. ⊣

Problems for Section 2A

In the first five problems we formulate precisely some simple properties of the semantics of recursive programs which follow easily from the definitions.

x2A.1. **Problem** (Change of order in the body). Prove that for any Φ-structure **A**, any extended Φ-program $E(\vec{x})$ with E as in (2A-1) and any permutation $\pi : \{1,\dots,K\} \rightarrowtail\!\!\!\!\twoheadrightarrow \{1,\dots,K\}$, if

$$E' \equiv E_0 \text{ where } \Big\{\mathsf{p}_{\pi(1)}(\vec{\mathsf{x}}_{\pi(1)}) = E_{\pi(1)},\dots,\mathsf{p}_{\pi(K)}(\vec{\mathsf{x}}_{\pi(K)}) = E_{\pi(K)}\Big\},$$

then for all $\vec{x}$, $\mathrm{den}(\mathbf{A}, E(\vec{x})) = \mathrm{den}(\mathbf{A}, E'(\vec{x}))$.

x2A.2. **Problem** (Localization). Prove that for any Φ-structure **A** and extended Φ-program $E(\vec{x})$,

$$\mathbf{A} \models E(\vec{x}) = w \iff \mathbf{G}_\infty(\mathbf{A},\vec{x}) \models E(\vec{x}) = w.$$

x2A.3. **Problem** (Invariance under isomorphisms). Prove that if $\pi : \mathbf{A} \rightarrowtail\!\!\!\!\twoheadrightarrow \mathbf{B}$ is an isomorphism between two Φ-structures and $E(\vec{x})$ is an extended Φ-program, then

$$\mathbf{A} \models E(\vec{x}) = w \iff \mathbf{B} \models E(\pi(\vec{x})) = \pi(w) \quad (\vec{x} \in A^n).$$

x2A.4. **Problem.** Prove the missing case in the proof of Theorem 2A.1, where some part other the head in (2A-10) is not explicit.

x2A.5. **Problem** (Transitivity). Prove that if g is **A**-recursive and f is $(\mathbf{A}, g)$-recursive, then f is **A**-recursive. It follows that if $(\mathbf{A}, \Psi)$ is an expansion of **A** by partial functions which are **A**-recursive, then

$$\mathbf{Rec}(\mathbf{A},\Psi) = \mathbf{Rec}(\mathbf{A}).$$

x2A.6. **Problem.** Prove that every structure **A** has an expansion by a total, unary relation $(\mathbf{A}, R)$ such that $\mathbf{Rec}(\mathbf{A}) = \mathbf{Rec}(\mathbf{A}, R)$ and $\mathbf{Expl}(\mathbf{A}, R)$ is closed under substitution (1A-12). HINT: Use Problem 1E.4.

x2A.7. **Problem.** Let $\mathbf{A} = (A, 0, \cdot, \text{eq}_0)$ be a structure with $\cdot$ binary, and define x^n for $x \in A$ and $n \geq 1$ by $x^1 = x$, $x^{n+1} = x \cdot x^n$. Define $f : A^2 \rightharpoonup A$ by

$$f(x, y) = x^k \text{ where } k = \text{ the least } n \geq 1 \text{ such that } y^n = 0,$$

and prove that f is **A**-recursive.

x2A.8. **Problem.** Prove that the following functions on L^* (from (1A-17), (1C-4) and (1C-5)) are recursive in the Lisp structure $\mathbf{L}^*$ defined in (1D-5):

$$u * v, \quad \text{half}_1(u), \quad \text{half}_2(u),$$

and for each $i \in \mathbb{N}$, $u \mapsto (u_i)$, set to nil if $i \geq |u|$. Infer that for every ordering $\leq$ of L, the functions merge(u, v) and sort(u) are $(\mathbf{L}^*, \leq)$-recursive.

x2A.9. **Problem.** Prove that $S \notin \mathbf{Rec}(\mathbb{N}, 0, \text{Pd}, \text{eq}_0)$. HINT: Use Problem x2A.2.

It is also true that $\text{Pd} \notin \mathbf{Rec}(\mathbb{N}, 0, S, \text{eq}_0)$, but (perhaps) this is not so immediate at this point, see Problem x2D.6.

x2A.10. **Problem.** True or false: S and Pd are recursive in the Euclidean structure $\mathbf{N}_\varepsilon = (\mathbb{N}, \text{rem}, \text{eq}_0, \text{eq}_1)$ or its expansion $(\mathbf{N}_\varepsilon, 0, 1)$ with the constants 0 and 1.

In the next few problems we consider the special cases of recursion in the structures $\mathbf{N}_u, \mathbf{N}_b, \mathbf{N}_{k\text{-}ary}, \mathbf{N}$ and their expansions by total functions, for which elementary recursion was first developed,

x2A.11. **Problem.** Prove that arithmetic subtraction $x \dot{-} y$ is $\mathbf{N}_u$-recursive, by verifying the identity

$$x \dot{-} y = \text{if } (y = 0) \text{ then } x \text{ else Pd}(x \dot{-} \text{Pd}(y)). \tag{2A-15}$$

x2A.12. **Problem.** Prove that

$$\mathbf{Rec}(\mathbf{N}_u) = \mathbf{Rec}(\mathbf{N}_b) = \mathbf{Rec}(\mathbf{N}_{k\text{-}ary}) = \mathbf{Rec}(\mathbb{N}, 0, S, =) = \mathbf{Rec}(\mathbf{N}),$$

where k-ary arithmetic $\mathbf{N}_{k\text{-}ary}$ and the standard (Peano) structure $\mathbf{N}$ are defined in (1D-3) and (1D-4).

x2A.13. **Problem** (Primitive recursion). Suppose $\mathbf{A} = (\mathbf{N}_u, \Phi)$ is an expansion of $\mathbf{N}_u$, $g : \mathbb{N}^n \rightharpoonup \mathbb{N}$ and $h : \mathbb{N}^{n+2} \rightharpoonup \mathbb{N}$ are **A**-recursive, and $f : \mathbb{N}^n \rightharpoonup \mathbb{N}$ satisfies the following two equations:

$$\begin{aligned} f(0, \vec{x}) &= g(\vec{x}), \\ f(y+1, \vec{x}) &= h(f(y, \vec{x}), y, \vec{x}). \end{aligned} \tag{2A-16}$$

Prove that f is **A**-recursive. Verify also that $f(y, \vec{x})\downarrow \implies (\forall i < y)[f(i, \vec{x})\downarrow]$.

Recall that the class of *primitive recursive functions* on $\mathbb{N}$ is the smallest class of (total) functions on $\mathbb{N}$ which contains the successor S, the n-ary constant functions $C_0^n(\vec{x}) = 0$ and the projection functions $P_i^n(\vec{x}) = x_i$, and which is closed under composition and primitive recursion. By the last two problems, *every primitive recursive function* $f : \mathbb{N}^n \to \mathbb{N}_s$ *is* $\mathbf{N}_u$*-recursive.*

Primitive recursive functions have been studied extensively, especially because of their applications to proof theory—the most basic of these stemming from the fact that they are all *provably recursive* in Peano arithmetic.

x2A.14. **Problem** (Minimalization). Prove that if $g : \mathbb{N}^{n+1} \rightharpoonup \mathbb{N}$ is recursive in some expansion $\mathbf{A} = (\mathbf{N}_u, \Psi)$ of $\mathbf{N}_u$, then so is the partial function

$$\text{(2A-17)} \quad f(\vec{x}) = \mu y[g(y,\vec{x}) = 0]$$
$$=_{\text{df}} \text{the least } y \text{ such that } (\forall i < y)(\exists w)[g(i,\vec{x}) = w + 1 \;\&\; g(y,\vec{x}) = 0].$$

HINT: Check out Problem x1B.4.

Combined with classical results (for example in Kleene [1952]), these last two problems imply easily that *a partial function* $f : \mathbb{N}^n \rightharpoonup \mathbb{N}$ *is* $\mathbf{N}_u$*-recursive exactly when it is Turing-computable.* We will formulate a version of this for functionals in Proposition 2C.3.

x2A.15*. **Problem** (Rózsa Péter). A function $f : \mathbb{N}^{n+1} \to \mathbb{N}$ is defined by *nested recursion* from g, h and $\tau_1, \dots, \tau_n$ if it satisfies the following equations:

$$\text{(2A-18)} \quad \begin{aligned} f(0,\vec{y}) &= g(\vec{y}), \\ f(x+1,\vec{y}) &= h(f(x,\tau_1(x,y),\dots,\tau_n(x,\vec{y})),x,\vec{y}). \end{aligned}$$

(1) Prove that if f is defined by nested recursion from $(\mathbf{N}_u, \mathbf{\Phi})$-recursive functions, then it is $(\mathbf{N}_u, \mathbf{\Phi})$-recursive. (This is easy.)

(2) Prove that if f is defined from primitive recursive functions by nested recursion, then it is primitive recursive. (This is not so easy.)

x2A.16. **Problem** (Double recursion). A function $f : \mathbb{N}^{2+n} \to \mathbb{N}$ is defined by *double recursion* from g, h_1, σ, h_2 if it satisfies the following equations for all $x, y, \vec{z}$:

$$\text{(2A-19)} \quad \begin{aligned} f(0,y,\vec{z}) &= g(y,\vec{z}), \\ f(x+1,0,\vec{z}) &= h_1(f(x,\sigma(x,\vec{z}),\vec{z}),x,\vec{z}), \\ f(x+1,y+1,\vec{z}) &= h_2(f(x+1,y,\vec{z}),x,y,\vec{z}). \end{aligned}$$

Prove that if f is defined by double recursion from $(\mathbf{N}_u, \mathbf{\Psi})$-recursive functions, then it is $(\mathbf{N}_u, \mathbf{\Psi})$-recursive.

x2A.17*. **Problem** (The Ackermann-Péter function). Consider the system of equations

$$\begin{aligned} A(0,x) &= x+1 \\ A(n+1,0) &= A(n,1) \\ A(n+1,x+1) &= A(n,A(n+1,x)), \end{aligned} \tag{2A-20}$$

on a function $A : \mathbb{N}^2 \to \mathbb{N}$.

(1) Verify that this defines $A(n,x)$ by double recursion.

(2) Prove that the Ackermann-Péter function is not primitive recursive.

HINT: For (2), prove first that every *Ackermann section*

$$A_n(x) = A(n,x)$$

is primitive recursive and then show that *for every primitive recursive function* $f(\vec{x})$ *there is some* m *such that*

$$f(\vec{x}) < A_m(\max \vec{x}) \quad (\vec{x} \in \mathbb{N}^n). \tag{2A-21}$$

This requires establishing some basic inequalities about these functions, including

$$A_n(x) \geq 1, \quad x < y \Longrightarrow A_n(x) < A_n(y),$$
$$n < m \Longrightarrow A_n(x) < A_m(x), \quad A_n(A_n(x)) < A_{n+2}(x)$$

which are also needed for the punchline—that $A(n,x)$ is not primitive recursive.

2B. Simple fixed points and tail recursion

One would expect that various restrictions on the general form (2A-10) of recursive definitions yield **A**-recursive functions and functionals with special properties, and there are many results of this kind in what is called *schematology*, cf. Greibach [1975] (and references given there) and the earlier Péter [1951] for recursion on $\mathbb{N}$. This is not our topic, but it is worth discussing here two notions of restricted recursion which are important for the theory of recursive programs.

Simple fixed points. A partial function $f : A^n \rightharpoonup A$ is a *simple fixed point* of **A** if it is the canonical (least) solution of a single equation

$$p(\vec{x}) = g(\vec{x}, p)$$

with an **A**-explicit $g(\vec{x}, p)$, so

$$f(\vec{x}) = p(\vec{x}) \text{ where } \left\{ p(\vec{x}) = g(\vec{x}, p) \right\}. \tag{2B-1}$$

Addition, for example, is a simple fixed point of $\mathbf{N}_u$ by (1B-10) as is arithmetic subtraction by (2A-15) in Problem x2A.11. One might think that every **A**-recursive partial function $f : A^n \rightharpoonup A$ is a simple fixed point and this is almost true—but not exactly, even in $\mathbf{N}_u$:

2B.1. **Proposition** (Moschovakis [1984]). *If* $\mathbf{A} = (\mathbf{N}_u, \mathbf{\Phi})$ *is an expansion of the unary numbers by any finite set* $\mathbf{\Phi} = (\phi_1, \dots, \phi_k)$ *of total functions, then there exists a total function* $f : \mathbb{N} \to \mathbb{N}$ *which is* **A**-*recursive but is not a simple fixed point of* **A**.

A proof of this is outlined in Problems x2B.3 - x2B.5.

On the other hand, the **A**-recursive partial functions are very close to the simple fixed points of **A**, as follows:

Pointed structures. An element $a \in A$ in the universe of a Φ-structure **A** is *strongly explicit* if some closed Φ-term denotes a and the equality-with-a relation

(2B-2) $$\mathrm{eq}_a(x) \iff x = a$$

is **A**-explicit; and **A** is *pointed* if it has at least two strongly explicit elements. For example, $\mathbf{N}_u$, $\mathbf{N}_b$ and every field are pointed with $a = 0, b = 1$ and every expansion of a pointed structure is also pointed.

2B.2. **Proposition.** *If* **A** *is pointed and* $f : A^n \rightharpoonup A_s$, *then* f *is* **A**-*recursive if and only if there is a simple fixed point* $g : A^{m+n} \rightharpoonup A_s$ *of* **A** *such that*

(2B-3) $$f(\vec{x}) = g(\vec{a}, \vec{x}) \quad (\vec{x} \in A^n, \vec{a} = \underbrace{a, \dots, a}_{m \text{ times}}).$$

When this equation holds, we say that f is a *section of* g by explicit constants.

PROOF. In a simple case when

$$f(\vec{x}) = f_0(\vec{x}, p_1, p_2) \text{ where } \Big\{ p_1(\vec{x}_1) = f_1(\vec{x}_1, p_1, p_2), p_2(\vec{x}_2) = f_2(\vec{x}_2, p_1, p_2) \Big\},$$

let $\overline{r}$ be the solution of the recursive equation

$$\begin{aligned} r(s, t, \vec{x}_1, \vec{x}_2, \vec{x}) = {} & \text{if } (s = t = a) \\ & \quad \text{then } f_0(\vec{x}, (\lambda\vec{x}_1)r(a, a, \vec{x}_1, \vec{x}_2, \vec{x}), (\lambda\vec{x}_2)r(a, b, \vec{x}_1, \vec{x}_2, \vec{x})) \\ \text{else if } & (s = a \;\&\; t = b) \\ & \quad \text{then } f_1(\vec{x}_1, (\lambda\vec{x}_1)r(a, a, \vec{x}_1, \vec{x}_2, \vec{x}), (\lambda\vec{x}_2)r(a, b, \vec{x}_1, \vec{x}_2, \vec{x})) \\ & \qquad \text{else } f_2(\vec{x}_2, (\lambda\vec{x}_1)r(a, a, \vec{x}_1, \vec{x}_2, \vec{x}), (\lambda\vec{x}_2)r(b, a, \vec{x}_1, \vec{x}_2, \vec{x})) \end{aligned}$$

and check by an easy fixed-point-argument that for all $\vec{x}, \vec{x}_1$ and $\vec{x}_2$,

$$f_1(\vec{x}_1) = \overline{r}(a, b, \vec{x}_1, \vec{x}_2, \vec{x}_1), \quad f_2(\vec{x}_2) = \overline{r}(b, a, \vec{x}_1, \vec{x}_2, \vec{x}),$$
$$\text{and so } f(\vec{x}) = \overline{r}(a, a, \vec{x}_1, \vec{x}_2, \vec{x}).$$

It follows that if $\vec{a}_1, \vec{a}_2$ are sequences of respective lengths those of $\vec{x}_1$ and $\vec{x}_2$, then

$$f(\vec{x}) = \overline{r}(a, a, \vec{a}_1, \vec{a}_2, \vec{x}).$$

For recursive definitions with $K + 1$ parts, we replace (s, t) by a suitably long $(s_1, \ldots, s_k)$ so that $2^k \geq K + 1$ and then use tuples $\vec{a}$ and $\vec{b}$ of length k to combine the equations in the system in one. ⊣

The *minimal* ("free") *pointed extension of* $\mathbf{A}$ is the structure

$$\text{(2B-4)} \qquad \mathbf{A}[a, b] = (A \cup \{a, b\}, \mathbf{\Phi}^{a,b}, a, b, \mathrm{eq}_a, \mathrm{eq}_b),$$

where a, b are distinct objects not in A, for each $f : A^n \rightharpoonup A_s$,

$$\text{(2B-5)} \qquad f^{a,b}(x) = \begin{cases} f(x), & \text{if } x \in A^n, \\ \uparrow & \text{otherwise} \end{cases} \qquad (x \in A[a, b]^n),$$

and $\mathbf{\Phi}^{a,b} = \{\phi^{a,b} : \phi \in \mathbf{\Phi}\}$.

2B.3. **Proposition.** *For every structure* $\mathbf{A}$ *and* $f : A^n \rightharpoonup A_s$,

$$f \in \mathbf{Rec}(\mathbf{A}) \iff f^{a,b} \in \mathbf{Rec}(\mathbf{A}[a, b]).$$

Taken together, these two Propositions say that every $\mathbf{A}$-recursive partial function is a section of a fixed point, except that to realize this, we may need to add two strongly explicit points to $\mathbf{A}$. This is an important fact about abstract recursion, but we will not need it, and so we leave its (technically messy) proof and a variation of it for Problems x2B.1*, x2B.2.

McColm [1989] has also shown that *multiplication is not a simple fixed point of* $\mathbf{N}_u$, along with several other results in this classical case. The general problem of characterizing in a natural way the *simple fixed points* of a structure $\mathbf{A}$ is largely open and it is not clear what (if any) their algebraic or foundational significance might be, cf. Problem x2B.6.

Tail recursion. A partial function $\overline{p} : A^k \rightharpoonup A_s$ is defined by *tail recursion* from test $: A^k \rightharpoonup \mathbb{B}$, output $: A^k \rightharpoonup A_s$ and $\sigma : A^k \rightharpoonup A^k$ if it is the (canonical, least) solution of the recursive equation

$$\text{(2B-6)} \qquad p(\vec{u}) = \text{if test}(\vec{u}) \text{ then output}(\vec{u}) \text{ else } p(\sigma(\vec{u})).$$

Typical is the definition of $\gcd(x, y)$ in (1C-7) by tail recursion from

$$\mathrm{test}(x, y) = \mathrm{eq}_0(\mathrm{rem}(x, y)),\ \mathrm{output}(x, y) = y,\ \sigma(x, y), = (y, \mathrm{rem}(x, y)).$$

Tail recursion captures the notion of *explicit iteration*, perhaps best expressed by this

2B.4. **Proposition.** *The canonical, least solution of the tail recursive equation* (2B-6) *is given by*

$$\text{(2B-7)}\quad \overline{p}(\vec{x}) = \text{output}(\sigma^m(\vec{x}))\ \textit{where}\ m = \mu k\ \text{test}(\sigma^k(\vec{x}))$$
$$=_{\text{df}} \text{output}(\sigma^m(\vec{x}))\ \textit{where m is least such that}$$
$$(\forall i < m)[\text{test}(\sigma^i(\vec{x})){\downarrow}]\ \&\ \text{test}(\sigma^m(\vec{x})).$$

PROOF. We use the construction of $\overline{p}(\vec{x})$ in the proof of the Fixed Point Lemma 1B.1, which starts with the totally undefined $\overline{p}^0(\vec{x})$ and sets

$$\overline{p}^{k+1}(\vec{x}) = \text{if test}(\vec{x})\ \text{then output}(\vec{x}))\ \text{else}\ \overline{p}^m(\sigma^k(\vec{x})).$$

By induction on k, we get

$$\overline{p}^{k+1}(\vec{x}) = \begin{cases} \text{output}(\sigma^m(\vec{x})) & \text{where } m \leq k \text{ is least such that test}(\sigma^m(\vec{x})), \\ \uparrow & \text{if no such } m \leq k \text{ exists,} \end{cases}$$

and then $\overline{p} = \lim_k \overline{p}^k$ easily satisfies (2B-7). ⊣

Tail recursive programs and functions. An extended Φ-program is *tail recursive* if it is of the form

$$\text{(2B-8)}\quad E(\vec{x}) \equiv \mathsf{p}(\text{input}(\vec{x}))$$
$$\text{where}\ \big\{\mathsf{p}(\vec{u}) = \text{if test}(\vec{u})\ \text{then output}(\vec{u})\ \text{else}\ \mathsf{p}(\sigma(\vec{u}))\big\}$$

with Φ-explicit terms input, test, output, σ of suitable sorts, arities and co-arities; and a partial function $f : A^n \rightharpoonup A_s$ is *tail recursive* in a Φ-structure **A** if it is computed in **A** by a tail recursive program, i.e., if

$$\text{(2B-9)}\quad f(\vec{x}) = p(\text{input}(\vec{x}))$$
$$\overline{\text{where}}\ \big\{p(\vec{u}) = \text{if test}(\vec{u})\ \text{then output}(\vec{u})\ \text{else}\ p(\sigma(\vec{u}))\big\},$$

with **A**-explicit input, test, output, σ. We set

$$\textbf{Tailrec}^0(\mathbf{A}) = \{f : A^n \rightharpoonup A_s : n \geq 0 \text{ and } f \text{ is tail recursive in } \mathbf{A}\}.$$

In general, this is not a well-behaved class, cf. Problem x2B.7. If **A** is pointed, however, it is a very natural, well-structured class of partial functions, especially because it is then closed under definitions by

Mutual tail recursion. A system of *mutual tail recursion* in a structure **A** is a set of equations

$$\text{(2B-10)}\qquad E : \big\{p_1(\vec{x}_1) = f_1(\vec{x}_1, \vec{p}), \dots, p_K(\vec{x}_K) = f_K(\vec{x}_K, \vec{p})\big\}$$

where each $f_i(\vec{x}_i, \vec{p})$ is in one of the following two *forms*, with **A**-explicit $\text{test}_i(\vec{x}_i)$, output_i, $\sigma_{ij}(\vec{x}_i)$ and $\tau_{ik}(\vec{x}_i)$:

$$\text{(tail call)}\quad f_i(\vec{x}_i) = \text{if test}_i(\vec{x}_i)\ \text{then output}_i(\vec{x}_i)\ \text{else}\ p_j(\sigma_{ij}(\vec{x}_i))$$
$$\text{(branching call)}\quad f_i(\vec{x}_i) = \text{if test}_i(\vec{x}_i)\ \text{then}\ p_j(\sigma_{ij}(\vec{x}_i))\ \text{else}\ p_k(\tau_{ik}(\vec{x}_i)).$$

2B.5. **Theorem.** *If* **A** *is pointed and* $\overline{p}_1, \dots, \overline{p}_n$ *are the solutions of a system of mutual tail recursion* (2B-10), *then each* $\overline{p}_i$ *is tail recursive in* **A**.

PROOF. If all the equations in (2B-10) are branching calls, then their solutions are the partial functions with empty domains of convergence; and if they are all tail calls, then the proof is simple and we leave it for Problem x2B.10.

So we may assume that there is an L such that $1 \le L < K$, every $f_i(\vec{x}_i)$ is a tail call for $i \le L$ and every $f_i(\vec{x}_i)$ is a branching call when $L < i \le K$. We will also simplify notation by

$$\text{writing } x \text{ for } \vec{x}_1, \dots, \vec{x}_n$$

and pretending for the moment that **A** has K strongly explicit constants, $1, \dots, K$. With these conventions, we put

$$\begin{aligned} \text{test}(i,x) &\iff_{\text{df}} \textstyle\bigvee_{1\le j\le L}[i = j \;\&\; \text{test}_i(\vec{x}_i)] \\ &\iff [i = 1 \;\&\; \text{test}_1(\vec{x}_i] \vee \cdots \vee [i = L \;\&\; \text{test}_L(\vec{x}_L], \end{aligned}$$

$$\text{output}(i,x) \;=_{\text{df}} \begin{cases} \text{output}_1(\vec{x}_1) & \text{if } i = 1, \\ \quad\vdots & \\ \text{output}_L(\vec{x}_L) & \text{otherwise, if } i \ge L. \end{cases}$$

We look for a similar definition by cases of an explicit $\rho(i,x)$ of appropriate co-arity so that the single tail recursive equation

$$r(i,x) = \text{if test}(i,x) \text{ then output}(i,x) \text{ else } r(\rho(i,x)) \tag{2B-11}$$

is equivalent to E, in the following strong sense: for each $i = 1, \dots, K$ and all $x = \vec{x}_1, \dots, \vec{x}_K$,

$$\overline{p}_i(\vec{x}_i) = \overline{r}(i,x) = \overline{r}(i, \vec{x}_1, \dots, \vec{x}_K). \tag{2B-12}$$

This gives $\overline{p}_i(\vec{x}_i) = \overline{r}(\vec{1}, \vec{x}_i, \vec{2})$ with suitably long sequences $\vec{1}$ and $\vec{2}$, and so $\overline{p}_i$ is tail recursive.

With the notation conventions in (2B-10) and the specification of tail and branching calls below it, we set first

$$\rho(i,x) = (j, x\{\vec{x}_i := \sigma_{ij}(\vec{x}_i)\}) \quad (i \le L),$$

where $x\{\vec{x}_j := \sigma_{ij}(\vec{x}_i)\}$ is obtained from x by replacing $\vec{x}_j$ by $\sigma_{ij}(\vec{x}_i)$. For $i > L$, we put

$$\rho(i,x) = \text{if test}(i,x) \text{ then } (j, x\{\vec{x}_j := \sigma_{ij}(\vec{x}_i)\}) \text{ else } (k, x\{\vec{x}_k := \tau_{ik}(\vec{x}_i)\}).$$

We now prove (2B-12) by showing that for all i, x,

$$\overline{r}^t(i,x) = \overline{p}_i^t(\vec{x}_i)$$

for all the iterates that build up the solutions $\overline{p}_i, \overline{r}$ in the proof of the Fixed Point Lemma, Theorem 1B.1.

This is certainly true at $t = 0$, when all these partial functions have empty domain of convergence.

In the induction step, first for $i \leq L$,

$$\begin{aligned}
\overline{r}^{t+1}(i,x) &= \text{if test}(i,x) \text{ then output}(i,x) \text{ else } \overline{r}^t(\rho(i,x))\\
&= \text{if test}_i(\vec{x}_i) \text{ then output}_i(\vec{x}_i) \text{ else } \overline{r}^t(j, x\{\vec{x}_j := \sigma_{ij}(\vec{x}_i)\})\\
&= \text{if test}_i(\vec{x}_i) \text{ then output}_i(\vec{x}_i) \text{ else } \overline{p}^t_j(\sigma_{ij}(\vec{x}_i)) \quad \text{(ind. hyp.)}\\
&= \overline{p}_i^{t+1}(\vec{x}_i).
\end{aligned}$$

For $i > L$, test(i, x) is never true, so:

$$\begin{aligned}
\overline{r}^{t+1}(i,x) &= \overline{r}^t(\rho(i,x))\\
&= \text{if test}(i,x) \text{ then } \overline{r}^t(j, x\{\vec{x}_j := \sigma_{ij}(\vec{x}_i)\}) \text{ else } \overline{r}^t(k, x\{\vec{x}_k := \tau_{ik}(\vec{x}_i)\})\\
&= \text{if test}(i,x) \text{ then } \overline{p}^t_j(\sigma_{ij}(\vec{x}_i)) \text{ else } \overline{p}^t_k(\tau_{ik}(\vec{x}_i)) \quad \text{(ind. hyp.)}\\
&= \overline{p}_i^{t+1}(\vec{x}_i).
\end{aligned}$$

To remove the assumption that **A** has K strongly explicit constants, we replace $1, \ldots, K$ by sufficiently long sequences of two constants, as in the proof of Proposition 2B.2. ⊣

Problems x2B.9 – x2B.17 develop (with generous hints) the basic properties of $\mathbf{Tailrec}^0(\mathbf{A})$ on pointed structures, including the fact that *for every expansion* $(\mathbf{N}_u, \mathbf{\Phi})$ *of the unary numbers by total functions,*

$$\mathbf{Rec}^0(\mathbf{N}_u, \mathbf{\Phi}) = \mathbf{Tailrec}^0(\mathbf{N}_u, \mathbf{\Phi}). \tag{2B-13}$$

This is an important fact about (classical) recursion on the natural numbers, a version of the *Normal Form Theorem*, Kleene [1952][Theorem IX]. It holds for many "rich" structures but not generally: there are interesting examples of total, pointed structures in which $\mathbf{Rec}^0(\mathbf{A}) \neq \mathbf{Tailrec}^0(\mathbf{A})$ and others where every **A**-recursive function can be computed by a tail recursion but at a cost in "efficiency". We will describe some of these examples in Section 2G and on page 124.

Relativization. A functional $f(\vec{x}, \vec{r})$ is a *simple fixed point* of a structure **A** if

$$f(\vec{x}, \vec{r}) = p(\vec{x}) \text{ where } \Big\{ p(\vec{x}) = g(\vec{x}, p, \vec{r}) \Big\} \tag{2B-14}$$

with an **A**-explicit functional $g(\vec{x}, p, \vec{r})$, i.e., if f satisfies the *relativized version* of (2B-1) in which we allow the given g to depend on the arbitrary *parameters* $\vec{r}$. In the same way, $f(\vec{x}, \vec{r})$ is *tail recursive* in **A** if for suitable **A**-explicit

$h_1, \dots, h_k$, test, output,

$$\text{(2B-15)} \quad f(\vec{x}, \vec{r}) = p(h_1(\vec{x}, \vec{r}), \dots, h_k(\vec{x}, \vec{r})) \text{ where}$$
$$\Big\{ p(\vec{u}) = \text{if test}(\vec{u}, \vec{r}) \text{ then output}(\vec{u}, \vec{r}) \text{ else } p(\sigma_1(\vec{u}), \dots, \sigma_k(\vec{u})) \Big\},$$

and **Tailrec**(**A**) is the family of **A**-tail recursive functionals. Most of the basic properties of simple fixed points and tail recursive partial functions can be extended to functionals, basically by *relativizing their proofs*, inserting the parameters $\vec{r}$ wherever this makes sense.

The process of relativizing definitions and proofs is a standard tool of recursion theory and we will sometimes appeal to it to simplify arguments, when it can be applied routinely—which is most of the time.

Problems for Section 2B

x2B.1*. **Problem** (Adding points, 1). Prove Proposition 2B.3: that for every **A** and every $f : A^n \rightharpoonup A_s$,

$$f \in \mathbf{Rec}(\mathbf{A}) \iff f^{a,b} \in \mathbf{Rec}(\mathbf{A}[a, b]),$$

with $\mathbf{A}[a, b]$ and $f^{a,b}$ defined by (2B-4) and (2B-5). HINT: For the non-trivial direction ($\Longleftarrow$), we need to code an arbitrary $\mathbf{A}[a, b]$-explicit recursive equation by a (long) system of **A**-explicit equations; it is clear that this can be done in some way, and the trick is to formulate a precise lemma which gives it and which can be proved "by induction", i.e., by appealing to Proposition 1E.1.

x2B.2. **Problem** (Adding points, 2). Define $\mathbf{A}[a, b]^*$ like $\mathbf{A}[a, b]$, except that each $\phi \in \Phi$ is now interpreted by

$$\phi^*(x_1, \dots, x_n) = \text{if } x_1, \dots, x_n \in A \text{ then } \phi(x_1, \dots, x_n) \text{ else } a,$$

so that if **A** is a total structure, then so is $\mathbf{A}[a, b]^*$. Prove that for any partial function $f : A^n \rightharpoonup A_s$, $f \in \mathbf{Rec}(\mathbf{A}) \iff f^{a,b} \in \mathbf{Rec}(\mathbf{A}[a, b]^*$. HINT: Use Problem x2B.1*, do not repeat its proof.

The next three problems lead to a proof of Proposition 2B.1.

x2B.3. **Problem.** Suppose $F(x, p)$ is a continuous functional whose fixed point $\overline{p} : \mathbb{N} \to \mathbb{N}$ is a total, unary function, and let

$$\text{(2B-16)} \qquad \text{stage}(x) = \text{stage}_F(x) = \text{the least } k \text{ such that } \overline{p}^k(x)\downarrow \ -1$$

in the notation of Lemma 1B.1. Prove that for infinitely many x,

$$\text{stage}(x) \le x.$$

x2B.4. **Problem.** Suppose $\psi : \mathbb{N} \to \mathbb{N}$ is strictly increasing, i.e.,

$$x < y \Longrightarrow \psi(x) < \psi(y),$$

and set by recursion on $\mathbb{N}$,

$$\psi^0(x) = x, \quad \psi^{n+1}(x) = \psi(\psi^n(x)).$$

A unary partial function $f : \mathbb{N} \rightharpoonup \mathbb{N}$ is *n-bounded* (relative to ψ, for $n > 0$) if

$$f(x)\downarrow \Longrightarrow f(x) \leq \psi^n(x);$$

and a functional $F(x, p)$ (with p a variable over unary partial functions) is *ℓ-bounded* (relative to ψ), if for all p and $n \geq 1$,

$$\text{if } p \text{ is } n\text{-bounded, then for all } x,\ F(x, p) \leq \psi^{\ell n}(x).$$

Suppose $\mathbf{A} = (\mathbb{N}, \{\phi^{\mathbf{A}}\}_{\phi \in \Phi})$ is a total Φ-structure and every primitive $\phi^{\mathbf{A}} : \mathbb{N}^n \to \mathbb{N}$ is bounded by some fixed ψ as above, in the sense that

$$\phi(\vec{x}) \leq \psi(\max \vec{x}).$$

Prove that for every term $E(\mathsf{x}, \mathsf{p})$ in the vocabulary $\Phi \cup \{\mathsf{p}\}$, there is an ℓ such that the functional

$$F(x, p) = \text{den}(E(x, p))$$

is ℓ-bounded. HINT: You will need to verify that $\psi(x) \geq x$, because ψ is increasing, and hence, for all ℓ, ℓ'

$$\ell \leq \ell' \Longrightarrow \psi^{\ell}(x) \leq \psi^{\ell'}(x).$$

(This is also needed in the next problem.)

x2B.5. **Problem.** Prove Proposition 2B.1.

x2B.6. **Open problem.** Prove that if $\mathbf{A} = (\mathbf{N}_u, \mathbf{\Phi})$ is an expansion of the unary numbers with any set $\mathbf{\Phi} = (\phi_1, \dots, \phi_k)$ of total, $\mathbf{N}_u$-recursive functions, then there exists a total relation $R : \mathbb{N} \to \mathbb{B}$ which is recursive but not a simple fixed point of $\mathbf{A}$.

x2B.7. **Problem.** As in Problem x1E.9, let $\mathbf{A} = (A, \phi)$ where A is any non-empty set, $\phi : A \rightharpoonup A$ is not total, and for some ordering $\leq$ on A,

$$\phi(x)\downarrow \Longrightarrow x < \phi(x).$$

Prove that $f : A \rightharpoonup \mathbb{B}$ is tail recursive in $\mathbf{A}$ if either $f(x)$ is explicit or $f(x)\uparrow$ for every x. Infer that $\mathbf{Tailrec}^0(\mathbf{A})$ is not closed under composition.

x2B.8. **Problem.** Prove that $f : A^n \rightharpoonup A_s$ satisfies (2B-9) in a Φ-structure $\mathbf{A}$ exactly when it is computed by the procedure

set $\vec{u} := \text{input}(\vec{x})$;
  while $\neg\text{test}(\vec{u})$, set $\vec{u} := \sigma(\vec{u})$;
return $\text{output}(\vec{u})$.

as this is customarily understood.

In Problems x2B.9 – x2B.17 we outline briefly the elementary theory of tail recursion (mostly) in a pointed structure; this includes closure properties of **Tailrec**(**A**) for pointed **A** which mirror those for **Rec**(**A**) in Problems x2A.5 – x2A.14 but are not quite as easy to prove.

x2B.9. **Problem.** Prove that $\mathbf{Tailrec}^0(\mathbf{A})$ is closed under mangling and explicit substitutions, $f(\vec{x}) = g(h(\vec{x}))$ with **A**-explicit $h(\vec{x})$ of suitable co-arity.

x2B.10. **Problem.** Prove Theorem 2B.5 for systems of mutual tail recursion in which all the equations are tail calls. HINT: The needed construction is a "reduct" of the argument we gave for the general case.

x2B.11*. **Problem.** Prove that if **A** is pointed, then $\mathbf{Tailrec}^0(\mathbf{A})$ is closed under substitutions

$$f(\vec{x}) = g(h(\vec{x})) = g(h_1(\vec{x}), \dots, h_m(\vec{x})).$$

HINT: With $m = 1$, for simplicity, we are given representations

$$g(u) = \overline{p}_g(\text{input}_g(u)), \quad h(\vec{x}) = \overline{p}_h(\text{input}_h(\vec{x})),$$

where $\overline{p}_g(\vec{v}), \overline{p}_h(u)$ are the solutions of the two recursive equations

$$p_g(\vec{v}) = \text{if test}_g(\vec{v}) \text{ then output}_g(\vec{v}) \text{ else } p_g(\sigma_g(\vec{v}))$$
$$p_h(u) = \text{if test}_h(u) \text{ then output}_h(u) \text{ else } p_h(\sigma_h(u)).$$

Replace the second equation by

$$q_h(u) = \text{if test}_h(u) \text{ then } p_g(\text{input}_g(\text{output}_h(u)) \text{ else } q_h(\sigma_h(u))$$

and prove that for the resulting system of mutual tail recursion,

$$\overline{q}_h(u) = g(\overline{p}_h(u)),$$

which then yields $g(h(\vec{x})) = g(\overline{p}_h(\text{input}_h(\vec{x}))) = \overline{q}_h(\text{input}_h(\vec{x}))$.

x2B.12*. **Problem.** Prove that if **A** is pointed, then $\mathbf{Tailrec}^0(\mathbf{A})$ is closed under branching

$$f(\vec{x}) = \text{if } f_1(\vec{x}) \text{ then } f_2(\vec{x}) \text{ else } f_3(\vec{x}).$$

HINT: This is quite simple when $f_1(\vec{x})$ is explicit. In the general case, we are given three tail recursive definitions

$$f_i(\vec{x}) = p_i(\text{input}_i(\vec{x}))$$
$$\text{where } \Big\{ p_i(u_i) = \text{if test}_i(u_i) \text{ then output}_i(u_i) \text{ else } p_i(\sigma_i(u_i)) \Big\}$$

for $i = 1, 2, 3$, where u_i varies over some product A^{k_i} and all the twelve partial functions in them are explicit in **A**.

Let $\overline{q}_1, \overline{q}_{11}, \overline{r}, \overline{q}_2, \overline{q}_{21}, \overline{q}_3, \overline{q}_{31}$ be the solutions of the following system E of mutual tail recursion:

$$
\begin{aligned}
q_1(\vec{x}) &= q_{11}(\vec{x}, \text{input}_1(\vec{x}))\\
q_{11}(\vec{x}, u) &= \text{if } \text{test}_1(u) \text{ then } r(\vec{x}, u) \text{ else } q_{11}(\vec{x}, \sigma_1(u))\\
r(\vec{x}, u) &= \text{if } \text{output}_1(u) \text{ then } q_2(\vec{x}) \text{ else } q_3(\vec{x})\\
q_2(\vec{x}) &= q_{21}(\vec{x}, \text{input}_2(\vec{x}))\\
q_{21}(\vec{x}, v) &= \text{if } \text{test}_2(v) \text{ then } \text{output}_2(v) \text{ else } q_{21}(\vec{x}, \sigma_2(\vec{x}))\\
q_3(\vec{x}) &= q_{31}(\vec{x}, \text{input}_3(\vec{x}))\\
q_{31}(\vec{x}, w) &= \text{if } \text{test}_3(w) \text{ then } \text{output}_3(w) \text{ else } q_{31}(\vec{x}, \sigma_3(w)).
\end{aligned}
$$

By Proposition 2B.4,

$$\overline{q}_{21}(\vec{x}, v) = \text{output}_2(\sigma_2^m(v)) \text{ with } m \text{ least such that } \text{test}_2(\sigma_2^m(v))$$

and so

$$
\begin{aligned}
\overline{q}_2(\vec{x}) = \text{output}_2(&\sigma_2^m(\text{input}_2(\vec{x}))\\
&\text{with } m \text{ least such that } \text{test}_2(\sigma_2^m(\text{input}_2(\vec{x}))) = f_2(\vec{x}),
\end{aligned}
$$

and similarly $\overline{q}_3(\vec{x}) = f_3(\vec{x})$. It takes just a bit more work to show that

$$\overline{q}_1(\vec{x}) = \text{if } f_1(\vec{x}) \text{ then } f_2(\vec{x}) \text{ else } f_3(\vec{x}) = f(\vec{x}).$$

x2B.13*. **Problem.** Prove that if **A** is pointed, then $\textbf{Tailrec}^0(\textbf{A})$ is closed under tail recursion.

HINT: By Problem x2B.11*, it is enough to prove that if

$$f(x) = p(x) \ \underline{\text{where}} \left\{ p(x) = \text{if } \text{test}(x) \text{ then } \text{output}(x) \text{ else } p(\sigma(x)) \right\}$$

and for some **A**-explicit $\text{input}_i, \text{test}_i, \text{output}_i, \sigma_i$ on suitable products of A

$$
\begin{aligned}
\text{test}(x) &= p_1(\text{input}_1(x))\\
&\qquad \underline{\text{where}} \left\{ p_1(u) = \text{if } \text{test}_1(u) \text{ then } \text{output}_1(u) \text{ else } p_1(\sigma_1(u)) \right\},\\
\text{output}(x) &= p_2(\text{input}_2(x))\\
&\qquad \underline{\text{where}} \left\{ p_2(v) = \text{if } \text{test}_2(v) \text{ then } \text{output}_2(v) \text{ else } p_2(\sigma_2(v)) \right\},\\
\sigma(x) &= p_3(\text{input}_3(x))\\
&\qquad \underline{\text{where}} \left\{ p_3(w) = \text{if } \text{test}_3(w) \text{ then } \text{output}_3(w) \text{ else } p_3(\sigma_3(w)) \right\}.
\end{aligned}
$$

Let $\overline{q}_1, \overline{q}_{11}, \overline{r}, \overline{q}_2, \overline{q}_{21}, \overline{q}_3, \overline{q}_{31}$ be the solutions of the following system E of mutual tail recursion:

$$\begin{aligned}
q_1(x) &= q_{11}(x, \text{input}_1(x))\\
q_{11}(x,u) &= \text{if test}_1(u) \text{ then } r(x,u) \text{ else } q_{11}(x, \sigma_1(u))\\
r(x,u) &= \text{if output}_1(u) \text{ then } q_2(x) \text{ else } q_3(x)\\
q_2(x) &= q_{21}(x, \text{input}_2(x))\\
q_{21}(x,v) &= \text{if test}_2(v) \text{ then output}_2(v) \text{ else } q_{21}(x, \sigma_2(x))\\
q_3(x) &= q_{31}(x, \text{input}_3(x))\\
q_{31}(x,w) &= \text{if test}_3(w) \text{ then } q_1(\text{output}_3(w)) \text{ else } q_{31}(x, \sigma_3(w)).
\end{aligned}$$

The key fact about this system is that

$$\overline{q}_1(x) = \text{if test}(x) \text{ then } \overline{q}_2(x) \text{ else } \overline{q}_3(x),$$

which is easy to check using Proposition 2B.4; together with the simpler

$$\overline{q}_2(x) = \text{output}(x), \quad \overline{q}_3(x) = \overline{q}_1(\sigma(x)),$$

it gives

$$\begin{aligned}
\overline{q}_1(x) &= \text{if test}(x) \text{ then output}(x) \text{ else } \overline{q}_1(\sigma(x))\\
&= \text{output}(\sigma^m(x)) \text{ where } m \text{ is least such that test}(\sigma^m(x)) = f(x).
\end{aligned}$$

x2B.14. **Problem** (**Tailrec^0** transitivity). Prove that if **A** is pointed and $\psi : A^k \rightharpoonup A_s$ is tail recursive in **A**, then $\mathbf{Tailrec}^0(\mathbf{A}, \psi) = \mathbf{Tailrec}^0(\mathbf{A})$.

x2B.15. **Problem.** Prove that if $f : \mathbb{N}^{n+2} \to \mathbb{N}$ is defined by primitive recursion from $\mathbf{N}_u$-tail recursive functions, then f is tail recursive in $\mathbf{N}_u$. Infer that every primitive recursive function on $\mathbb{N}$ is $\mathbf{N}_u$-tail recursive.

HINT: It suffices to prove that if $f(y, \vec{x}, w)$ is defined by

$$f(0, \vec{x}, w) = w, \quad f(y+1, \vec{x}, w) = h(f(y, \vec{x}, w), y, \vec{x})$$

with $h(u, y, \vec{x})$ tail recursive in **A** then $f(y, \vec{x}, w)$ is tail recursive in **A**. Let $\overline{p}(i, j, \vec{x}, w)$ be the canonical solution to the recursive equation

$$p(i, y, \vec{x}, w) = \text{if } (i = 0) \text{ then } w \text{ else } p(i \mathbin{\dot{-}} 1, y+1, \vec{x}, h(w, y, \vec{x})).$$

This is tail recursive in $(\mathbf{N}_u, h)$ and hence tail recursive in $\mathbf{N}_u$, so it suffices to prove that $f(y, \vec{x}, w) = \overline{p}(0, y, \vec{x}, w)$. Skipping the parameters $\vec{x}$ which do not enter the argument, this follows from

$$\overline{p}(i, y, f(y, w)) = f(i + y, w).$$

x2B.16. **Problem.** Prove that if $g(y, \vec{x})$ is $\mathbf{N}_u$-tail recursive, then so is

$$f(\vec{x}) = \mu y[g(y, \vec{x}) = 0],$$

cf. (2A-17).

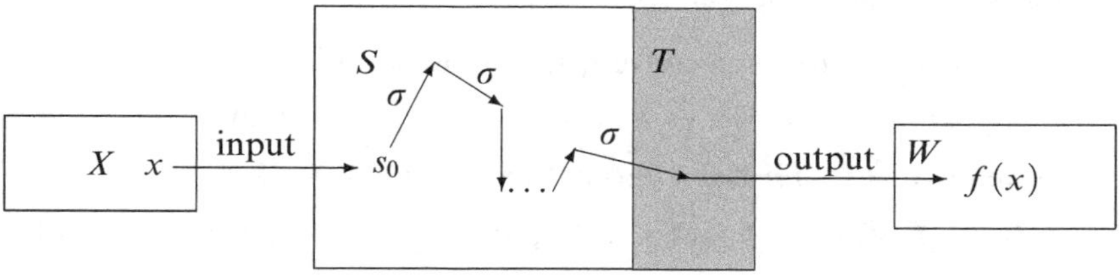

FIGURE 1. Total iterator computing $f : X \rightharpoonup W$.

x2B.17. **Problem.** Prove that if $(\mathbf{N}_u, \mathbf{\Phi})$ is an expansion of $\mathbf{N}_u$ by total functions, then $\mathbf{Rec}^0(\mathbf{N}_u, \mathbf{\Phi}) = \mathbf{Tailrec}^0(\mathbf{N}_u, \Phi)$.

HINT: By the classical Normal Form Theorem of Kleene,

$$f \in \mathbf{Rec}(\mathbf{N}_u, \mathbf{\Phi}) \Longrightarrow f(\vec{x}) = U(\mu y[g(y, \vec{x}) = 0])$$

with suitable U, g which are primitive recursive in $\mathbf{\Phi}$. This takes some computation to prove, which cannot be avoided here.

2C. Iterators (sequential machines, computation models)

All the standard deterministic models of computation for partial functions $f : X \rightharpoonup W$ on one set to another are captured by the following, well-known, general notion[16]: for any two sets X and W, a (partial) *iterator* or *sequential machine*

$$\mathrm{i} : X \rightsquigarrow W$$

is a quintuple (input, S, σ, T, output), satisfying the following conditions:

(I1) S is a non-empty set, the *set of states* of i;
(I2) input : $X \rightharpoonup S$ is the *input function* of i;
(I3) $\sigma : S \rightharpoonup S$ is the *transition function* of i;
(I4) $T \subseteq S$ is the set of *terminal states* of i, and $s \in T \Longrightarrow \sigma(s) = s$;
(I5) output : $T \rightharpoonup W$ is the *output function* of i.

Most often—and most usefully—the iterator i is *total*, i.e., input(x), $\sigma(s)$ and output(s) are all total, but it is useful to have around the general case.

A *partial computation* of i is any finite sequence $(s_0, \dots, s_n)$ of states such that for all $i < n$, s_i is not terminal and $\sigma(s_i) = s_{i+1}$, and it is *convergent* if, in addition, $s_n \in T$. Note that (with $n = 0$), this includes every one-term

[16]Cf. van Emde Boas [1990, 1.2].

sequence (s), and (s) is convergent if $s \in T$. We write

(2C-1) $s \to_{\mathrm{i}}^{*} s'$ there is a convergent computation $(s_0, \dots, s_n)$
with $s_0 = s, s_n = s'$,

and we say that i *computes* a partial function $f : X \rightharpoonup W$ if

(2C-2) $f(x) = w \iff (\exists s \in T)[\mathrm{input}(x) \to_{\mathrm{i}}^{*} s \;\&\; \mathrm{output}(s) = w]$.

It is clear that there is at most one convergent computation starting from any state s_0, and so exactly one partial function $\bar{\mathrm{i}} : X \rightharpoonup W$ is computed by i. *The computation* of i on x is the finite sequence

(2C-3) $\mathrm{Comp}_{\mathrm{i}}(x) = (\mathrm{input}(x), s_1, \dots, s_n, \mathrm{output}(s_n)) \quad (x \in X, \bar{\mathrm{i}}(x)\downarrow)$,

such that $(\mathrm{input}(x), s_1, \dots, s_n)$ is a convergent computation, and its length

(2C-4) $\mathrm{Time}_{\mathrm{i}}(x) = n + 2$

is the natural *time complexity* of i.

There is little structure to this definition of course, and the important properties of specific computation models derive from the judicious choice of the set of states and the transition function, but also the input and output functions. The first two depend on what operations (on various *data structures*) are assumed as given (primitive) and regulate how the iterator calls them, while the input and output functions often involve *representing* the members of X and W in some specific way, taking for example numbers in unary or binary notation if $X = W = \mathbb{N}$.

We will review the definitions of some of the most commonly used computation models in Section 2F.

Reduction of iteration to tail recursion. Fix an iterator i and let

$$A_{\mathrm{i}} = X \uplus W \uplus S$$

be the *disjoint union* (as in (1D-7)) of its input set, its output set and the set of its states. We identify, as usual, each $Z \subseteq A_{\mathrm{i}}$ with its characteristic function $Z : A_{\mathrm{i}} \to \mathbb{B}$ and we set

(2C-5) $\mathbf{A}_{\mathrm{i}} = (A_{\mathrm{i}}, X, W, S, \mathrm{input}, \sigma, T, \mathrm{output})$,

where X, W, S are viewed as subsets of A_{i} and $\mathrm{input}(x), \sigma(s), \mathrm{output}(s)$ are the functions of the iterator, viewed now as partial functions on A_{i} which return their argument when it is not in the appropriate input set. This is the *structure of* i, it is a total structure when i is a total iterator, and the *extended tail recursive program associated with* i is

(2C-6) $E_{\mathrm{i}}(\mathsf{x}) \equiv \mathsf{q}(\mathsf{input}(\mathsf{x}))$ where
$\Big\{\mathsf{q}(\mathsf{s}) = \text{if } T(\mathsf{s}) \text{ then } \mathrm{output}(\mathsf{s}) \text{ else } \mathsf{q}(\sigma(\mathsf{s}))\Big\}$.

2C.1. **Theorem.** *For all* $x \in X$, $\bar{\mathfrak{i}}(x) = \text{den}(\mathbf{A}_{\mathfrak{i}}, E_{\mathfrak{i}}(x))$.

In particular, the partial function computed by an iterator $\mathfrak{i} : X \rightsquigarrow W$ *is tail recursive in the associated structure* $\mathbf{A}_{\mathfrak{i}}$.

This follows easily from Proposition 2B.4 and we leave its proof for Problem x2C.2. Its significance is that it reduces *computability*, as it is captured by any specific computation model $\mathfrak{i}$, to *tail recursiveness* relative to the data structures and primitives of $\mathfrak{i}$. It is sometimes more interesting in the following version, "internal" to a structure **A**.

Explicit representation. A bijection $\pi : S \rightarrowtail\!\!\!\!\twoheadrightarrow A^k$ is an *explicit representation* of an iterator

$$\mathfrak{i} = (\text{input}, S, \sigma, T, \text{output}) : A^n \rightharpoonup A_s$$

in a Φ-structure **A** with universe A, if there are (necessarily unique) **A**-explicit partial functions

$$\text{input}_\pi : A^n \rightharpoonup A^k, \ \sigma_\pi : A^k \rightharpoonup A^k, \ T_\pi : A^k \rightharpoonup \mathbb{B}, \ \text{output}_\pi : A^k \rightharpoonup A_s$$

such that

$$\text{input}_\pi(\vec{x}) = \pi(\text{input}(\vec{x})), \ \sigma_\pi(\pi(s)) = \pi(\sigma(s)),$$
$$T_\pi(\pi(s)) = T(s), \ \text{output}_\pi(\pi(s)) = \text{output}(s).$$

2C.2. **Theorem.** *For any structure* **A**, *a partial function* $f : A^n \rightharpoonup A_s$ *is tail recursive in* **A** *if and only if it is computed by an iterator* $\mathfrak{i}$ *which is explicitly representable in* **A**.

This, too is an easy consequence of Proposition 2B.4 and we leave it for Problem x2C.3, but it is the main tool for relating computability and complexity theory developed (as is usual) using computation models to abstract recursion. For example:

2C.3. **Proposition** (Turing computability and recursion). *Every deterministic Turing machine which operates on natural numbers using their binary expansions is explicitly representable in the structure* $\mathbf{N}_b = (\mathbb{N}, 0, \text{parity}, \text{iq}_2, \text{em}_2, \text{om}_2, \text{eq}_0)$ *of binary arithmetic.*

As a consequence, a functional $f(\vec{x}, \vec{r})$ *on* $\mathbb{N}$ *is computable by a deterministic Turing machine if and only if it is tail recursive in* $\mathbf{N}_b$—which is equivalent to f being recursive in $\mathbf{N}_b$ or $\mathbf{N}_u$.

OUTLINE OF PROOF. The second claim follows from the general theory of tail recursion we developed in Section 2B and some unavoidable Turing machine programming.

For a detailed proof of the main claim, we would need to settle on one of a myriad of precise specifications of Turing machines and do a lot of coding, which is not of the moment. We explain, instead, the simple idea which is needed using a toy example and leave for Problem x2C.4 the argument for a more general, reasonably realistic case.

Suppose a Turing machine M has only one two-way infinite tape, only one symbol in its alphabet, 1, (internal) states $Q_0, \dots, Q_k$ (with Q_0 declared *initial*), and computes a unary partial function $f : \mathbb{N} \rightharpoonup \mathbb{N}$ by operating on the unary (tally) representation of numbers

$$\overline{n} = \underbrace{1 \cdots 1}_{n+1},$$

so that $\overline{0} = 1, \overline{1} = 11$, etc. If we use 0 to denote the blank square, then the "complete configuration" of M at a stage in a computation is a triple (Q_j, τ, i), where $\tau : \mathbb{Z} \to \{0, 1\}$, $\tau(l) = 0$ for all but finitely many l's, and $i \in \mathbb{Z}$ is the location of the scanned cell; and if we then write τ as a pair of sequences emanating from the scanned cell y_0

$$\cdots x_3 x_2 x_1 x_0 \underset{\uparrow}{y_0} y_1 y_2 y_3 \cdots$$

one "growing" to the left and the other to the right, we can code (τ, i) by the pair of numbers

$$(x, y) = (\textstyle\sum_k x_k 2^k, \sum_k y_k 2^k)$$

and code the complete configuration of M by the triple of numbers (j, x, y). So we can identify the set S of states of the iterator defined by M with $\mathbb{N}^3$, and then $\text{output}(j, x, y) = y$ and the set T of terminal states is defined by cases on j, keeping in mind that the numbers $0, \dots, k$ are all strongly explicit in $\mathbf{N}_b$—as are all numbers.

To check that the transition function $\sigma : \mathbb{N}^3 \to \mathbb{N}^3$ is also explicit in $\mathbf{N}_b$, notice first that $y_0 = \text{parity}(y)$ so that the scanned symbol can be computed from x and y by $\mathbf{N}_b$-operations. The input configuration for the number n is coded by the triple $(Q_0, 0, y)$ where $\overline{n}$ is the binary expansion of y, i.e., $y = 1 + 2 + \cdots + 2^n = 2^{n+1} - 1$; and all machine operations correspond to simple $\mathbf{N}_b$-explicit functions on these codes. For example:

$$\text{move to the right} : x \mapsto 2x + \text{parity}(y), \quad y \mapsto \text{iq}_2(y),$$

$$\text{move to the left} : x \mapsto \text{iq}_2(x), \quad y \mapsto 2y + \text{parity}(x),$$

$$\text{print 1 on the scanned square} : x \mapsto x, \quad y \mapsto 1 + \text{em}(\text{iq}_2(y)) = \text{om}(\text{iq}_2(y)),$$

where, with the notation of (1D-2),

$$2x + \text{parity}(y) = \text{if } (\text{parity}(y) = 0) \text{ then } \text{em}_2(x) \text{ else } \text{om}_2(x).$$

Using these functions, it is quite simple to construct an explicit definition of σ (setting $\sigma(j, x, y) = (j, x, y)$ if $j > k$).

If M has two symbols 1 and a and we want it to receive the input in binary notation, which is typical, then the scanned symbol can be any one of blank,

0, 1 or a, we code these by the four pairs 00, 10, 01 and 11 respectively, and we "pad" the tapes so that they have even length and they can be read as sequences of these symbols. If, for example, the input is 4, which is 001 in binary, then the initial configuration is $(0, 0, y)$ where the binary expansion of y is 101001 so $y = 1 + 2^2 + 2^5 = 1 + 4 + 32 = 37$. It is again easy to check that the elementary operations of M are explicit in $\mathbf{N}_b$—and the same well-known technique of using binary sequences of fixed length to code any finite number of symbols and padding the tapes works in general. ⊣

About implementations (I). An *implementation of a "program" or "algorithm" which computes a partial function* $f : X \rightharpoonup W$ *is*—at a minimum—*an iterator* $\mathfrak{i} : X \rightsquigarrow W$ *which also computes* f, it is *correct*. They come in many varieties and are classified in many ways, especially by whether they are *sequential* or *parallel* and by what *resources* (e.g., *space*) they use.

None of this is precise, of course, as there is no general agreement on what "programs" (in general) or "algorithms" are, and it is certainly not enough: no one would argue that the merge-sort and the insert-sort have the same implementations, even though they both compute the sorting function from the same primitives. On the other hand, these terms are intuitively understood quite well, they are widely used and there are few arguments about whether a particular iterator is an implementation of a specific algorithm, whether it is parallel or sequential, etc.

We will define a (classical) sequential implementation of recursive programs in the next section and we will also discuss briefly the (difficult) problem of making these notions and claims about them precise in Section 2H; other than that, we will use these terms loosely, with their customary, intuitive meaning when they help explain or motivate what we want to do or have done. For example, we can read Theorem 2C.1 as claiming (in part) that an iterator $\mathfrak{i}$ implements the tail recursive program

$$E_{\mathfrak{i}} \ : \ \mathsf{p}(\mathrm{input}(\vec{\mathsf{x}})) \text{ where } \Big\{\mathsf{p}(\vec{\mathsf{u}}) = \text{if test}(\vec{\mathsf{u}}) \text{ then output}(\vec{\mathsf{u}}) \text{ else } \mathsf{p}(\sigma(\vec{\mathsf{u}}))\Big\}$$

which expresses it in $\mathbf{A}_{\mathfrak{i}}$, but it is clear from the proof that much more than that is established about the connection between $\mathfrak{i}$ and $E_{\mathfrak{i}}$, cf. Problem x3A.3.

Problems for Section 2C

x2C.1. **Problem.** True or false: every partial function $f : X \rightharpoonup W$ is computable by a total iterator $\mathfrak{i} : X \rightsquigarrow W$.

x2C.2. **Problem** (Theorem 2C.1). Prove that the partial function computed by an iterator $\mathfrak{i} : X \rightsquigarrow W$ is tail recursive in $\mathbf{A}_{\mathfrak{i}}$.

x2C.3. **Problem** (Theorem 2C.2). Prove that $f : A^n \rightharpoonup A_s$ is tail recursive in a structure **A** exactly when it is computed by an iterator $\mathfrak{i}$ which is explicitly representable in **A**.

x2C.4. **Problem.** Outline a proof of Proposition 2C.3 for a Turing machine M which computes a functional $f(x, y, p)$ on $\mathbb{N}$ with p ranging over unary partial functions, has K symbols in its alphabet and uses two two-way tapes—one of them to call the "oracle" for p.

2D. The recursive machine

We associate here with each Φ-structure **A** and each extended Φ-program $E(\vec{x})$ an iterator which computes the denotation of $E(\vec{x})$ in **A**. This is one of the classical *implementations of recursion* and an important tool for studying the connection between recursion and computation.

An $(\mathbf{A}, E)$-*term* is a closed term

$$M \equiv N(y_1, \dots, y_m), \tag{2D-1}$$

where $N(\vec{y})$ is an extended term such that N is a subterm of one of the parts E_i of E and $y_1, \dots, y_m \in A$. These are voc(E)-terms with parameters from A, but not all such: the $(\mathbf{A}, E)$-terms are constructed by substituting parameters from A into the *finitely many* subterms of E.

The states of $\mathfrak{i} = \mathfrak{i}(\mathbf{A}, E(\vec{x}))$ are all finite sequences s of the form

$$a_0 \cdots a_{m-1} : b_0 \cdots b_{n-1}$$

where the elements $a_0, \dots, a_{m-1}, b_0, \dots, b_{n-1}$ of s satisfy the following conditions:

- Each a_i is a function symbol in Φ, or one of $\mathsf{p}_1, \dots, \mathsf{p}_K$, or the special symbol ?, or an $(\mathbf{A}, E)$-term, and
- each b_j is a parameter from A or a truth value, i.e., $b_j \in A \cup \mathbb{B}$.

The special *separator symbol* ':' has exactly one occurrence in each state, and the sequences $\vec{a}, \vec{b}$ are allowed to be empty, so that the following sequences are states (with $x \in A \cup \mathbb{B}$):

$$x : \qquad : x \qquad :$$

The *terminal states* of $\mathfrak{i}$ are the sequences of the form

$$: w$$

i.e., those with no elements on the left of ':' and just one constant on the right; the *output* function of $\mathfrak{i}$ simply reads this constant w, i.e.,

$$\text{output}(: w) = w;$$

(pass)	$\vec{a}\,\underline{x :}\,\vec{b} \;\to\; \vec{a} : x\,\vec{b} \quad (x \in A)$
(e-call)	$\vec{a}\,\underline{\phi_i : \vec{x}}\,\vec{b} \;\to\; \vec{a} : \phi_i^{\mathbf{A}}(\vec{x})\,\vec{b}$
(i-call)	$\vec{a}\,\underline{\mathsf{p}_i : \vec{x}}\,\vec{b} \;\to\; \vec{a}\,E_i(\vec{x}, \vec{\mathsf{p}}) : \vec{b}$
(comp)	$\vec{a}\,\underline{h(F_1, \ldots, F_n) :}\,\vec{b} \;\to\; \vec{a}\,h\,F_1 \cdots F_n : \vec{b}$
(br) (br0) (br1)	$\vec{a}\,\underline{\text{if } F \text{ then } G \text{ else } H :}\,\vec{b} \;\to\; \vec{a}\,G\,H\,?\,F : \vec{b}$ $\vec{a}\,\underline{G\,H\,? : \mathrm{tt}}\,\vec{b} \;\to\; \vec{a}\,G : \vec{b}$ $\vec{a}\,\underline{G\,H\,? : \mathrm{ff}}\,\vec{b} \;\to\; \vec{a}\,H : \vec{b}$

- The underlined words are those which trigger a transition and are changed by it.
- In (pass), $x \in A \cup \mathbb{B}$.
- In the *external call* (e-call), $\vec{x} = x_1, \ldots, x_n$, $\phi_i \in \Phi$, and arity$(\phi_i) = n$.
- In the *internal call* (i-call), p_i is an n-ary recursive variable of E defined by the equation $\mathsf{p}_i(\vec{\mathsf{x}}) = E_i(\vec{\mathsf{x}}, \vec{\mathsf{p}})$.
- In the *composition transition* (comp), h is a (constant or variable) function symbol in voc(E) with arity$(h) = n$.

TABLE 1. Transition Table for the recursive machine $\mathrm{i}(\mathbf{A}, E)$.

and the input function uses the head term $E_0(\vec{\mathsf{x}})$ of $E(\vec{\mathsf{x}})$,

$$\text{input}(\vec{x}) \equiv E_0(\vec{x}) :$$

The transition function of i is defined by the seven cases in the Transition Table 1, i.e.,

$$\sigma(s) = \begin{cases} s', & \text{if } s \to s' \text{ is a special case of some line in Table 1,} \\ s, & \text{otherwise,} \end{cases}$$

and it is a partial function, because for a given s (clearly) at most one transition $s \to s'$ is *activated* by s. Notice that only the external calls depend on the structure $\mathbf{A}$ and only the internal calls depend on the program E—and so, in particular, all programs with the same body share the same transition function.

An illustration of how these machines compute is given in Figure 2 on page 77.

The next result is a trivial but very useful observation:

2D.1. **Lemma** (Transition locality). *If $s_0, s_1, \ldots, s_n$ is a partial computation of $\mathrm{i}(\mathbf{A}, E)$ and $\vec{a}^*, \vec{b}^*$ are such that the sequence $\vec{a}^*\, s_0\, \vec{b}^*$ is a state, then the sequence*

$$\vec{a}^*\, s_0\, \vec{b}^*, \vec{a}^*\, s_1\, \vec{b}^*, \ldots, \vec{a}^*\, s_n\, \vec{b}^*$$

is also a partial computation of $\mathrm{i}(\mathbf{A}, E)$.

2D.2. **Theorem** (Implementation correctness). (1) *Suppose $\mathbf{A}$ is a Φ-structure, $E(\vec{\mathsf{x}})$ is an extended Φ-program with recursive variables $\mathsf{p}_1, \ldots, \mathsf{p}_K$, the partial functions $\overline{p}_1, \ldots, \overline{p}_K$ are the mutual fixed points in $\mathbf{A}$ of the system in the body of $E(\vec{\mathsf{x}})$*

$$p_i(\vec{x}_i) = \mathrm{den}((\mathbf{A}, p_1, \ldots, p_K), E_i(\vec{x}_i)) \quad (i = 1, \ldots, K) \tag{2D-2}$$

and M is an $(\mathbf{A}, E)$-term. Then for every $w \in A \cup \mathbb{B}$,

$$\mathrm{den}((\mathbf{A}, \overline{p}_1, \ldots, \overline{p}_K), M) = w \iff M : \to^*_{\mathrm{i}(\mathbf{A}, E(\vec{\mathsf{x}}))} : w. \tag{2D-3}$$

In particular, with $M \equiv E_0(\vec{x})$,

$$\mathrm{den}(\mathbf{A}, E(\vec{x})) = w \iff E_0(\vec{x}) \to^*_{\mathrm{i}(\mathbf{A}, E(\vec{\mathsf{x}}))} : w,$$

and so the extended program $E(\vec{\mathsf{x}})$ and the recursive machine $\mathrm{i}(\mathbf{A}, E(\vec{\mathsf{x}}))$ compute the same partial function in $\mathbf{A}$.

(2) *Similarly for functionals: if $E(\vec{\mathsf{x}})$ is an extended Φ program with parameters $\vec{\mathsf{r}} \equiv \mathsf{r}_1, \ldots, \mathsf{r}_k$ which computes a functional $f(\vec{x}, \vec{r})$ on A, then, for every $\vec{r}$, the recursive machine $\mathrm{i}((\mathbf{A}, \vec{r}), E(\vec{\mathsf{x}}))$ associated with $E(\vec{\mathsf{x}})$ and $\vec{r}$ computes $f(\vec{x}, \vec{r})$.*

OUTLINE OF PROOF. With $\mathbf{A}$ and $E(\vec{\mathsf{x}})$ fixed, we write $\mathrm{i} = \mathrm{i}(\mathbf{A}, E(\vec{\mathsf{x}}))$.

(1) First we define the partial functions computed by i,

$$\widetilde{p}_i(\vec{x}_i) = w \iff \mathsf{p}_i(\vec{x}_i) : \to^*_{\mathrm{i}} : w \quad (i = 1, \ldots, K)$$

and show by an easy induction on the closed term F that

$$\mathrm{den}((\mathbf{A}, \widetilde{p}_1, \ldots, \widetilde{p}_K), F) = w \iff F : \to^*_{\mathrm{i}} : w. \tag{2D-4}$$

When we apply this to the terms $E_i(\vec{x}_i)$, we get

$$\mathrm{den}((\mathbf{A}, \widetilde{p}_1, \ldots, \widetilde{p}_K), E_i(\vec{x}_i)) = w \iff E_i(\vec{x}_i) \to^*_{\mathrm{i}} w \iff \widetilde{p}_i(\vec{x}_i) = w;$$

which then means that the partial functions $\widetilde{p}_1, \ldots, \widetilde{p}_K$ satisfy the system (2D-2).

Next we show that for any closed term F as above and any system $p_1, \ldots, p_K$ of solutions of (2D-2),

$$F : \to^*_{\mathrm{i}} w \Longrightarrow \mathrm{den}((\mathbf{A}, p_1, \ldots, p_K), F) = w.$$

$$
\begin{array}{rcll}
f(2,3) & : & & (\text{comp}) \\
f\ 2\ 3 & : & & (\text{pass, pass}) \\
f & : & 2\ 3 & (\text{i-call}) \\
\text{if } (2=0) \text{ then } 3 \text{ else } S(f(\text{Pd}(2),3)) & : & & (\text{br}) \\
3\ S(f(\text{Pd}(2),3))\ ?\ \text{eq}_0(2) & : & & (\text{comp, pass, e-call}) \\
3\ S(f(\text{Pd}(2),3))\ ? & : & \mathrm{ff} & (\text{br2}) \\
S(f(\text{Pd}(2),3)) & : & & (\text{comp}) \\
S\ f(\text{Pd}(2),3) & : & & (\text{comp}) \\
S\ f\ \text{Pd}(2)\ 3 & : & & (\text{pass}) \\
S\ f\ \text{Pd}(2) & : & 3 & (\text{comp}) \\
S\ f\ \text{Pd}\ 2 & : & 3 & (\text{pass}) \\
S\ f\ \text{Pd} & : & 2\ 3 & (\text{e-call}) \\
S\ f & : & 1\ 3 & (\text{i-call}) \\
S \text{ if } (1=0) \text{ then } 3 \text{ else } S(f(\text{Pd}(1),3)) & : & & (\text{br}), (\text{comp many times}) \\
S\ S\ f\ \text{Pd}(1)\ 3 & : & & (\text{pass}) \\
S\ S\ f\ \text{Pd}(1) & : & 3 & (\text{comp}) \\
S\ S\ f\ \text{Pd}\ 1 & : & 3 & (\text{pass}) \\
S\ S\ f\ \text{Pd} & : & 1\ 3 & (\text{e-call}) \\
S\ S\ f & : & 0\ 3 & (\text{i-call}) \\
S\ S \text{ if } (0=0) \text{ then } 3 \text{ else } S(f(\text{Pd}(0),3)) & : & & (\text{br}), (\text{comp many times}), (\text{pass}) \\
S\ S\ 3\ S\ f(\text{Pd}(0),3)\ ?\ \text{eq}_0(0) & : & & \\
S\ S\ 3\ S\ f(\text{Pd}(0),3)\ ? & : & \mathrm{tt} & (\text{br0}) \\
S\ S\ 3 & : & & (\text{pass}) \\
S\ S & : & 3 & (\text{e-call}) \\
S & : & 4 & (\text{e-call}) \\
& : & 5 &
\end{array}
$$

FIGURE 2. The computation of $2+3$ by the program $f(i,x) = \text{if eq}_0(i) \text{ then } x \text{ else } S(f(\text{Pd}(i),x))$.

This is done by induction of the length of the computation which establishes the hypothesis; and when we apply it with $F \equiv \mathsf{p}_i(\vec{x}_i)$ $(i = 1,\dots,K)$, it yields $\widetilde{p}_1 \sqsubseteq p_1, \dots, \widetilde{p}_K \sqsubseteq p_K$. It follows that $\widetilde{p}_1, \dots, \widetilde{p}_K$ are the least solutions of (2D-2),i.e., $\widetilde{p}_i = \overline{p}_i$, which completes the proof.

Both arguments appeal repeatedly to the simple but basic Lemma 2D.1.

(2) is proved by "relativizing" the proof of (1), i.e., by entering in every step the parameters $\vec{r}$ which are treated like primitives and do not enter the argument. ⊣

Reduction of recursion to iteration. Theorem 2D.2 reduces recursion to iteration, much as Theorem 2C.1 reduces iteration to (tail) recursion. Notice, however, that the recursive machine $\mathfrak{i} = \mathfrak{i}(\mathbf{A}, E(\vec{x}))$ runs in the structure $\mathbf{A}_{\mathfrak{i}}$ which is richer than **A**—its universe is a set of finite sequences (which contains all sequences from A) and it has additional primitives needed to manipulate these *stacks*. In many cases, $\mathbf{A}_{\mathfrak{i}}$ can be suitably "interpreted" in **A** and then $\mathbf{Rec}(\mathbf{A}) = \mathbf{Tailrec}(\mathbf{A})$, cf. Problems x2B.17 for the classical case and Problem x2D.4* below for a reasonably general result. We will consider structures where $\mathbf{Tailrec}(\mathbf{A}) \subsetneq \mathbf{Rec}(\mathbf{A})$ in Section 2G.

We should also mention here that rigorous semantics of recursive definitions were first given in terms of implementations (like the recursive machine).

Symbolic computation. The *symbolic recursive machine* $\mathfrak{i}_s = \mathfrak{i}_s(\Phi, E)$ associated with a vocabulary Φ and a Φ-program E is defined as follows.

The states of $\mathfrak{i}_s$ are all finite sequences s of the form

$$a_0 \ \dots \ a_{m-1} : b_0 \ \dots \ b_{n-1}$$

where the elements $a_0, \dots, a_{m_1}, b_0, \dots, b_{n-1}$ of s satisfy the following conditions:

- Each a_i is a function symbol in Φ or one of $\mathsf{p}_1, \dots, \mathsf{p}_K$, or a pure voc$(E)$-term, or the special symbol ?, and
- each b_j is a pure, algebraic Φ-term.

The transitions of $\mathfrak{i}_s$ are those listed for the recursive machine in Table 1, except that the following three are modified as follows:

(e-call)	$\vec{a}\ \underline{\phi_i : \vec{x}}\ \vec{b}$	$\to$	$\vec{a}\ \underline{: \phi_i(\vec{x})}\ \vec{b}$	
(br0)	$\vec{a}\ \underline{G\ H\ ? : b_0}\ \vec{b}$	$\to$	$\vec{a}\ \underline{G :}\ \vec{b}$	(if $b_0 = \mathrm{tt}$)
(br1)	$\vec{a}\ \underline{G\ H\ ? : b_0}\ \vec{b}$	$\to$	$\vec{a}\ \underline{H :}\ \vec{b}$	(if $b_0 = \mathrm{ff}$)

In the last two commands, b_0 is a pure, algebraic Φ-term (perhaps with variables in it), and the conditions $b_0 = \mathrm{tt}$ or $b_0 = \mathrm{ff}$ cannot be checked, unless b_0 is one of tt or ff. The computations of $\mathfrak{i}_s$ are defined relative to an *environment*, a set of boolean conditions

$$\mathcal{E} = \{\mathrm{tt} = \mathrm{tt}, P_0 = \mathrm{tt}, P_1 = \mathrm{tt}, \dots, P_{m-1} = \mathrm{tt},$$
$$\mathrm{ff} = \mathrm{ff}, N_0 = \mathrm{ff}, N_1 = \mathrm{ff}, \dots, N_{n-1} = \mathrm{ff}\},$$

where the P_i and N_j are pure, algebraic Φ-terms of boolean sort. We say that $\mathcal{E}$ *activates* (or *justifies*) the transition (br0) if $(b_0 = \mathrm{tt}) \in \mathcal{E}$, and $\mathcal{E}$ *activates* (br1) if $(b_0 = \mathrm{ff}) \in \mathcal{E}$. A computation relative to an environment $\mathcal{E}$ is a sequence of states $s_0, s_1, \dots, s_n$, where for each $i < n$ the Table and the environment justifies the transition $s_i \to s_{i+1}$.

Take, for example, the program which computes $2x$ in $\mathbf{N}_u$,

$$E \equiv p(\mathsf{u},\mathsf{u}) \text{ where } \{p(\mathsf{u},\mathsf{v}) = \text{if } \mathrm{eq}_0(\mathsf{v}) \text{ then } \mathsf{u} \text{ else } S(p(\mathsf{u},\mathrm{Pd}(\mathsf{v})))\}$$

and consider the symbolic computation starting with the head $p(\mathsf{u},\mathsf{u})$:

$$\begin{aligned} p(\mathsf{u},\mathsf{u}) : \to\ & \text{if } \mathrm{eq}_0(\mathsf{u}) \text{ then } \mathsf{u} \text{ else } S(p(\mathsf{u},\mathrm{Pd}(\mathsf{u}))) : \\ & \to\ \mathsf{u}\ S(p(\mathsf{u},\mathrm{Pd}(\mathsf{u})))\ ?\ \mathrm{eq}_0(\mathsf{u}) : \\ \to\ \mathsf{u}\ S(p(\mathsf{u},\mathrm{Pd}(\mathsf{u})))\ ?\ \mathrm{eq}_0\ \mathsf{u} : \to\ & \mathsf{u}\ S(p(\mathsf{u},\mathrm{Pd}(\mathsf{u})))\ ?\ \mathrm{eq}_0\ : \mathsf{u} \\ & \to\ \mathsf{u}\ S(p(\mathsf{u},\mathrm{Pd}(\mathsf{u})))\ ? : \mathrm{eq}_0(\mathsf{u}) \end{aligned}$$

If the environment does not *decide* the term $\mathrm{eq}_0(\mathsf{u})$, then the computation cannot go any further, it *stalls*. If the environment has the condition $\mathrm{eq}_0(\mathsf{u}) = \mathrm{ff}$, then (br1) is activated and we continue:

$$\begin{aligned} & \mathsf{u}\ S(p(\mathsf{u},\mathrm{Pd}(\mathsf{u}))\ ? : \mathrm{eq}_0(\mathsf{u}) \to S(p(\mathsf{u},\mathrm{Pd}(\mathsf{u}))) : \to\ S\ p(\mathsf{u},\mathrm{Pd}(\mathsf{u})) : \\ & \quad \to\ S\ p\ \mathsf{u},\mathrm{Pd}(\mathsf{u}) : \to\ S\ p\ \mathsf{u},\mathrm{Pd}\ \mathsf{u} : \\ & \qquad \to\ S\ p\ \mathsf{u},\mathrm{Pd} : \mathsf{u} \to\ S\ p\ \mathsf{u} : \mathrm{Pd}(\mathsf{u}) \to\ S\ p : \mathsf{u}\ \mathrm{Pd}(\mathsf{u}) \\ & \qquad\quad \to\ S \text{ if } \mathrm{eq}_0(\mathrm{Pd}(\mathsf{u})) \text{ then } \mathsf{u} \text{ else } S(p(\mathsf{u},\mathrm{Pd}^2(\mathsf{u}))) : \to \cdots \end{aligned}$$

The next time that ? will show up, we will need to have one of the two conditions

$$\mathrm{eq}_0(\mathrm{Pd}(\mathsf{u})) = \mathrm{tt} \text{ or } \mathrm{eq}_0(\mathrm{Pd}(\mathsf{u})) = \mathrm{ff}$$

in the environment to continue, etc. The computation will go on forever unless the environment has a condition $\mathrm{eq}_0(\mathrm{Pd}^n(\mathsf{u})) = \mathrm{tt}$ for some n, which will then turn it around so that eventually it stops in the state

$$: S^n(\mathsf{u})$$

which gives the correct answer for $\mathsf{u} = n$.

We will not do much with symbolic computation and we have left for Problem x2D.5 its basic—quite obvious—property; but it is one of the basic notions of computation theory and it has some (perhaps ineluctable) applications, e.g., Problem x2D.6.

Problems for Section 2D

x2D.1. **Problem.** Consider the following three extended $\mathbf{N}_u$-programs:

$$E_1(x) \equiv p(x) \text{ where } \{p(\mathsf{x}) = S(p(x))\},$$
$$E_2(x) \equiv p(x) \text{ where } \{p(x) = p(q(x)), q(x) = x\},$$
$$E_3(x) \equiv p(x,y) \text{ where } \{p(x,y) = q(p(x,y),y), q(x,y) = x\}.$$

Determine the partial functions computed by them and discuss how their computations by the recursive machine differ.

x2D.2. **Problem.** Let **A** be a Φ-structure where Φ contains the binary function constant ϕ and the unary function constant ψ which are interpreted by total functions in **A**. Let

$$f(x) = \phi^{\mathbf{A}}(\psi^{\mathbf{A}}(x), \psi^{\mathbf{A}}(x)) \quad (x \in A).$$

(1) Check that the recursive machine for the (explicit) program

$$E \equiv \phi(\psi(\mathsf{x}), \psi(\mathsf{x}))$$

which computes $f(x)$ in **A** will make two calls to ψ in its computations.

(2) Construct a better recursive program E which computes $f(x)$ in **A** using only one call to ψ.

x2D.3*. **Problem** (Stack discipline). (1) Prove that for every program E in a *total* structure **A**, and every $(\mathbf{A}, E)$-term M, there is no computation of $\mathrm{i}(\mathbf{A}, E)$ of the form

$$M : \to s_1 \to \cdots \to s_m \tag{2D-5}$$

which is *stuck*, i.e., the state s_m is not terminal and there is no s' such that $s \to s'$.

(2) Prove that if **A** is a partial structure, M is an $(\mathbf{A}, E)$-term and the finite computation (2D-5) is stuck, then its last state s_m is of the form

$$\vec{a}\ \phi_j : y_1, \dots, y_{n_j}\ \vec{b}$$

where ϕ_j is a primitive function of **A** of arity n_j and $\phi_j(y_1, \dots, y_{n_j})\uparrow$.

HINT (Clinton Conley): Prove that if (2D-5) holds and there is no s' such that $s_m \to s'$, then either s_m is terminal or the conclusion of (2) holds.

x2D.4*. **Problem** (Pairing schemes). A *pairing scheme* on a pointed structure **A** is a triple of total functions

$$\text{pair} : A^2 \to A, \quad \text{first} : A \to A, \quad \text{second} : A \to A \tag{2D-6}$$

such that

(2D-7) pair : $A^2 \rightarrowtail A \setminus \{0\}$ is an injection, and

$$\text{first}(\text{pair}(x,y)) = x, \quad \text{second}(\text{pair}(x,y)) = y \quad (x, y \in A).$$

(1) Prove that if a pointed structure **A** admits an explicit pairing scheme, then $\mathbf{R}(\mathbf{A}) = \mathbf{Tailrec}(\mathbf{A})$. HINT: Consider the system of mutual tail recursion

$$\Big\{p(x) = \text{if } (x = 0) \text{ then } \mathrm{tt} \text{ else } q(x),$$
$$q(x) = \text{if } (\text{first}(x) = 0) \text{ then } p(\text{second}(x)) \text{ else } q(x)\Big\}$$

whose solutions $\overline{p}, \overline{q}$ are tail recursive by Theorem 2B.5, and set

$$\mathbb{N}^{\mathbf{A}} = \{x \in A : \overline{p}(x)\downarrow\} = \{0, \text{pair}(0,0), \text{pair}(0,(0,0)), \dots\}.$$

This is an infinite set and if $S^{\mathbf{A}}, \text{Pd}^{\mathbf{A}}$ are the restrictions to $\mathbb{N}^{\mathbf{A}}$ of the functions pair$(0, x)$ and second(x), then $(\mathbb{N}^{\mathbf{A}}, 0, S^{\mathbf{A}}, \text{Pd}^{\mathbf{A}}, \text{eq}_0)$ is isomorphic with $\mathbf{N}_u$. Next use the techniques of solving Problems x2B.15 and x2B.16 to prove that **Tailrec**(**A**) is closed under definitions by primitive recursion and minimalization, with n varying over $\mathbb{N}^{\mathbf{A}}$; and finally, use Theorem 2D.2 to adapt any of the classical proofs of Kleene's Normal Form Theorem to prove that every **A**-recursive partial function satisfies an equation of the form

$$f(\vec{x}) = U(\mu n \in \mathbb{N}^{\mathbf{A}}[\tau(n, \vec{x}) = 0])$$

with suitable tail recursive U and τ.[17]

(2) Infer that $\mathbf{R}(\mathbf{A}) = \mathbf{Tailrec}(\mathbf{A})$ if **A** admits a *tail recursive pairing scheme*, i.e., one in which the functions tail(x, y), first(z) and second(z) are all tail recursive. HINT: Use Problem x2B.14.

x2D.5. **Problem** (Symbolic Computation). Fix a Φ-structure and a Φ-program E, and suppose that

$$N(x_1, \dots, x_n) : \to s_1 \to \cdots \to : w$$

is a computation of the recursive machine of E which computes the value of the $(\mathbf{A}, E)$ term $N(x_1, \dots, x_n)$ with the indicated parameters. Make precise and prove the following: there is an environment $\mathcal{E}$ in the distinct variables $\mathsf{x}_1, \dots, \mathsf{x}_n$ which is *sound for* $x_1, \dots, x_n$ in **A**, such that the given computation is obtained from the symbolic computation relative to $\mathcal{E}$ and starting with $N(\mathsf{x}_1, \dots, \mathsf{x}_n)$ by replacing each x_i in it by x_i.

There are many applications of symbolic computation, including the following simple fact for which there is no obvious, more elementary proof:

x2D.6. **Problem.** Prove that Pd(x) is not $(\mathbb{N}, 0, S, \text{eq}_0)$-recursive.

---

[17]The argument is fussy and perhaps not worth working out, but it is interesting to note what (I think) are the minimal, known hypotheses on **A** which guarantee that $\mathbf{R}(\mathbf{A}) = \mathbf{Tailrec}(\mathbf{A})$—they are satisfied, for example, when **A** is a proper, elementary extension of **A**. They also imply reasonable versions of the Enumeration and S^m_n theorems, so that a decent part of classical recursion theory can be developed for these structures. (And as far as I know, it was Jon Barwise who first noticed this, for the corresponding problem in the theory of *inductive definability* on an arbitrary **A**.)

2E. Finite nondeterminism

Much of the material in Section 2C can be extended easily to (finitely) *non-deterministic computation models*, in which the (typically total) transition function allows a finite number of choices of *the next state*. We will not study nondeterminism in any serious way, but it is important to give the precise definition of *nondeterministic recursive programs*, as they are covered by the intrinsic complexity lower bound results in Part II.

Beyond this, our main aim in this section is to define and analyze Pratt's nondeterministic algorithm for the gcd in Theorem 2E.2 which is relevant to the Main Conjecture on page 2.

A (finitely) *nondeterministic iterator*[18] $\mathfrak{i} : X \rightsquigarrow W$ is a tuple

$$\mathfrak{i} = (\text{input}, S, \sigma_1, \dots, \sigma_k, T, \text{output})$$

which satisfies (I1) – (I5) in Section 2C except that (I3) is replaced by the obvious

(I3′) for every $i = 1, \dots, k$, $\sigma_i : S \rightharpoonup S$.

So $\mathfrak{i}$ has k transition functions and a *partial computation* of $\mathfrak{i}$ is any finite sequence $(s_0, \dots, s_n)$ of states such that for every $i < n$, s_i is not terminal and for some $j = 0, \dots, k$, $s_{i+1} = \sigma_j(s_i)$, and it is *convergent* if $s_n \in T$. This allows the possibility that the machine may produce more than one value on some input, and we must be careful in specifying what it means for $\mathfrak{i}$ to compute some $f : X \rightharpoonup W$. The formal definitions are as before: we set

$$\text{(2E-1)} \quad s \to_{\mathfrak{i}}^* s' \iff \text{there is a convergent computation } (s_0, \dots, s_n) \text{ with } s_0 = s, s_n = s'$$

and we say that $\mathfrak{i}$ *computes* $f : X \rightharpoonup W$ if

$$\text{(2E-2)} \qquad f(x) = w \iff (\exists s \in T)[\text{input}(x) \to_{\mathfrak{i}}^* s \ \&\ \text{output}(s) = w],$$

but they must be read more carefully now: $\mathfrak{i} : X \rightsquigarrow W$ *computes* f *if whenever* $f(x)\downarrow$, *then at least one convergent computation starting with* input(x) *produces the value* $f(x)$ *and no convergent computation from* input(x) *produces a different value*. Divergent computations are disregarded.

Nondeterministic recursive programs are defined exactly as before, except that we allow multiple definitions for each recursive variable. For example, in

[18]**Note.** By "iterator", we always mean "deterministic iterator", while by "nondeterministic iterator" we mean some $\mathfrak{i}$ which may be deterministic, and the same for programs; in other words, "deterministic" is the default, even if sometimes, for emphasis, we refer to a "deterministic" iterator or program.

$(\mathbb{N}, 0, S, \phi)$, we might have

$$\text{(2E-3)} \qquad E^* \equiv \phi(\mathsf{p}(\vec{x}), \vec{x}) \text{ where } \Big\{\mathsf{p}(\vec{x}) = 0, \mathsf{p}(\vec{x}) = S(\mathsf{p}(\vec{x}))\Big\}.$$

The recursive machine $\mathrm{i}(\mathbf{A}, E(\vec{x}))$ associated with a nondeterministic extended program is now nondeterministic: if

$$\mathsf{p}(\vec{x}) = E^1(\vec{x}) \text{ and } \mathsf{p}(\vec{x}) = E^2(\vec{x})$$

are both in the body of E, then $\mathrm{i}(\mathbf{A}, E(\vec{x}))$ allows both transitions

$$\mathsf{p} : \vec{x} \ \rightarrow \ E^1(\vec{x}) : \ \text{and } \mathsf{p} : \vec{x} \ \rightarrow \ E^2(\vec{x}) :$$

And, again, we say that $E(\vec{x})$ *defines* or *computes* $f : A^n \rightharpoonup A_s$ in a Φ-structure $\mathbf{A}$ if (2E-2) holds for the iterator $\mathrm{i}(\mathbf{A}, E(\vec{x}))$, and then we write

$$\text{(2E-4)} \qquad \mathbf{A} \models E(\vec{x}) = w \iff_{\text{df}} f(\vec{x}) = w \iff E_0(\vec{x}) : \to^*_{\mathrm{i}(\mathbf{A},E)} : w.$$

We also put

$$\text{(2E-5)} \quad c^s(\Phi_0)(\mathbf{A}, E(\vec{x}))$$
$$= \min\Big(\text{number of external calls to } \phi \in \Phi_0$$
$$\text{in any convergent computation of } \mathrm{i}(\mathbf{A}, E) \text{ on the input } \vec{x}\Big),$$

where a call to ϕ in a computation by the recursive machine is a state of the form $\vec{a}\ \phi : w_1 \ \cdots \ w_n\ \vec{b}$. This is the only complexity measure on nondeterministic programs that we will need (for now).

The definitions extend naturally to *nondeterministic programs with function parameters* which compute functionals, we set

$$\mathbf{Rec}_{\text{nd}}(\mathbf{A}) = \text{the set of nondeterministically } \mathbf{A}\text{-recursive functionals}$$

and as for deterministic recursion, we let $\mathbf{Rec}^0_{\text{nd}}(\mathbf{A})$ be the set of all partial functions in $\mathbf{Rec}_{\text{nd}}(\mathbf{A})$.

It is well known that $\mathbf{Rec}^0(\mathbf{A}) = \mathbf{Rec}^0_{\text{nd}}(\mathbf{A})$ when $\mathbf{A}$ is a total expansion of $\mathbf{N}_u$, but this fails for expansions by partial functions—and for functionals, it even fails for $\mathbf{N}_u$. We have included in Problems x2E.7 – x2E.10* some basic, elementary results of this type.

Certificates and computations. The next Proposition makes precise the connection between certificates and computations to which we have alluded, still leaving the motivation for the term "certificate" for Section 4D.

2E.1. **Proposition.** ***If*** *$E(\vec{x})$ is an extended nondeterministic Φ-program,*

$$\text{(2E-6)} \qquad \mathbf{c} = (E_0(\vec{x}) :, \dots, : w)$$

is a convergent computation of the recursive machine for E in a Φ-structure $\mathbf{A}$ and $\mathbf{U}_c \subseteq_p \mathbf{A}$ is the structure with

$$\text{eqdiag}(\mathbf{U}_c) = \{(\phi, \vec{u}, v) : \textit{a transition } \vec{a}\ \phi : \vec{u}\ \vec{b} \to \vec{a} : v\ \vec{b} \textit{ occurs in } \mathbf{c}\},$$

then $(\mathbf{U}_{\mathfrak{c}}, \vec{x})$ *is a certificate, i.e.,* $\mathbf{U}_{\mathfrak{c}} = \mathbf{G}_m(\mathbf{U}, \vec{x})$ *for some* m.

Conversely*, every certificate* $(\mathbf{U}, \vec{x})$ *is* $(\mathbf{U}_{\mathfrak{c}}, \vec{x})$ *for some computation* $\mathfrak{c}$ *of a nondeterministic program* $E(\vec{\mathsf{x}})$ *as above.*

PROOF. The universe U of $\mathbf{U}$ comprises all the parameters from A which occur in eqdiag($\mathbf{U}$); so if $s_0 = E_0(\vec{x})\ :\ ,\dots, s_m =\ :\ w$ enumerates the states in $\mathfrak{c}$, then by an easy induction on $k \leq m$, every parameter which occurs in $s_0, \dots, s_k$ is in $G_\infty(\mathbf{U}, \vec{x})$; so $\mathbf{U} = \mathbf{G}_\infty(\mathbf{U}, \vec{x})$, and since it is finite, $\mathbf{U} = \mathbf{G}_m(\mathbf{U}, \vec{x})$ for some m.

For the converse, given $(\mathbf{U}, \vec{x})$, choose by Problem x1D.8 an enumeration

$$\text{eqdiag}(\mathbf{U}) = \Big((\phi_0, \vec{u}_0, w_0), \dots, (\phi_m, \vec{u}_m, w_m)\Big),$$

so each t in $\vec{u}_0$ is x_j for some j, and for each $s < m$, each t in the tuple $\vec{u}_{s+1}$ occurs in the sequence $\vec{x}, \vec{u}_0, w_0, \vec{u}_1, w_1, \dots, \vec{u}_s, w_s$. Define the program

$$E \equiv p_0(\vec{\mathsf{x}}, \phi_0(\vec{\mathsf{u}}_0)) \text{ where } \Big\{p_0(\vec{\mathsf{x}}, y) = p_1(\vec{\mathsf{x}}, \phi_1(\vec{\mathsf{u}}_1)),$$
$$p_1(\mathsf{x}, \mathsf{y}) = p_2(\vec{\mathsf{x}}, \phi_2(\vec{\mathsf{u}}_2)), \dots, p_m(\vec{\mathsf{x}}, \mathsf{y}) = \phi_m(\vec{\mathsf{u}}_m)\Big\}$$

with suitably chosen tuples of variables $\vec{\mathsf{u}}_i$ so that if we set $\vec{\mathsf{x}} := \vec{x}$, then $\vec{\mathsf{u}}_0 = \vec{u}_0$, $\vec{\mathsf{u}}_1 = \vec{u}_1, \dots, \vec{\mathsf{u}}_m = \vec{u}_m$; and finally check that if $\mathfrak{c}$ is the computation of the recursive machine for E on the input $E_0(\vec{x}) = p_0(\vec{x}, \phi_0(\vec{u}))$, then $\mathbf{U} = \mathbf{U}_{\mathfrak{c}}$. ⊣

Fixed point semantics for nondeterministic programs. It is also useful to characterize the denotations of nondeterministic programs using a natural least-fixed-point operation, if for no other reason than to verify that the definitions we gave do not depend on the specific choice of the recursive machine as a "preferred" implementation—as there are many others. We define here the notions we need and leave for Problems x2E.14* and x2E.15* the precise formulation and (easy) proofs of the relevant correctness results.

A *partial multiple valued* (pmv) *function* $f : X \rightharpoonup_{\text{mv}} W$ assigns to each $x \in X$ a (possibly empty) subset of W. We write

$$\begin{aligned} f(x) \rightarrow_{\text{mv}} w &\iff w \in f(x), \\ f(x)\downarrow &\iff f(x) \neq \emptyset, \\ G_f(x, w) &\iff f(x) \rightarrow_{\text{mv}} w \quad \text{(the } graph \text{ of } f), \end{aligned}$$

and we put on the space $(X \rightharpoonup_{\text{mv}} W)$ of all pmv functions on X to W the partial ordering induced by their graphs,

$$f \subseteq g \iff G_f \subseteq G_g \iff (\forall x, w)[f(x) \rightarrow_{\text{mv}} w \Longrightarrow g(x) \rightarrow_{\text{mv}} w].$$

Every partial function $f : X \rightharpoonup W$ can be viewed as a pmv function whose values are empty or singletons, $f(x) = \{f(x)\} = \{w : f(x) = w\}$, and the basic operations on partial functions we introduced in Section 1A (easily and naturally) extend to pmv functions. For example: the *composition* and the *conditional* operations on pmv functions are defined by

$$f(g_1(x),\dots,g_m(x)) \rightharpoonup_{\text{mv}} w$$
$$\iff (\exists w_1,\dots,w_m)[g_1(x) \rightharpoonup_{\text{mv}} w_1 \;\&\; \cdots g_m(x) \rightharpoonup_{\text{mv}} w_m \;\&\; f(w_1,\dots,w_m) \rightharpoonup_{\text{mv}} w],$$

$$\text{if } f(x) \text{ then } g(x) \text{ else } h(x) \rightharpoonup_{\text{mv}} w$$
$$\iff [f(x) \rightharpoonup_{\text{mv}} \mathrm{tt} \;\&\; g(x) \rightharpoonup_{\text{mv}} w] \vee [f(x) \rightharpoonup_{\text{mv}} \mathrm{ff} \;\&\; h(x) \rightharpoonup_{\text{mv}} w].$$

The definitions extend directly to *pmv functionals*, cf. Problem x2E.14*.

For any vocabulary Φ, a *pmv Φ-structure* is a tuple $\mathbf{A} = (A, \{\phi^{\mathbf{A}}\}_{\phi\in\Phi})$, where each $\phi \in \Phi$ of arity n and sort $s \in \{\texttt{ind}, \texttt{boole}\}$ is interpreted by a pmv function $\phi^{\mathbf{A}} : A^n \rightharpoonup_{\text{mv}} A_s$; and, again, every Φ-structure can be viewed as a pmv Φ-structure.

Next we turn to the following important result of Vaughn Pratt.[19]

2E.2. **Theorem** (Pratt's nuclid algorithm). *Consider the following non-deterministic recursive program E_P in the structure $\mathbf{N}_\varepsilon$ of the Euclidean defined in* (1C-10):

$$E_P \equiv \text{nuclid}(a,b,a,b) \text{ where } \Big\{$$
$$\text{nuclid}(a,b,m,n) = \text{if } (n \neq 0) \text{ then nuclid}(a,b,n,\text{rem}(\text{choose}(a,b,m),n))$$
$$\text{else if } (\text{rem}(a,m) \neq 0) \text{ then nuclid}(a,b,m,\text{rem}(a,m))$$
$$\text{else if } (\text{rem}(b,m) \neq 0) \text{ then nuclid}(a,b,m,\text{rem}(b,m))$$
$$\text{else } m,$$
$$\text{choose}(a,b,m) = m, \quad \text{choose}(a,b,m) = a, \quad \text{choose}(a,b,m) = b\Big\}.$$

If $a \geq b \geq 1$, then $\mathbf{N}_\varepsilon \models E_P(a,b) = \gcd(a,b)$.

PROOF. Fix $a \geq b \geq 1$, and let

$$(m,n) \to (m',n')$$
$$\iff \Big(n \neq 0 \;\&\; m' = n \;\&\; [n' = \text{rem}(m,n) \vee n' = \text{rem}(a,n) \vee n' = \text{rem}(b,n)]\Big)$$
$$\vee \Big(n = 0 \;\&\; \text{rem}(a,m) \neq 0 \;\&\; m' = m \;\&\; n' = \text{rem}(a,m)\Big)$$
$$\vee \Big(n = 0 \;\&\; \text{rem}(b,m) \neq 0 \;\&\; m' = m \;\&\; n' = \text{rem}(b,m)\Big).$$

[19] This theorem and Problems x2E.16 – x2E.19 are in Pratt [2008] which has not been published. They are included here with Vaughan Pratt's permission.

This is the transition relation of the nondeterministic iterator associated with E_P (with a, b and the "housekeeping" details of the recursive machine suppressed in the notation) and it obviously respects the property $m > 0$. The terminal states are

$$T(a, b, m, n) \iff n = 0 \mathbin{\&} m \mid a \mathbin{\&} m \mid b,$$

and the output on a terminal $(a, b, m, 0)$ is m.

It is obvious that there is at least one computation which outputs $\gcd(a, b)$, because one of the choices at each step is the one that the Euclidean would make. To see that no convergent computation produces any other value, we observe that directly from the definition,

If x divides a, b, m and n and $(m, n) \to (m', n')$, then x divides m' and n'.

This implies that every common divisor of a and b divides every output m; and because of the conditions on the terminal state, every output m divides both a and b, so that the only output is $\gcd(a, b)$. ⊣

In fact, E_P does not have any divergent computations, see Problem x2E.16.

Pratt's algorithm allows at each stage to replace

the Euclidean's $(m, n) \to (n, \text{rem}(m, n))$ by

$$(m, n) \to (n, \text{rem}(a, n)) \text{ or } (m, n) \to (n, \text{rem}(b, n)),$$

which does not lose any common divisors of a and b, and then simply adds a check at the end which insures that the output is not some random divisor of (say) a which does not also divide b. The important thing about it is that in some cases this guessing can produce a much faster computation of $\gcd(a, b)$: see Problems x2E.17 – x2E.19 which outline a proof that for successive Fibonacci numbers it can compute $\gcd(F_{t+1}, F_t)$ using only

$$O(\log t) = O(\log\log(F_t))$$

calls to the remainder function, thus beating the Euclidean on its worst case. A complete analysis of the inputs on which it does better than the Euclidean does not appear to be easy.

We will discuss briefly the relevance of these results of Pratt for the Main Conjecture in Corollary 6C.7 on page 188.

Problems for Section 2E

x2E.1. **Problem.** Prove that the following are equivalent for a Φ-structure $\mathbf{A}$ and a nondeterministic extended Φ-program $E(\vec{x})$:

(a) $E(\vec{x})$ computes a partial function in $\mathbf{A}$.
(b) $E(\vec{x})$ computes a partial function in every substructure $\mathbf{U} \subseteq_p \mathbf{A}$.
(c) $E(\vec{x})$ computes a partial function in every finite substructure $\mathbf{U} \subseteq_p \mathbf{A}$.

x2E.2. **Problem** (The homomorphism property for $\mathbf{Rec}_{\mathrm{nd}}$). Prove that if $\mathbf{A}$ is a Φ-structure, $E(\vec{x})$ an n-ary nondeterministic extended program which computes a partial function in $\mathbf{A}$, $\mathbf{U}, \mathbf{V} \subseteq_p \mathbf{A}$ and $\pi : \mathbf{U} \to \mathbf{V}$ is a homomorphism, then

$$\mathbf{U} \models E(\vec{x}) = w \Longrightarrow \mathbf{V} \models E(\pi(\vec{x})) = \pi(w) \quad (\vec{x} \in U^n). \tag{2E-7}$$

HINT: For each computation $\mathfrak{c}$ of the recursive machine which proves that $\mathbf{U} \models E(\vec{x}) = w$ as in (2E-6), define a sequence of states $\pi(\mathfrak{c})$ by replacing every parameter $u \in A$ which occurs in $\mathfrak{c}$ by $\pi(u)$, and verify that $\pi(\mathfrak{c})$ is a computation of the recursive machine which proves that $\mathbf{V} \models E(\pi(\vec{x})) = \pi(w)$.

It is also useful to notice the version of this fact for isomorphisms, which is proved the same way:

x2E.3. **Problem.** Prove that if $\pi : \mathbf{A} \rightarrowtail\!\!\!\twoheadrightarrow \mathbf{B}$ is an isomorphism between two Φ-structures and $E(\vec{x})$ is an extended, nondeterministic Φ-program which computes a partial function in $\mathbf{A}$, then $E(\vec{x})$ computes a partial function in $\mathbf{B}$ and

$$\mathbf{A} \models E(\vec{x}) = w \iff \mathbf{B} \models E(\pi(\vec{x})) = \pi(w).$$

x2E.4. **Problem** (The finiteness property for $\mathbf{Rec}_{\mathrm{nd}}$). Prove that if $\mathbf{A}$ is a Φ-structure and $E(\vec{x})$ is an n-ary nondeterministic extended program which computes a partial function in $\mathbf{A}$, then

$$\mathbf{A} \models E(\vec{x}) = w \Longrightarrow (\exists m)[\mathbf{G}_m(\mathbf{A}, \vec{x}) \models E(\vec{x}) = w]. \tag{2E-8}$$

Moreover, if $E(\vec{x})$ is a program with empty body (a pure Φ-term), then

$$(\exists m)(\forall \vec{x}, w)[\mathbf{A} \models E(\vec{x}) = w \Longrightarrow \mathbf{G}_m(\mathbf{A}, \vec{x}) \models E(\vec{x}) = w].$$

x2E.5. **Problem.** Prove for $\mathbf{Rec}_{\mathrm{nd}}(\mathbf{A})$ the closure properties listed for $\mathbf{Rec}(\mathbf{A})$ in Corollary 2A.2.

x2E.6. **Problem.** Formulate for $\mathbf{Rec}_{\mathrm{nd}}(\mathbf{A})$ and prove the properties of $\mathbf{Rec}(\mathbf{A})$ in Problems x2A.5, x2A.13 and x2A.14.

x2E.7. **Problem.** Prove that if $\mathbf{\Phi}$ is a set of total functions on $\mathbb{N}$, then

$$\mathbf{Rec}^0_{\mathrm{nd}}(\mathbf{N}_u, \mathbf{\Phi}) = \mathbf{Rec}^0_{\mathrm{nd}}(\mathbf{N}_b, \mathbf{\Phi}) = \mathbf{Rec}^0(\mathbf{N}_u, \mathbf{\Phi}).$$

HINT: Appeal to the classical result, that a partial function $f(\vec{x})$ is recursive in $\mathbf{\Phi}$ if and only if its graph is *semirecursive* in $\mathbf{\Phi}$, i.e., if for some partial $g(\vec{x}, w)$ which is recursive in $\mathbf{\Phi}$,

$$f(\vec{x}) = w \iff g(\vec{x}, w)\downarrow .$$

x2E.8*. **Problem.** Prove that if $\mathbf{A}$ is pointed and admits a tail recursive pairing scheme (as in Problem x2D.4*), then $\mathbf{Rec}^0(\mathbf{A}) = \mathbf{Rec}^0_{\mathrm{nd}}(\mathbf{A})$. HINT: The tedious argument for Problem x2D.4* works for this too—it being an abstract version of Kleene's Normal Form Theorem which was originally proved for the nondeterministic Gödel-Herbrand-Kleene programs.

x2E.9. **Problem.** Prove that the functional

$$f(p) = \text{if } (p(0)\downarrow \text{ or } p(1)\downarrow) \text{ then } 0 \text{ else } \mu y[y = y+1]$$

is nondeterministically $\mathbf{N}_u$-recursive but not $\mathbf{N}_u$-recursive. HINT: Consider the computation of some deterministic $\mathbf{N}_u$-program with parameter p which might compute $f(p)$.

x2E.10*. **Problem.** Give an example of a partial function $\phi : \mathbb{N}^2 \rightharpoonup \mathbb{N}$ such that

$$\mathbf{Rec}^0(\mathbf{N}_u, \phi) \subsetneq \mathbf{Rec}^0_{\text{nd}}(\mathbf{N}_u, \phi).$$

x2E.11. **Problem.** For the program E^* defined in (2E-3), prove that

$$\text{den}_{E^*}(\vec{x}) = w \iff (\exists n)[\phi^{\mathbf{A}}(n, \vec{x}) = w]$$

provided that

$$(\forall n, m, u, v)[\phi^{\mathbf{A}}(n, \vec{x}) = u \;\&\; \phi^{\mathbf{A}}(m, \vec{x}) = v] \Longrightarrow u = v\};$$

if this condition does not hold, then E^* does not compute a partial function in $(\mathbb{N}, 0, S, \phi)$. Define also a related nondeterministic program E^{**} which computes in the same structure $(\mathbb{N}, 0, S, \phi)$ a partial function $\text{den}_{E^{**}}$ such that

$$\text{den}_{E^{**}}(\vec{x})\downarrow \iff (\exists n)[\phi^{\mathbf{A}}(n, \vec{x})\downarrow].$$

x2E.12. **Problem.** Prove that the basic fact in Problem x2A.9 holds for any $\mathbf{A}$ and any $f \in \mathbf{Rec}_{\text{nd}}(\mathbf{A})$, and infer that $S \notin \mathbf{Rec}_{\text{nd}}(\mathbb{N}, 0, \text{Pd}, \text{eq}_0)$.

x2E.13. **Problem** (Symbolic computation). Prove a version of Problem x2D.5 for nondeterministic programs and use it to show that

$$\text{Pd} \notin \mathbf{Rec}_{\text{nd}}(\mathbb{N}, 0, S, \text{eq}_0).$$

x2E.14*. **Problem.** Define what it means for a *pmv functional*

$$f : X \times (Y_1 \rightharpoonup_{\text{mv}} W_1) \times \cdots (Y_k \rightharpoonup_{\text{mv}} W_k) \rightharpoonup_{\text{mv}} W \tag{2E-9}$$

to be *monotone* and *continuous* by extending the definitions starting on page 10; formulate and prove the pmv version of the Fixed Point Lemma 1B.1—that systems of equations of continuous pmv functionals have least fixed points.

x2E.15*. **Problem** (nd-implementation correctness). Suppose $\mathbf{A}$ is a pmv Φ-structure, $E(\vec{\mathsf{x}})$ is an extended nondeterministic Φ-program with recursive variables $\mathsf{p}_1, \ldots, \mathsf{p}_K$, the pmv functions $\overline{p}_1, \ldots, \overline{p}_K$ are the mutual fixed points in $\mathbf{A}$ of the system in the body of $E(\vec{\mathsf{x}})$

$$p_i(\vec{x}_i) = \text{den}((\mathbf{A}, p_1, \ldots, p_K), E_i(\vec{x}_i)) \quad (i = 1, \ldots, K),$$

and M is an $(\mathbf{A}, E)$-term. Prove that for every $w \in A \cup \mathbb{B}$,

$$\text{den}((\mathbf{A}, \overline{p}_1, \ldots, \overline{p}_K), M) \rightarrow_{\text{mv}} w \iff M : \rightarrow^*_{\mathrm{i}(\mathbf{A}, E(\vec{\mathsf{x}}))} : w.$$

Infer that, with $M \equiv E_0(\vec{x})$,

$$\mathrm{den}(\mathbf{A}, E(\vec{x})) \to_{\mathrm{mv}} w \iff E_0(\vec{x}) \to^*_{\mathrm{i}(\mathbf{A}, E(\vec{x}))} : w,$$

and so the nondeterministic extended program $E(\vec{x})$ and the nondeterministic recursive machine $\mathrm{i}(\mathbf{A}, E(\vec{x}))$ compute the same pmv function in $\mathbf{A}$. HINT: Follow closely the argument in the proof of Theorem 2D.2, making all the necessary (technical and notational) adjustments.

The results in the remaining problems in this section are due to Vaughan Pratt.

x2E.16. **Problem.** Prove that the program E_P in Theorem 2E.2 has no divergent (infinite) computations. HINT: Check the claim by hand for the cases where one of a, b is 0. For $a, b \geq 1$, prove convergence of the main loop by induction on $\max(m, n)$ and within this by induction on n.

The complexity estimate for Pratt's algorithm depends on some classical identities that relate the Fibonacci numbers.

x2E.17. **Problem.** Prove that for all $t \geq 1$ and $m \geq t$,

$$F_m(F_{t+1} + F_{t-1}) = F_{m+t} + (-1)^t F_{m-t}. \tag{2E-10}$$

HINT: Prove in sequence, by direct computation, that

$$\varphi\hat{\varphi} = -1; \quad \varphi + \frac{1}{\varphi} = \sqrt{5}; \quad \hat{\varphi} + \frac{1}{\hat{\varphi}} = -\sqrt{5}; \quad F_{t+1} + F_{t-1} = \varphi^t + \hat{\varphi}^t.$$

x2E.18. **Problem.** (1) Prove that for all odd $t \geq 2$ and $m \geq t$,

$$\mathrm{rem}(F_{m+t}, F_m) = F_{m-t}. \tag{2E-11}$$

(2) Prove that for all even $t \geq 2$ and $m \geq t$,

$$\mathrm{rem}(F_{m+t}, F_m) = F_m - F_{m-t}, \tag{2E-12}$$

$$\mathrm{rem}(F_m, (\mathrm{rem}(F_{m+t}, F_m))) = F_{m-t}. \tag{2E-13}$$

HINT: For (2E-13), check that for $t \geq 2$, $2F_{m-t} < F_m$.

x2E.19. **Problem.** Fix $t \geq 2$. Prove that for every $s \geq 1$ and every u such that $u \leq 2^s$ and $t - u \geq 2$, there is a computation of Pratt's algorithm which starts from $(F_{t+1}, F_t, F_{t+1}, F_t)$ and reaches a state $(F_{t+1}, F_t, ?, F_{t-u})$ doing no more than $2s$ divisions.

Infer that for all $t \geq 3$,

$$c^s(\mathrm{rem})(\mathbf{N}_\varepsilon, E_P(F_{t+1}, F_t)) \leq 2\lceil \log(t-2) \rceil + 1 = O(\log\log(F_t)). \tag{2E-14}$$

HINT: The measure $c^s(\Phi_0)(\mathbf{A}, E(\vec{x}))$ for nondeterministic recursive programs is defined in (2E-5). For (2E-14), you also need Problem x1C.7.

2F. Some standard models of computation

We collect here the definitions of some computation models—other than Turing machines—which are commonly used to express and analyze algorithms, especially in arithmetic and algebra.[20]

Finite register machines. These are the simplest "structured" iterators which capture tail recursion in pointed structures in a useful way.

A (finite) *register program* P in the vocabulary Φ, of sort s and $k > 0$ registers is a sequence of *lines* or *commands* consecutively numbered from 0 to some $L \geq 0$ and satisfying the following conditions.

(1) Each line is in one of the following forms, where

$$P_l(\vec{\mathsf{u}}) \equiv P_l(\mathsf{u}_1, \dots, \mathsf{u}_k)$$

is a pure, extended Φ-term:

$l.$ $\mathsf{u}_i := P_l(\vec{\mathsf{u}})$, goto l' (Assignment)
$l.$ if $P_l(\vec{\mathsf{u}})$ then goto to l' else goto l'' (Branching)
$l.$ halt, return $P_l(\vec{\mathsf{u}})$. (Halting)

(2) If l is an assignment, then $\text{sort}(P_l) = \texttt{ind}$; if l is a branching command, then $\text{sort}(P_l) = \texttt{boole}$; and if l is a halting command, then P_l is a term of the fixed sort s of P.

Each finite register program P defines an iterator

$$\mathfrak{i}_P = (\text{input}, S, \sigma, T, \text{output}) : A^k \rightharpoonup A_s$$

on every Φ-structure $\mathbf{A}$ in the obvious way:

$$S = \{l, \vec{u}) : l \leq L, \vec{u} \in A^k\}, \quad \text{input}(\vec{u}) = (0, \vec{u}),$$

$$(l, \vec{u}) \in T \iff l \text{ is a halting command}, \quad \text{output}(l, \vec{u}) = \text{den}(\mathbf{A}, P_l(\vec{u})),$$

and the transition function is defined by cases on the form of each line:

$$\begin{aligned} &\text{if } l \text{ is an assignment}, \sigma(l, \vec{u}) = (l', \vec{u}\{u_i := \text{den}(\mathbf{A}, P_l(\vec{u}))\}), \\ &\text{if } l \text{ is a branching}, \sigma(l, \vec{u}) = \text{if } P_l(\vec{u}) \text{ then } (l', \vec{u}) \text{ else } (l'', \vec{u}), \\ &\text{if } l \text{ is a halt}, \sigma(l, \vec{u}) = (l, \vec{u}). \end{aligned}$$

A partial function $f : A^n \rightharpoonup A_s$ is *register computable* in $\mathbf{A}$ if

(2F-1) $$f(\vec{x}) = \overline{\mathfrak{i}_P}(g(\vec{x}))$$

for some register program P and with some $\mathbf{A}$-explicit $g : A^n \rightharpoonup A^k$.

2F.1. **Proposition.** *If* $\mathbf{A}$ *is pointed, then* $f : A^n \rightharpoonup A_s$ *is register computable exactly when it is tail recursive.*

This is an easy Corollary of Theorem 2C.2 and we leave it for Problem x2F.1.

[20] We will stick with deterministic models, for clarity, but all these schemes have natural nondeterministic versions which are defined as in Section 2E.

Primitive register machines. We chose this definition of register programs because it makes it easy to prove register computability—first to check this Proposition, and then to develop the basic theory of tail recursiveness that we outlined in Problems x2B.9 –x2B.17 using much easier arguments, at least for those who are adept at imperative programming. To prove that some specific function or relation *is not* register computable, we often need to restrict severely the terms P_l of the program, as follows:

A register program P is *primitive* if all the terms in its specification are prime, algebraic terms in the variables $\mathsf{u}_1, \dots, \mathsf{u}_k$,

$$P_l :\equiv \mathrm{tt} \mid \mathrm{ff} \mid \mathsf{u}_i \mid \phi(\mathsf{u}_{\pi(1)}, \dots, \mathsf{u}_{\pi(n_i)})$$

where $\phi \in \Phi$ and $\pi : \{1, \dots, n\} \to \{0, \dots, k\}$; and $f : A^n \rightharpoonup A_s$ is computable by a primitive P in a pointed $\mathbf{A}$ if (2F-1) holds with

$$\text{(2F-2)} \qquad g(x_1, \dots, x_n) = (x_1, \dots, x_n, a_{n+1}, \dots, a_k) \quad (k \geq n)$$

with some a_i which are strongly explicit in $\mathbf{A}$.

Every $f : A^n \rightharpoonup A_s$ which is register computable in a pointed $\mathbf{A}$ is computable in $\mathbf{A}$ by a primitive register program, cf. Problem x2F.3.

Decision trees. These are finite register programs which satisfy two, additsional restrictions:

(1) All the goto clauses point to lines further down in the program, i.e.,

if $\underline{l.\ \mathsf{u}_i := M_l(\vec{\mathsf{u}}), \text{goto } l'}$ is a line, then $l' > l$,

if $\underline{l.\ \text{if } M_l(\vec{\mathsf{u}}) \text{ then goto to } l' \text{ else goto } l''}$ is a line, then $l', l'' > l$.

(2) Every line other than 0 is pointed to by a goto from exactly one (earlier) line.

The terminology is chosen because the goto relation defines a tree on the set of all symbolic computations of i_P and each actual computation for a specific input in a structure $\mathbf{A}$ follows (exactly) one of the branches of this tree. These computations are now all finite, of length $\leq L$, and the partial functions computed on A by i_P are all explicit. Decision trees are mostly used in arithmetic or algebraic complexity theory, where the explicit terms are polynomials, rational functions, inequalities, etc. They also come in many varieties, e.g., in some cases the most interesting questions are best expressed using **primitive decision trees**.

Simplest among these are the **straight line programs** (or **computation sequences**), primitive decision trees with no branching commands; they are used extensively in arithmetic and algebraic complexity.

Random Access Machines (RAMs). We give a very general and rather abstract definition of these computation models, quite far removed from both the original definition in Cook and Reckhow [1973] and most of the numerous variations discussed in van Emde Boas [1990]; there are good reasons for the abstraction, which we will discuss in Section 4G, and we will soon explain how the Cook-Reckhow version is covered by this definition.

Fix an infinite, pointed Φ-structure $\mathbf{A} = (A, \mathbf{\Phi})$. An n-ary *random access machine* (RAM) *over* $\mathbf{A}$ of sort $s \in \{\texttt{ind}, \texttt{boole}\}$ and $k > n$ *accumulators* is an iterator

$$\mathfrak{i} = (\text{input}, S, \sigma, T, \text{output}) : A^n \rightharpoonup A_s$$

such that the following conditions hold with some $L \in \mathbb{N}$.

(1) The set of states of $\mathfrak{i}$ is

$$= \{(l, \vec{u}, R) : l \leq L, \vec{u} \in A^k, R : A \to A\}.$$

(2) $\text{input}(\vec{x}) = (0, (0, x_1, \dots, x_n, 0, \dots, 0), R_0)$ where $R_0 = (\lambda t)0$.

(3) If the sort s of the RAM is $\texttt{ind}$, then $\text{output}(l, \vec{u}, R) = u_0$, and if $s = \texttt{boole}$, then $\text{output}(l, \vec{u}, R) = \text{eq}_0(u_0)$.

(4) The transition function and the terminal states of $\mathfrak{i}$ are defined by a RAM *program* over $\mathbf{A}$, a sequence of *lines* or *commands* numbered from 0 to L of the following form, where

$$P_l(\vec{\mathsf{u}}) \equiv P_l(\mathsf{u}_1, \dots, \mathsf{u}_k) \text{ and } C_l(\mathsf{t})$$

are pure, extended Φ-terms of sorts $\texttt{ind}$ and $\texttt{boole}$ respectively, $0 \leq i, j \leq k$ and R ranges over $(A \to A)$:

l. $\mathsf{u}_i := P_l(\vec{\mathsf{u}})$, goto l' (Accumulator assignment)

l. if $C_l(\mathsf{u}_i)$ then $\mathsf{u}_j := \mathsf{R}(\mathsf{u}_i)$, goto l', else halt (Application)

l. if $C_l(\mathsf{u}_i)$ then $\mathsf{R}(\mathsf{u}_i) := \mathsf{u}_j$, goto l', else halt (Indirect application)

l. If $P_l(\vec{\mathsf{u}})$ then goto l' else goto l'' (Branching)

l. Halt

We leave for the problems the precise specification of the transition function of a RAM iterator as well as the trivial proof of the following (half) analog of Proposition 2F.1:

2F.2. **Proposition.** *If a partial function $f : A^n \rightharpoonup A_s$ is tail recursive in an infinite, pointed structure $\mathbf{A}$, then f is tail recursive in the two-sorted structure*

$$\text{(2F-3)} \qquad \mathsf{RAM}(\mathbf{A}) = (\mathbf{A}, (A \to A), \chi_A, R_0, \text{ap}, \text{indap}),$$

where χ_A is the characteristic function of A by (1A-6) *and* (with $u, v, t, \in A$ and $R : A \to A$)

$$R_0(t) = 0, \quad \text{ap}(R, u) = R(u),$$
$$\text{indap}(R, u, v)(t) = \text{if } (t = u) \text{ then } R(v) \text{ else } R(t).$$

The last primitive indap : $(A \rightharpoonup A) \times A^2 \to (A \rightharpoonup A)$ is used to make precise the Indirect application command: its effect on a state $(l, \vec{u}, R)$ is either to declare it *halting*, if $\neg C_l(u_i)$, or to move it to the state $(l', \vec{u}, R')$ where

$$R'(t) = \text{indap}(R, u_i, u_j)(t) = \begin{cases} u_j, & \text{if } [t = u_i], \\ R(t), & \text{otherwise.} \end{cases}$$

As we described it here, a RAM is basically a machine with finitely many registers, what we have called now *accumulators*, enriched by the addition of potentially infinite memory coded in some function $R : A \rightharpoonup A$; notice that the direct and indirect application commands allow us to implement commands of the form

$$\mathsf{R}(\mathsf{u}_i) := \mathsf{R}(\mathsf{R}(\mathsf{u}_j)) \text{ or } \mathsf{R}(\mathsf{R}(\mathsf{u}_i)) := \mathsf{R}(\mathsf{u}_j)$$

by setting (for the first of these) in sequence

$$\mathsf{u}_j := \mathsf{R}(\mathsf{u}_j), \mathsf{u}_j := \mathsf{R}(\mathsf{u}_j), \mathsf{R}(\mathsf{u}_i) := \mathsf{u}_j.$$

The tests $C_l(u_j)$ can be used to restrict the values of $R(t)$ which are relevant: if $C_l(t) = \mathrm{tt}$ for all $t \in A$, then $R : A \rightharpoonup A$ is unrestricted; if, on the other hand,

$$C_l(t) = \chi_N(t) \text{ for some explicit } N \subset A, \text{ e.g., } \mathbb{N},$$

then the computation halts when it is asked either to read $R(u_i)$ or to update $R(u_i)$ for some $u_i \notin N$. This is the case with the simplest RAMs defined by Cook and Reckhow [1973], which (as we set things up) are over the structure

$$\text{(2F-4)} \qquad \mathbf{A}_{\mathrm{CR}} = (\mathbb{Z}, \mathbb{N}, 0, 1, \mathrm{eq}_0, +, -).$$

Problems for Section 2F

x2F.1. **Problem.** Prove Proposition x2F.1, that if $\mathbf{A}$ is pointed, then the register computable partial functions are exactly the tail recursive ones.

x2F.2. **Problem.** Prove that if $f : A^n \rightharpoonup A_s$ is register computable in a pointed structure $\mathbf{A}$, then there is a register program P with $k \geq n$ such that

$$f(\vec{x}) = \overline{\mathrm{i}_P}(x_1, \dots, x_n, a_{n+1}, \dots, a_k)$$

with some a_i which are strongly explicit in $\mathbf{A}$.

x2F.3. **Problem.** Prove that if $f : A^n \rightharpoonup A_s$ is computed in a pointed structure $\mathbf{A}$ by a register program, then it is also computed in $\mathbf{A}$ by a primitive one. HINT: Apply Problem x2F.2, and then use induction on the number of terms in P which are not prime; you will need to do some elementary but fussy "register machine programming".

x2F.4*. **Problem** (Bounded stacks characterization). Prove that if a structure **A** is pointed, then a partial function $f : A^n \rightharpoonup A_s$ is tail recursive in **A** if and only if there is an extended recursive program $E(\vec{x})$ which computes f and satisfies the following *bounded stacks condition*: there is a number K such that every state

$$a_0 \dots a_{m-1} : b_0 \cdots b_{n-1}$$

which occurs in a computation of the recursive machine $\mathfrak{i}(\mathbf{A}, E(\vec{x}))$ on input$(\vec{x})$ has length $m + n + 1 \le K$.

HINT: For the more interesting "if" direction, define an iterator whose states are the states of the recursive machine $\mathfrak{i}(\mathbf{A}, E(\vec{x}))$ of length $K + 1$ and those a_i which are not elements of A (tt, ff, terms, ?, etc.) are coded by (sequences of) constants.

x2F.5. **Problem.** Specify a straight line program which computes the value of the polynomial $a_0 + a_1 x + a_2 x^2$ in a field $\mathbf{F} = (F, 0, 1, +, -, \cdot, \div, =)$ from the input a_0, a_1, a_2, x using Horner's rule on page 25.

x2F.6. **Problem.** Specify a primitive decision tree which decides the nullity relation (1C-12) for degree $n = 2$ in $\mathbf{F} = (F, 0, 1, +, -, \cdot, \div, =)$ using Horner's rule.

x2F.7. **Problem.** Suppose $\mathbf{F} = (F, 0, 1, +, -, \cdot, \div, =)$ is a field of characteristic $\neq 2$. Specify a primitive decision tree which decides the nullity relation $N_F(a_0, a_1, x) \iff a_0 + a_1 x = 0$ of degree 1 in the reduct $(F, 0, 1, \cdot, =)$, i.e., using no additions, subtractions or divisions.

x2F.8. **Problem.** Prove Proposition 2F.2.

x2F.9. **Problem.** Prove that for every pointed structure $\mathbf{A} = (A, \Phi)$ and every **A**-explicit $N \subseteq A$, the equality relation eq_N on N is decidable by a RAM machine over **A** in which $C_l(t) = \chi_N(t)$ in every direct or indirect application.

Infer that the converse of Proposition 2F.2 is not true for all pointed **A**.[21]

HINT: To decide if $u = v$, set $R(u) := 1, t := R(v)$ and check if $t = 0$.

2G. Full vs. tail recursion (I)

We discuss here some of the many important results on the relation between full and tail recursion. It is an interesting and difficult subject, but not central

[21]This does not affect the Cook-Reckhow RAMs over the structure $\mathbf{A}_{\mathrm{CR}}$ in (2F-4), in which $\mathrm{eq}_{\mathbb{N}}$ is obviously explicit.

to our concerns, and so we will focus on just a few, characteristic facts and mostly skip proofs.[22]

Examples where Tailrec^0(A) $\subsetneq$ Rec^0(A). Perhaps the first example of a structure where some recursive function is not tail recursive is the following

2G.1. **Theorem** (Patterson and Hewitt [1970], Lynch and Blum [1979]). *There is a total, pointed, countable structure*

$$\mathbf{T} = (T, 0, 1, s, d, \mathrm{leaf}, \mathrm{red}, \mathrm{eq}_0, \mathrm{eq}_1)$$

with $s, d : T \to T$ *and* $\mathrm{leaf}, \mathrm{red}, \mathrm{eq}_0, \mathrm{eq}_1 : T \to \mathbb{B}$, *such that*

$$\mathbf{Tailrec}^0(\mathbf{T}) \subsetneq \mathbf{Rec}^0(\mathbf{T}),$$

in fact some total relation on T *is recursive but not tail recursive.*

PROOF. For each $n \geq 0$, let

$$T_n = \{(n, t_0, \ldots, t_{i-1}) : i \leq n \ \&\ (\forall j \leq (i-1))[t_j \leq 1]\},$$

viewed as the complete binary tree of sequences of length $\leq (n+1)$, with the root labelled n and all the other nodes labelled 0 or 1. It carries the obvious *son*, *daughter* and *leaf* functions

$$s(u) = u * \{0\}, \quad d(u) = u * \{1\} \quad (|u| < n+1),$$
$$\mathrm{leaf}(u) \iff |u| = n+1,$$

where $|(n, t_0, \ldots, t_{i-1})| = i + 1$ as in (1A-17) and $s(u) = d(u) = u$ if $\mathrm{leaf}(u)$. It has 2^n leaves and $2^{n+1} - 1$ nodes.

A (restricted) *coloring* of T_n assigns to each node *white* or *red*, so that all the nodes which are not leaves are white and all but at most one leaf are painted red; so there are $2^n + 1$ colored trees of depth n, say

$$T_{n,1}, \ldots, T_{n,2^n+1} \text{ with } T_{n,1} \text{ the tree where all leaves red.}$$

The (disjoint) union

$$T = \bigcup_{n,1 \leq j \leq 2^n+1} T_{n,j}$$

is the universe of $\mathbf{T}$, and the functions and relations on it are those inherited from the individual trees, with *the constants* 0 and 1 respectively naming the leaves $(1, 0), (1, 1)$ of the tree $T_{1,1}$.

The unary (total) relation

$$\text{(2G-1)} \qquad \mathrm{Red}(u) \iff \textit{all the leaves below } u \textit{ are red}$$

is recursive in $\mathbf{T}$—in fact a simple fixed point—since it is (easily) the unique solution of the recursive equation

$$p(u) = \text{if } \mathrm{leaf}(u) \text{ then } \mathrm{red}(u) \text{ else } \Big(p(u * \{0\}) \ \&\ p(u * \{1\})\Big).$$

[22] I am indebted to Siddharth Bhaskar for advising me on what to include from this area in which I am anything but expert—but, of course, the blame for what is left out is entirely mine.

To complete the proof, it suffices to prove that $\text{Red}(u)$ is not tail recursive, or equivalently by Proposition 2F.1 and the discussion on page 91, that $\text{Red}(u)$ is not decided by a primitive register program; so assume towards a contradiction that *some primitive register program P of* **T** *with k registers and $L+1$ commands decides* $\text{Red}(u)$, so that by (2F-2)

$$\text{Red}(u) \iff P \text{ starting on } (0, u, \vec{a}) \text{ returns tt}$$

with suitable constants $\vec{a} = a_2, \dots, a_k$. We consider the computations of P on inputs of the form

(2G-2) $\qquad u = (n) \in T_{n,j}, \quad \text{where } n \geq 2 \;\&\; 1 \leq j \leq 2^n + 1,$

and we leave for Problem x2G.1 the proofs of three simple lemmas about them:

Lemma 1. If a state $(l, x_1, \dots, x_k)$ occurs in the computation of P starting with (2G-2)*, then every x_i is either a constant or a node in T_n.*

Call two states

$$s = (l, x_1, \dots, x_k) \text{ and } s' = (l', x'_1, \dots, x'_k)$$

in a computation by P *equivalent* if the following conditions hold:

(i) $l = l'$.

(ii) For $i = 1, \dots, k$,

$$x_i = 0 \iff x'_i = 0, \; x_i = 1 \iff x'_i = 1, \; \text{red}(x_i) \iff \text{red}(x'_i).$$

(iii) For $i = 1, \dots, k$, $|x_i| =$ the length of the node $x_i = |x'_i|$.

Notice that by (i),

$$\text{if } s \text{ is equivalent with } s', \text{ then } [s \text{ is halting} \iff s' \text{ is halting}].$$

Lemma 2. If s and s' are equivalent and the transition function of P takes s to $\overline{s}$ and s' to $\overline{s}'$, then $\overline{s}$ and $\overline{s}'$ are also equivalent.

Lemma 3. The number of equivalence classes of states in a computation of P starting with (2G-2) *is bounded by $(L+1)2^{3k}(n+1)^k$.*

Granting these lemmas, consider now the computation

$$(0, (n), \vec{a}), (l_1, \vec{x}_1), \dots, (l_m, \vec{x}_m)$$

of P with *shape* (sequence of addresses) $(0, \dots, l_m)$ where

$$2^n > k(L+1)2^{3k}(n+1)^k$$

and (n) is the root of $T_{n,1}$, the copy of T_n in which all leaves are red. The state $(l_m, \vec{x}_m)$ is halting and returns tt, and no previous state is halting; this means, by Lemma 2, that

$$m < (L+1)2^{3k}(n+1)^k < 2^n,$$

since otherwise some $(l_i, \vec{x}_i)$ would be equivalent with some $(l_{i+k}, \vec{x}_{i+k})$, and then

$$l_i = l_{i+k} = l_{i+2k}, \dots,$$

and the computation would diverge. It follows that at least one leaf of $T_{n,1}$ does not occur in any of the registers during the computation, since the number of nodes which are "seen" is $\leq k(L+1)2^{3k}(n+1)^k < 2^n$; and this means that "the same" computation is executed by P if we start with (n) the root of the tree $T_{n,i}$ in which this leaf is white, which then returns the wrong value. ⊣

Patterson and Hewitt [1970] was written at a time when a central problem was the *logic of programs*, the search for natural conditions which insure that two *program schemas* compute the same partial function on all structures of a fixed vocabulary, cf. Walker and Strong [1973]. This influenced the formulation of results, including Theorem 2 of Patterson and Hewitt [1970] which claims only that a specific recursive program is not logically equivalent to any finite register program. Lynch and Blum [1979] define a specific structure in which some recursive function is not tail recursive as in Theorem 2G.1, which we proved here by a slight modification of the Patterson-Hewitt construction.

Many better examples were given later, including Stolboushkin and Taitslin [1983] which starts with a *Burnside group*, not quite trivial to construct and especially Tiuryn [1989], which also gives many references to the extensive literature on the problem and related results on the difference between the expressive power between deterministic and nondeterministic register programs. The Tiuryn example is simple to describe and it includes the identity relation among the primitives—which makes the proof of the version of Lemma 3 that is needed much more difficult. We will discuss some of this work in Problems x2G.2 – x2G.4, but we will not reproduce Tiuryn's proof.

Examples where $\mathbf{Tailrec}^0(\mathbf{A})$ *should be* $\subsetneq$ $\mathbf{Rec}^0(\mathbf{A})$. Jones [1999] proves that

$$\mathbf{Tailrec}^0(\mathbf{L}_b^-) \subsetneq \mathbf{Rec}^0(\mathbf{L}_b^-) \iff \text{LOGSPACE} \neq \text{PTIME},$$

where $\mathbf{L}_b^- = (\{0,1\}^*, \text{nil}, \text{eq}_{\text{nil}}, \text{head}, \text{tail})$ is the *cons-free* reduct of the Lisp structure $\mathbf{L}^*$ on binary strings. He proves many other things in this seminal paper, and in the subsequent Jones [2001] he obtains similar characterizations of the distinction between full and tail recursion for some higher type structures.[23]

Bhaskar [2018] uses one of Jones' methods to show that

$$\mathbf{Tailrec}^0(\mathbf{N}_{\text{Pd}}) \subsetneq \mathbf{Rec}^0(\mathbf{N}_{\text{Pd}}) \iff \text{EXPTIME} \neq \text{LINSPACE},$$

where *predecessor arithmetic* $\mathbf{N}_{\text{Pd}} = (\mathbb{N}, 0, 1, \text{Pd}, \text{eq}_0)$ is the *successor-free* reduct of unary arithmetic $\mathbf{N}_u$. A more interesting (and substantially more difficult) result in this paper is the equivalence

$$\mathbf{Tailrec}^0(\overline{\mathbf{F}}_p) \subsetneq \mathbf{Rec}^0(\overline{\mathbf{F}}_p) \iff \text{EXPTIME} \neq \text{LINSPACE} \quad (p \text{ prime}),$$

[23] Some results of this type were also cited in Tiuryn [1989] about the related distinction between deterministic and nondeterministic dynamic logic.

where $\overline{\mathbf{F}}_p$ is the algebraic closure of the finite, prime field of characteristic p. Bhaskar's analysis uses the crucial fact that both $\mathbf{N}_{\mathrm{Pd}}$ and $\overline{\mathbf{F}}_p$ are *locally finite* and suggests that it may be possible to characterize usefully the difference between full and tail recursion in a structure $\mathbf{A}$ in terms of model-theoretic properties of $\mathbf{A}$, cf. Open Problem x2G.5.

Problems for Section 2G

x2G.1. **Problem.** Prove the three lemmas in the proof of Theorem 2G.1. HINT: The commands in the primitive register program P are all in one of the forms

$$x_i := x_j;\ \text{goto}\ l',\ x_i := s(x_j);\ \text{goto}\ l',\ x_i := d(x_j);\ \text{goto}\ l'$$

$$\text{if test}(x_i)\ \text{goto}\ l'\ \text{else}\ \ \text{goto}\ l'',\qquad \text{halt, return output}(x_i),$$

where test and output are leaf, red, eq_0, or eq_1. Taking cases on this list verifies easily Lemmas 1 and 2, and Lemma 3 is also easily proved by a counting argument.

x2G.2. **Problem.** Let for each $n \geq 0$

$$T_n = \{(n, t_0, \dots, t_{i-1}) : i \leq n \ \&\ (\forall j \leq (i-1))[t_j \leq 1]\},$$

as in the proof of Theorem 2G.1, and consider the *forest*

$$\mathbf{T}^s = (\bigcup_n T_n, 0, 1, s, d, =)$$

as before but without the colors and with the addition of $=$. Prove that the relation

$$\text{(2G-3)}\qquad B(u, v) \iff u \sqsubseteq v \quad (v \text{ is equal to or below } u)$$

is recursive in $\mathbf{T}^s$.

Tiuryn [1989] proves that $B(u, v)$ is not tail recursive in $\mathbf{T}^s$.

x2G.3. **Problem.** Prove that in the structure $\mathbf{T}$ of Theorem 2G.1, the (partial) function

$$\text{White}(u) = \text{if there exists a white leaf below } u \text{ then } \mathrm{tt} \text{ else } \uparrow$$

is computed by a nondeterministic register program but not by any deterministic one. HINT: Use a variation of the proof of Theorem 2G.1.

The corresponding problem for total functions is (as far as I know) open:

x2G.4. **Open problem.** Define a total, pointed structure $\mathbf{A}$ and a total function $f : A^n \to \mathbb{B}$ which is computed by a nondeterministic register program but not by any deterministic one.

x2G.5. **Open problem** (Bhaskar). Find model-theoretic conditions on a total structure **A** which characterize the strict inclusion $\mathbf{Tailrec}^0(\mathbf{A}) \subsetneq \mathbf{Rec}^0(\mathbf{A})$.

x2G.6. **Open problem.** For $\mathbf{A} = (\mathbf{N}_\varepsilon, 0, 1) = (\mathbb{N}, \text{rem}, \text{eq}_0, \text{eq}_1, 0, 1)$, the pointed expansion of the Euclidean structure, is every recursive function tail recursive? A negative answer would produce an interesting Tiuryn-type example (with a binary primitive), while a positive one might have a bearing on the Main Conjecture on page 2.

2H. What is an algorithm?

We have already mentioned in the Introduction the basic foundational problem of *defining algorithms*[24] and placed it outside the scope of this book. It is, however, worth including here a brief discussion of the question, mostly to point to the (few) publications that have tried to deal with it.

With the notation of Section 1E, we will focus on algorithms that compute partial functions and relations

$$f : A^n \rightharpoonup A_s \quad (s \in \{\texttt{ind}, \texttt{boole}\}) \tag{2H-1}$$

from the primitives of a (partial) Φ-structure

$$\mathbf{A} = (A, \{\phi^{\mathbf{A}}\}_{\phi \in \Phi}).$$

The most substantial part of this restriction is that it leaves out algorithms with side effects and interaction, cf. the footnote on page 3 and the relevant Section 3B in Moschovakis [1989a].

Equally important is the restriction to *algorithms from specified primitives*, especially as the usual formulations of the *Church-Turing Thesis* suggest that the primitives of a Turing machine are in some sense “absolutely computable” and need not be explicitly assumed or justified. We have noted in the Introduction (and several other places) why this is not a useful approach when we want to formulate, derive and justify robust lower complexity bounds for mathematical problems; but in trying to understand *computability* and the meaning of the Church-Turing Thesis, it is natural to ask whether there are absolutely computable primitives and what those might be. See Sections 2 and 8 of Moschovakis [2014] for a discussion of the problem and references to relevant work, especially the eloquent analyses in Gandy [1980] and Kripke [2000].

[24]Using imprecise formulations of the *Church-Turing Thesis* and vague references to Church [1935], [1936] and Turing [1936], it is sometimes claimed naively that *algorithms are Turing machines*. This does not accord with the original formulations of the Church-Turing Thesis, cf. the discussion in Section 1.1 of Moschovakis [2014] (which repeats points well known and understood by those who have thought about this matter); and as we mentioned in the Introduction, it is not a useful assumption.

There is also the restriction to *first-order primitives*, partial functions and relations. This is necessary for the development of a conventional theory of complexity, but recursion and computability from higher type primitives have been extensively studied: see Kleene [1959], Kechris and Moschovakis [1977] and Sacks [1990] for the higher-type recursion which extends directly the first-order notion we have adopted, and Longley and Normann [2015] for a near-complete exposition of the many and different approaches to the topic.[25]

Once we focus on algorithms which compute partial functions (2H-1) from the primitives of a Φ-structure, then the problem of defining them rigorously comes down basically to choosing between *iterative algorithms* specified by computation models as in Section 2C and *elementary recursive algorithms* expressed directly by recursive programs; at least this is my view, which I have explained and defended as best I can in Section 3 of Moschovakis [1998].

The first of these choices—that algorithms are iterative processes—is *the standard view*, explicitly or implicitly adopted (sometimes with additional restrictions) by most mathematicians and computer scientists, including Knuth in Section 1.1 of his classic Knuth [1973]. More recently (and substantially more abstractly, on arbitrary structures), this standard view has been developed, advocated and defended by Gurevich and his collaborators, cf. Gurevich [1995], [2000] and Dershowitz and Gurevich [2008]; see also Tucker and Zucker [2000] and Duží [2014].

I have made the second choice, that *algorithms* (which compute partial functions from the primitives of a Φ-structure $\mathbf{A}$) *are directly expressed by recursive programs*, and I have developed and defended this view in several papers, most carefully in Moschovakis [1998].

With the notation in Section 2A, my understanding of *the algorithm expressed by an extended recursive Φ-program*

$$E(\vec{x}) \equiv E_0(\vec{x}) \text{ where } \Big\{\mathsf{p}_1(\vec{x}_1) = E_1, \dots, \mathsf{p}_K(\vec{x}_K) = E_K\Big\}$$

in a Φ-structure $\mathbf{A}$ is that it calls for solving in $\mathbf{A}$ the system of recursive equations in the body of $E(\vec{x})$ and then plugging the solutions in its head to compute for each $\vec{x} \in A^n$ the value $\text{den}(\mathbf{A}, E(\vec{x}))$; *how* we find the canonical solutions of this system is not part of *the elementary recursive algorithm* expressed by $E(\vec{x})$ in this view, and this raises some obvious questions:

About implementations (II). To compute $\text{den}(\mathbf{A}, E(\vec{x}))$ in specific cases, we might use the method outlined in the proof of the Fixed Point Lemma 1B.1, or the recursive machine defined in Section 2C or any one of several well known and much studied *implementations of recursion*. These are iterative algorithms which (generally) use fresh primitives and should satisfy additional properties,

[25]See also Moschovakis [1989a]—which is about recursion on structures with arbitrary monotone functionals for primitives—and the subsequent Moschovakis [1989b] where the relevant notion of *algorithm from higher-type primitives* is modeled rigorously.

cf. the discussion on page 73; so to specify *the admissible implementations of a recursive program* is an important (and difficult) part of this approach to the foundations of the theory of algorithms, cf. Moschovakis [1998] and (especially) Moschovakis and Paschalis [2008] which includes a precise definition and some basic results.

Imperative vs. functional programming. This "incompleteness" of elementary recursive algorithms is surely a weakness of this view and an important advantage of the standard view that asks for a specific iterative process rather than the (more difficult) identification of all admissible implementations of a recursive program. Some of this advantage disappears, however, when we want to understand algorithms which are not expressed by tail recursive programs (page 61): for example, what is *the* iterative program which expresses the merge-sort as it is specified in Proposition 1C.2? There are many, and it seems that we can understand "the merge-sort" and reason about it better by looking at this proposition rather than by focussing on choosing a specific implementation of it.

Moreover, there are also some advantages to this incompleteness of elementary recursive algorithms:

Proofs of correctness. In Proposition 1C.2, we claimed the correctness of the merge-sort—that it sorts—by just saying that

The sort function satisfies the equation . . .

whose proof was too trivial to deserve more than the single sentence

The validity of (1C-6) *is immediate, by induction on* $|u|$.

To prove the correctness of an iterator that "expresses the merge-sort", you must first design a specific one and then explain how you can extract from all the "housekeeping" details necessary for the specification of iterators a proof that what is actually being computed is the sorting function; most likely you will trust that a formal version of (1C-6) is implemented correctly by some compiler or interpreter of whatever higher-level language you are using, as we did for the recursive machine.

Simply put, whether correct or not, the view that algorithms are faithfully expressed by systems of recursive equations separates proofs of their correctness, which involve only purely mathematical facts from the relevant subject and standard results of fixed-point-theory, from proofs of *implementation correctness* for programming languages which are ultimately necessary but have a very different flavor.

What we will not do in this book is to give a precise (set-theoretic) definition of *the algorithm on a* Φ*-structure expressed by a* Φ*-recursive program*; several versions of this are given in the several papers already cited, it is neither simple nor as definitive as I would want it to be—and we do not need it.

CHAPTER 3

COMPLEXITY THEORY FOR RECURSIVE PROGRAMS

Suppose Π is a class of extended programs which compute (in some precise sense) partial functions and relations on a set A. In the most general terms, a *complexity measure* for Π associates with each n-ary $E \in \Pi$ which computes $f : A^n \rightharpoonup A_s$ an n-ary partial function

(3-2) $$C_E : A^n \rightharpoonup \mathbb{N}$$

which intuitively assigns to each $\vec{x}$ such that $f(\vec{x})\downarrow$ a *cost* of some sort of the computation of $f(\vec{x})$ by E.

We introduce here several natural complexity measures on the (deterministic) recursive programs of a Φ-structure $\mathbf{A}$, directly from the programs, i.e., without reference to the recursive machine. These somewhat abstract, "implementation-independent" (*logical*) definitions help clarify some questions about complexity, and they are also useful in the derivation of upper bounds for recursive programs and robust lower bounds for problems. Much of what we will say extends naturally to nondeterministic programs, but the formulas and arguments are substantially more complex and it is better to keep matters simple by dealing first with the more important, deterministic case.

3A. The basic complexity measures

Suppose $\mathbf{A}$ is a structure, $E(\vec{x})$ is an n-ary, extended, $\mathbf{A}$-program and M is an $(\mathbf{A}, E)$-term as these were defined in (2D-1), i.e.,

$$M \equiv N(y_1, \dots, y_m) \text{ with } N \text{ a subterm of } E \text{ and } y_1, \dots, y_m \in A.$$

Extending the notation of (2A-5), we set

(3A-1) $$\overline{M} = \text{den}(\mathbf{A}, E, M) = \text{den}((\mathbf{A}, \overline{p}_1, \dots, \overline{p}_K), M),$$

(3A-2) $$\text{Conv}(\mathbf{A}, E) = \{M : M \text{ is an } (\mathbf{A}, E)\text{-term and } \overline{M}\downarrow\},$$

where $\mathsf{p}_1, \dots, \mathsf{p}_K$ are the recursive variables of E and $\overline{p}_1, \dots, \overline{p}_K$ their mutual fixed points.

Normally we will use the simpler $\overline{M}$ rather than $\mathrm{den}((\mathbf{A}, \overline{p}_1, \dots, \overline{p}_K), M)$, since $\mathbf{A}$ and E will be held constant in most of the arguments in this section.

The tree-depth complexity $D_E^{\mathbf{A}}(M)$. With each convergent $(\mathbf{A}, E)$-term M, we will associate a *computation tree* $\mathcal{T}(M)$ which represents an abstract, parallel computation of $\overline{M}$ using E. The tree-depth complexity of M is the depth of $\mathcal{T}(M)$, but it is easier to define $D_E^{\mathbf{A}}(M)$ first and $\mathcal{T}(M)$ after that, by recursion on $D_E^{\mathbf{A}}(M)$.

3A.1. **Lemma.** *Fix a structure* $\mathbf{A}$ *and an* $\mathbf{A}$*-program* E. *There is exactly one function* $D = D_E^{\mathbf{A}}$ *which is defined for every* $M \in \mathrm{Conv}(\mathbf{A}, E)$ *and satisfies the following conditions*:

(D1) $D(\mathrm{tt}) = D(\mathrm{ff}) = D(x) = D(\phi) = 0$ (if $\mathrm{arity}(\phi) = 0$ and $\phi^{\mathbf{A}}\downarrow$).
(D2) $D(\phi(M_1, \dots, M_m)) = \max\{D(M_1), \dots, D(M_m)\} + 1$.
(D3) *If* $M \equiv \text{if } M_0 \text{ then } M_1 \text{ else } M_2$, *then*

$$D(M) = \begin{cases} \max\{D(M_0), D(M_1)\} + 1, & \textit{if } \overline{M}_0 = \mathrm{tt}, \\ \max\{D(M_0), D(M_2)\} + 1, & \textit{if } \overline{M}_0 = \mathrm{ff}. \end{cases}$$

(D4) *If* p *is a recursive variable of* E,[26] *then*

$$D(\mathsf{p}(M_1, \dots, M_m)) = \max\{D(M_1), \dots, D(M_m), d_{\mathsf{p}}(\overline{M}_1, \dots, \overline{M}_m)\} + 1,$$

where $d_{\mathsf{p}}(\vec{w}) = D(E_{\mathsf{p}}(\vec{w}))$.

The tree-depth complexity in $\mathbf{A}$ *of an extended program* $E(\vec{x})$ *is that of its head term*,

$$d_E(\vec{x}) = d(\mathbf{A}, E(\vec{x})) =_{\mathrm{df}} D(E_0(\vec{x})) \quad (\mathrm{den}(\mathbf{A}, E(\vec{x}))\downarrow). \tag{3A-3}$$

PROOF. If $\overline{M} = \mathrm{den}((\mathbf{A}, \overline{p}_1, \dots, \overline{p}_K), M)\downarrow$, then there is some k such that

$$\overline{M}^k = \mathrm{den}((\mathbf{A}, \overline{p}_1^k, \dots, \overline{p}_K^k), M)\downarrow, \tag{3A-4}$$

where $\overline{p}_1^k, \dots, \overline{p}_K^k$ are the iterates of the system that defines $\overline{M}$ as in the proof of the Fixed Point Lemma, Theorem 1B.1. We define $D(M)$ by recursion on

$$\mathrm{stage}(M) = \text{the least } k \text{ such that } \overline{M}^k\downarrow, \tag{3A-5}$$

and recursion on the length of terms within this. We consider cases on the form of M.

(D1) If M is tt, ff, $x \in A$ or a convergent, nullary constant, set $D(M) = 0$.
(D2) If $M \equiv \phi(M_1, \dots, M_m)$ for some $\phi \in \Phi$ and $\overline{M}\downarrow$, then

$$\mathrm{stage}(M) = \max\{\mathrm{stage}(M_1), \dots, \mathrm{stage}(M_m)\},$$

[26] If $\mathsf{p} \equiv \mathsf{p}_i$ is a recursive variable of E, we sometimes write $E_{\mathsf{p}} \equiv E_i$ for the term which defines p in E. It is a useful convention which saves typing double subscripts.

and these subterms are all smaller than M, so we may assume that $D(M_i)$ is defined for $i = 1, \dots, m$; we set

$$D(M) = \max\{D(M_1), \dots, D(M_m)\} + 1.$$

(D3) If $M \equiv$ if M_0 then M_1 else M_2 and $\overline{M}\downarrow$, then either $\overline{M}_0 = \mathrm{tt}$, $\overline{M}_1\downarrow$ and $\mathrm{stage}(M) = \max\{\mathrm{stage}(M_0), \mathrm{stage}(M_1)\}$ or the corresponding conditions hold with M_0 and M_2. In either case, the terms M_0, M_i are proper subterms of M, and we can assume that D is defined for them and define $D(M)$ appropriately as in case (D2).

(D4) If $M \equiv \mathsf{p}(M_1, \dots, M_m)$ with a recursive variable p of E, $\overline{M}\downarrow$ and $k = \mathrm{stage}(M)$, then

$$\overline{M}^k = \overline{p}^k(\overline{M}_1^k, \dots, \overline{M}_m^k)\downarrow,$$

and so $\mathrm{stage}(M_i) \leq k$ and we can assume that $D(M_i)$ is defined for $i = 1, \dots, n$, since these terms are smaller than M. Moreover, if $\overline{M}_1 = w_1, \dots, \overline{M}_m = w_m$, then

$$\overline{p}^k(w_1, \dots, w_m) = \mathrm{den}((\mathbf{A}, \overline{p}_1^{k-1}, \dots, \overline{p}_K^{k-1}), E_{\mathsf{p}}(w_1, \dots, w_m))\downarrow$$

by the definition of the iterates in the proof of Lemma 1B.1, and so

$$k \geq \mathrm{stage}(\mathsf{p}(w_1, \dots, w_m)) = \mathrm{stage}(E_{\mathsf{p}}(w_1, \dots, w_m)) + 1; \tag{3A-6}$$

so we may assume that $D(E_{\mathsf{p}}(w_1, \dots, w_m))$ is defined, and define $D(M)$ so that (D4) in the lemma holds.

The uniqueness of D is proved by a simple induction on $\mathrm{stage}(M)$, following the definition. ⊣

There is no reasonable way to implement recursive programs so that the number of steps required to compute $\overline{M}$ is $D(M)$. For example, to attain

$$D(\mathsf{p}(M)) = \max\{D(M), D(E_{\mathsf{p}}(\overline{M}))\} + 1,$$

we need to compute in parallel the value $\overline{M}$ of M and the value of $E_{\mathsf{p}}(\overline{M})$, but we cannot start on the second computation until we complete the first, so that we know $\overline{M}$. We can imagine a nondeterministic process which "guesses" the correct $\overline{M}$ and works with that; but if $\mathbf{A}$ is infinite, then this amounts to infinite nondeterminism, which is not a useful idealization.

In any case, our methods do not yield any interesting lower bounds for tree-depth complexity, but it is a very useful tool for defining rigorously and analysing many properties of recursive programs.

The computation tree. The *computation tree* $\mathcal{T}(M) = \mathcal{T}(\mathbf{A}, E, M)$ for $M \in \mathrm{Conv}(\mathbf{A}, E)$ is defined by recursion on $D(M)$ using the operation Top in (1A-20), see Figure 3. We take cases, corresponding to the definition of $D(M)$:

($\mathcal{T}$1) If M is tt, ff, $x \in A$ or a nullary constant ϕ, set $\mathcal{T}(M) = \{(M)\}$.

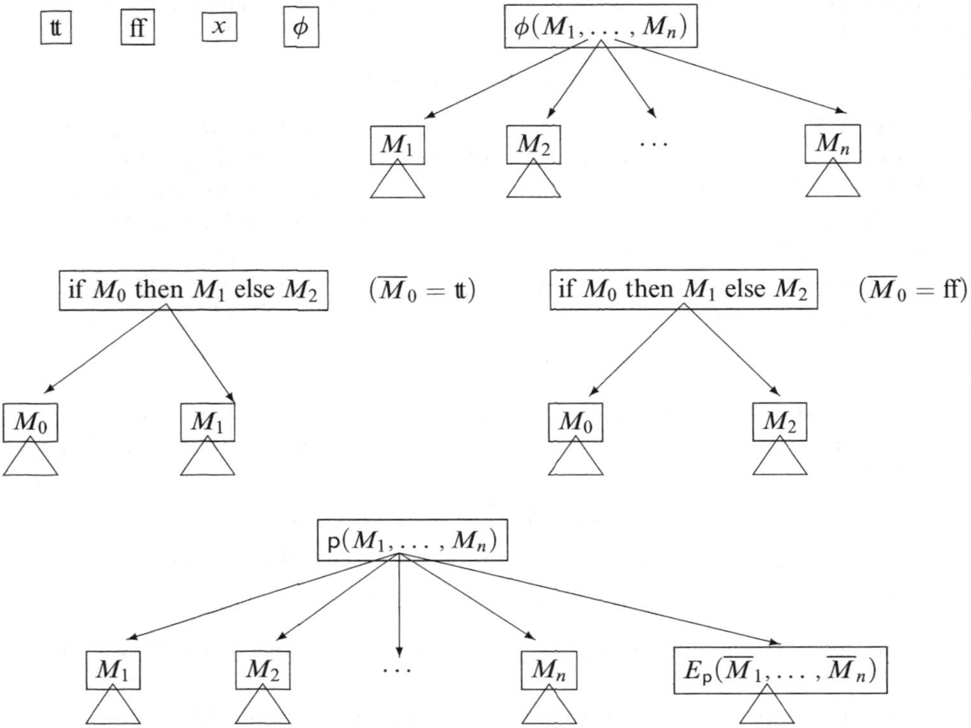

FIGURE 3. Computation trees.

($\mathcal{T}2$) If $M \equiv \phi(M_1, \dots, M_m)$, set $\mathcal{T}(M) = \text{Top}(M, \mathcal{T}(M_1), \dots, \mathcal{T}(M_m))$.

($\mathcal{T}3$) If $M \equiv$ if M_0 then M_1 else M_2, set

$$\mathcal{T}(M) = \begin{cases} \text{Top}(M, \mathcal{T}(M_0), \mathcal{T}(M_1)) & \text{if } \overline{M}_0 = \text{tt}, \\ \text{Top}(M, \mathcal{T}(M_0), \mathcal{T}(M_2)) & \text{if } \overline{M}_0 = \text{ff}. \end{cases}$$

($\mathcal{T}4$) If $M \equiv \mathsf{p}(M_1, \dots, M_m)$, set

$$\mathcal{T}(M) = \text{Top}(M, \mathcal{T}(M_1), \dots, \mathcal{T}(M_m), \mathcal{T}(E_{\mathsf{p}}(\overline{M}_1, \dots, \overline{M}_m))).$$

3A.2. **Proposition.** *For every* $M \in \text{Conv}(\mathbf{A}, E)$,

$$D(M) = \text{depth}(\mathcal{T}(M)).$$

PROOF is trivial, by induction on $D(M)$. ⊣

The sequential logical complexity $L^s(M)$ (time). For a fixed Φ-structure **A** and a Φ-program E, the *sequential logical complexity*

$$L^s(M) = L^s(\mathbf{A}, E, M) \quad (M \in \mathrm{Conv}(\mathbf{A}, E))$$

is defined by the following recursion on $D(M)$:

(L^s1) $L^s(\mathrm{tt}) = L^s(\mathrm{ff}) = L^s(x) = 0 \quad (x \in A)$ and $L^s(\phi) = 1$.

(L^s2) $L^s(\phi(M_1, \dots, M_n)) = L^s(M_1) + L^s(M_2) + \dots + L^s(M_n) + 1$.

(L^s3) If $M \equiv$ if M_0 then M_1 else M_2, then

$$L^s(M) = \begin{cases} L^s(M_0) + L^s(M_1) + 1 & \text{if } \overline{M}_0 = \mathrm{tt}, \\ L^s(M_0) + L^s(M_2) + 1 & \text{if } \overline{M}_0 = \mathrm{ff}. \end{cases}$$

(L^s4) If p is a recursive variable of E, then

$$L^s(\mathsf{p}(M_1, \dots, M_n)) = L^s(M_1) + \dots + L^s(M_n) + L^s(E_{\mathsf{p}}(\overline{M}_1, \dots, \overline{M}_n)) + 1.$$

The sequential logical complexity in **A** of an extended program $E(\vec{x})$ is that of its head term

(3A-7) $$l^s_E(\vec{x}) = l^s(\mathbf{A}, E(\vec{x})) =_{\mathrm{df}} L^s(E_0(\vec{x})) \quad (\mathrm{den}^{\mathbf{A}}_E(\vec{x})\downarrow).$$

Intuitively, $L^s(M)$ counts the number of steps required for the computation of $\overline{M}$ using "the algorithm expressed by" E and it should be a good approximation of the *time complexity* of any implementation of E. We cannot prove this in general absent a precise definition of implementations, but it can be verified for specific cases: for example, if

(3A-8) $$\mathrm{Time}_{\mathbf{A},E}(M) = \text{the length of the computation of the recursive machine for } \mathbf{A} \text{ and } E \text{ on } M:$$

then there is a number C (depending only on the program E) such that

(3A-9) $$L^s(M) \le \mathrm{Time}(M) \le L^s(M) + C \quad (\text{for all } M \in \mathrm{Conv}(\mathbf{A}, E)),$$

cf. Problem x3A.2*. This implies that

$$l^s_E(\vec{x}) \le \mathrm{Time}_{\mathbf{A},E}(\vec{x}) \le l^s_E(\vec{x}) + C \quad (\mathbf{A} \models E(\vec{x})\downarrow),$$

and we would expect that for every iterator $\mathfrak{i}$ which implements $E(\vec{x})$ on **A**, at least

$$l^s_E(\vec{x}) \le \mathrm{Time}_{\mathfrak{i}}(\vec{x}) = O(l^s_E(\vec{x}) + C)$$

for some constant C.

The parallel logical complexity $L^p(M)$. For a fixed Φ-structure $\mathbf{A}$ and a Φ-program E, the *parallel logical complexity*

$$L^p(M) = L^p(\mathbf{A}, E, M) \quad (M \in \mathrm{Conv}(\mathbf{A}, E))$$

is defined by the following recursion on $D(M)$:

(L^p1) $L^p(\mathrm{tt}) = L^p(\mathrm{ff}) = L^p(x) = 0 \quad (x \in A)$ and $L^p(\phi) = 1$.

(L^p2) $L^p(\phi(M_1, \dots, M_n)) = \max\{L^p(M_1), \dots, L^p(M_n)\} + 1$.

(L^p3) If $M \equiv$ if M_0 then M_1 else M_2, then

$$L^p(M) = \begin{cases} \max\{L^p(M_0), L^p(M_1)\} + 1, & \text{if } \overline{M}_0 = \mathrm{tt}, \\ \max\{L^p(M_0), L^p(M_2)\} + 1, & \text{if } \overline{M}_0 = \mathrm{ff}. \end{cases}$$

(L^p4) If p is a recursive variable of E, then

$$L^p(\mathsf{p}(M_1, \dots, M_n)) = \max\{L^p(M_1), \dots, L^p(M_n)\} + L^p(E_{\mathsf{p}}(\overline{M}_1, \dots, \overline{M}_n)) + 1;$$

and again, we put

(3A-10) $$l^p_E(\vec{x}) = l^p(\mathbf{A}, E(\vec{x})) =_{\mathrm{df}} L^p(E_0(\vec{x})) \quad (\mathrm{den}^{\mathbf{A}}_E(\vec{x})\downarrow).$$

Intuitively, $L^p(M)$ counts the minimal number of steps that must be executed in sequence in any computation of $\overline{M}$ by the algorithm expressed by E. No chance to prove this rigorously, of course, but it seems that the difference between $L^s(M)$ and $L^p(M)$ measures (in some vague sense) how "parallel" the algorithm expressed by E is. It is no more than exponential with base one more than the *total arity* of the program

(3A-11) $$\ell = \ell(E(\vec{\mathsf{x}})) = \max\{\mathrm{arity}(E(\vec{\mathsf{x}})), \max\{\mathrm{arity}(\phi) : \phi \in \Phi\}, \max\{\mathrm{arity}(\mathsf{p}_i) : i = 1, \dots, K\}\}.$$

3A.3. **Theorem.** *For every Φ-structure $\mathbf{A}$, every Φ-program E of total arity $\ell \geq 1$ and every $M \in \mathrm{Conv}(\mathbf{A}, E)$,*

(1) $L^s(M) \leq |\mathcal{T}(M))|$,
(2) $\mathrm{depth}(\mathcal{T}(M)) \leq L^p(M)$,
(3) $L^s(M) \leq (\ell + 1)^{L^p(M)+1} - 1$,
(4) $L^s(M) \leq (\ell + 1)^{L^p(M)}$,

and hence, for all $\vec{x}$ such that $\overline{p}_E(\vec{x})\downarrow$,

$$l^s_E(\vec{x}) \leq (\ell + 1)^{l^p_E(\vec{x})}.$$

PROOF. (1) and (2) are verified by simple inductions on $D(M)$, and (3) follows from (1) and (3) of Problem x1A.4, because $\mathrm{degree}(\mathcal{T}(M)) = \ell + 1$. (These inequalities are basically obvious from the tree pictures in Figure 3.)

(4) is somewhat stronger than (3) and simpler, which makes it (perhaps) worth proving, although this involves some seemingly unavoidable computations.

By Problem x3A.11 a stronger inequality holds when $\ell = 1$, so we assume that $\ell \geq 2$ and proceed again by induction on $D(M)$.

Case 1. $D(M) \leq 1$. In this case M is tt, ff, a parameter $x \in A$ or a nullary constant ϕ, so that $L^p(M) = L^s(M) \leq 1 \leq (\ell+1)^{L^p(M)}$.

Case 2, $M \equiv \phi(M_1, \dots, M_n)$. The induction hypothesis gives us the result for $M_1, \dots, M_n$, and we compute:

$$\begin{aligned}
L^s(M) &= L^s(M_1) + \cdots + L^s(M_n) + 1 \\
&\leq (\ell+1)^{L^p(M_1)} + \cdots + (\ell+1)^{L^p(M_n)} + 1 \\
&\leq \ell(\ell+1)^A + 1 \qquad\qquad (A = \max\{L^p(M_1), \dots, L^p(M_n)\}) \\
&\leq \ell(\ell+1)^A + (\ell+1)^A = (\ell+1)^{A+1} = (\ell+1)^{L^p(M)}.
\end{aligned}$$

Case 3, $M \equiv$ if M_0 then M_1 else M_2. The argument is very similar to Case 2 and we skip it.

Case 4, $M \equiv p(M_1, \dots, M_n)$. Now

$$L^s(M) = L^s(M_1) + \cdots + L^s(M_n) + L^s(E_p(\overline{M}_1, \dots, \overline{M}_n)) + 1.$$

If $L^p(M_1) = \cdots = L^p(M_n) = 0$, then each M_i is a truth value or a parameter x, so their sequential logical complexities are also $= 0$, and then, using the induction hypothesis:

$$\begin{aligned}
L^s(M) &= L^s(E_p(\overline{M}_1, \dots, \overline{M}_n)) + 1 \\
&\leq (\ell+1)^{L^p(E_p(\overline{M}_1, \dots, \overline{M}_n))} + 1 \\
&\leq (\ell+1)^{L^p(E_p(\overline{M}_1, \dots, \overline{M}_n))+1} = (\ell+1)^{L^p(M)}.
\end{aligned}$$

In the opposite case, setting

$$A = \max\{L^p(M_1), \dots, L^p(M_n)\} \geq 1, \quad B = L^p(E_p(\overline{M}_1, \dots, \overline{M}_n)),$$

we can compute as above:

$$\begin{aligned}
L^s(M) &= L^s(M_1) + \cdots + L^s(M_n) + L^s(E_i(\overline{M}_1, \dots, \overline{M}_n)) + 1 \\
&\leq (\ell+1)^{L^p(M_1)} + \cdots + (\ell+1)^{L^p(M_n)} + (\ell+1)^B + 1 \\
&\leq \ell(\ell+1)^A + (\ell+1)^{B+1},
\end{aligned}$$

and it suffices to prove that for all $A \geq 1$ and all $B \in \mathbb{N}$,

$$\ell(\ell+1)^A + (\ell+1)^{B+1} \leq (\ell+1)^{A+B+1};$$

but this inequality is equivalent to

$$\frac{\ell}{(\ell+1)^{B+1}} + \frac{1}{(\ell+1)^A} \leq 1,$$

which is obvious when $A \geq 1$. ⊣

Next we define two complexity measures on recursive programs which disregard the logical steps and count only calls to the primitives.

The number-of-calls complexity $C^s(\Phi_0)(M)$**.** Fix a Φ-structure $\mathbf{A}$, a subset $\Phi_0 \subseteq \Phi$ of the vocabulary and an extended Φ-program $E(\vec{x})$. The *number of calls to* Φ_0

$$C^s(\Phi_0)(M) = C^s(\Phi_0)(\mathbf{A}, E(\vec{x}), M)$$

of any $M \in \mathrm{Conv}(\mathbf{A}, E)$ is defined by the following recursion on $D(M)$:

(C^s1) $C^s(\Phi_0)(\mathrm{tt}) = C^s(\Phi_0)(\mathrm{ff}) = C^s(\Phi_0)(x) = 0 \quad (x \in A)$; and for nullary constants, $C^s(\Phi_0)(\phi) = 0$ if $\phi \notin \Phi_0$, and $C^s(\Phi_0)(\phi) = 1$ if $\phi \in \Phi_0$.

(C^s2) If $M \equiv \phi(M_1, \dots, M_n)$, then

$$C^s(\Phi_0)(M) = \begin{cases} C^s(\Phi_0)(M_1) + \cdots + C^s(\Phi_0)(M_n) + 1, & \text{if } \phi \in \Phi_0, \\ C^s(\Phi_0)(M_1) + \cdots + C^s(\Phi_0)(M_n), & \text{otherwise.} \end{cases}$$

(C^s3) If $M \equiv$ if M_0 then M_1 else M_2, then

$$C^s(\Phi_0)(M) = \begin{cases} C^s(\Phi_0)(M_0) + C^s(\Phi_0)(M_1), & \text{if } \overline{M}_0 = \mathrm{tt}, \\ C^s(\Phi_0)(M_0) + C^s(\Phi_0)(M_2), & \text{if } \overline{M}_0 = \mathrm{ff}. \end{cases}$$

(C^s4) If $M \equiv \mathsf{p}(M_1, \dots, M_n)$ with p a recursive variable of E, then

$$C^s(\Phi_0)(M) = C^s(\Phi_0)(M_1) + \cdots + C^s(\Phi_0)(M_n) + C^s(\Phi_0)(E_{\mathsf{p}}(\overline{M}_1, \dots, \overline{M}_n)).$$

The number of Φ_0-calls complexity in $\mathbf{A}$ of $E(\vec{x})$ is that of its head term,

(3A-12) $$c^s(\Phi_0)(\vec{x}) = c^s(\Phi_0)(\mathbf{A}, E(\vec{x})) =_{\mathrm{df}} C^s(\Phi_0)(\mathbf{A}, E(\vec{x}), E_0(\vec{x})),$$

and it is defined exactly when $\mathrm{den}_E^{\mathbf{A}}(\vec{x})\downarrow$. We also set

(C^s, c^s) $$C^s(M) = C^s(\Phi)(M), \quad c^s(\vec{x}) = c^s(\Phi)(\vec{x})$$

when we want to count the calls to all primitives.

This is a very natural complexity measure: $C^s(\Phi_0)(M)$ counts *the number of calls to the primitives in* Φ_0 which are required for the computation of $\overline{M}$ using "the algorithm expressed" by the program E and disregarding the "logical steps" (branching and recursive calls) as well as calls to primitives not in Φ_0. It does not distinguish between parallel and sequential implementations of E, although it is more directly relevant to the second—so we will sometimes refer to it as the *sequential calls complexity*.

Notice that E may (stupidly) call many times for the same value of one of the primitives, and all these calls will be counted separately by $C^s(\Phi_0)(M)$. Most of the lower bounds of algebraic problems that we will derive are about a somewhat smaller measure which counts only the number of distinct calls to the primitives in Φ_0.

$C^s(M)$ can be easily read off the computation tree $\mathcal{T}(M)$ and also off the computation of the recursive machine starting with M : , Problems x3A.13 and x3A.14.

The depth-of-calls complexity $C^p(\Phi_0)(M)$**.** Fix again a Φ-structure $\mathbf{A}$, a subset $\Phi_0 \subseteq \Phi$ of the vocabulary and an extended Φ-program $E(\vec{x})$. The *depth of calls to* Φ_0

$$C^p(\Phi_0)(M) = C^p(\Phi_0)(\mathbf{A}, E, M)$$

of any $M \in \mathrm{Conv}(\mathbf{A}, E)$ is defined by the following recursion on $D(M)$:

(C^p1) $C^p(\Phi_0)(\mathrm{tt}) = C^p(\Phi_0)(\mathrm{ff}) = C^p(\Phi_0)(x) = 0 \quad (x \in A)$; and for nullary constants, $C^p(\Phi_0)(\phi) = 0$ if $\phi \notin \Phi_0$, and $C^p(\Phi_0)(\phi) = 1$ if $\phi \in \Phi_0$.

(C^p2) If $M \equiv \phi(M_1, \dots, M_n)$, then

$$C^p(\Phi_0)(M) = \begin{cases} \max\{C^p(\Phi_0)(M_1), \dots, C^p(\Phi_0)(M_n)\} + 1, & \text{if } \phi \in \Phi_0, \\ \max\{C^p(\Phi_0)(M_1), \dots, C^p(\Phi_0)(M_n)\}, & \text{otherwise.} \end{cases}$$

(C^p3) If $M \equiv$ if M_0 then M_1 else M_2, then

$$C^p(\Phi_0)(M) = \begin{cases} \max\{C^p(\Phi_0)(M_0), C^p(\Phi_0)(M_1)\}, & \text{if } \overline{M}_0 = \mathrm{tt}, \\ \max\{C^p(\Phi_0)(M_0), C^p(\Phi_0)(M_2)\}, & \text{if } \overline{M}_0 = \mathrm{ff}. \end{cases}$$

(C^p4) If $M \equiv \mathsf{p}(M_1, \dots, M_n)$ of E with p a recursive variable of E, then

$$C^p(\Phi_0)(M) = \max\{C^p(\Phi_0)(M_1), \dots, C^p(\Phi_0)(M_n)\} \\ + C^p(\Phi_0)(E_{\mathsf{p}}(\overline{M}_1, \dots, \overline{M}_n)$$

The depth of calls to Φ_0 of E in $\mathbf{A}$ is that of its head term,

(3A-13) $$c^p(\Phi_0)(\vec{x}) = c^p(\Phi_0)(\mathbf{A}, E(\vec{x})) =_{\mathrm{df}} C^p(\Phi_0)(\mathbf{A}, E_0(\vec{x})),$$

and again, we skip the subscript when it is Φ,

(C^p, c^p) $$C^p(M) = C^p(\Phi))(M), \quad c^p(\vec{x}) = c^p(\Phi)(\vec{x}).$$

Intuitively, the number $C^p(\Phi_0)(M)$ counts *the maximal number of calls to the primitives in* Φ_0 *that must be executed in sequence* in any computation of $\overline{M}$ by (any implementation of) E. It is more directly relevant to parallel implementations—which is why we will sometimes call it the *parallel calls complexity*. It is not as easy to read it from $\mathcal{T}(M)$, however, which assumes not only parallelism but (potentially infinite) nondeterminism. In the key recursive calls $\mathsf{p}(M_1, \dots, M_n)$, for example, we put the children on the same level,

$$M_1, \dots, M_n, E_{\mathsf{p}}(\overline{M}_1, \dots, \overline{M}_n)$$

so that the depth of the tree is one more than the maximum of the depths of the trees for these terms; but $\overline{M}_1, \dots, \overline{M}_n$ must be computed *before* the computation of the rightmost child can be (realistically) started,[27] which is why we set

$$\begin{aligned} &C^p(\Phi_0)(\mathsf{p}(M_1, \dots, M_n)) \\ &= \max\{C^p(\Phi_0)(M_1), \dots, C^p(\Phi_0)(M_n)\} + C^p(\Phi_0)(E_{\mathsf{p}}(\overline{M}_1, \dots, \overline{M}_n)). \end{aligned}$$

$C^p(\Phi_0)(M)$ is also not easy to read off the computation of the recursive machine starting with M : , for a different reason: it is a measure of parallel complexity and the recursive machine is decidedly sequential.

The complexity measure $c^p(\vec{x})$ is majorized by all natural complexity measures of all reasonable implementations of recursive programs, and so lower bound results about it have wide applicability. Most of the lower bounds for problems in arithmetic we will derive in Part II are for a related "intrinsic" complexity measure, somewhat smaller than $c^p(\vec{x})$.

Problems for Section 3A

x3A.1. **Problem.** Make precise and prove that all five of the complexity functions we introduced for recursive programs are preserved under isomorphisms.

For example, if $\pi : \mathbf{A} \rightarrowtail\!\!\!\!\twoheadrightarrow \mathbf{B}$ is an isomorphism of one Φ-structure onto another, $N(\vec{y})$ is a pure voc(E)-term and $\vec{y} \in A^n$, then

$$\begin{aligned} \mathbf{A} \models N(\vec{y}) = w \Longrightarrow \Big(&\mathbf{B} \models N(\pi(\vec{y})) = \pi(w) \\ &\& \; C^s(\Phi_0)(\mathbf{A}, E, N(\vec{y})) = C^s(\Phi_0)(\mathbf{B}, E, N(\pi(\vec{y})))\Big). \end{aligned}$$

HINT: Prove this first for the tree-depth complexity $D(M)$.

x3A.2*. **Problem.** Prove (3A-9), that for each program E, there is a number $C = C_E$ such that for every $M \in \mathrm{Conv}(\mathbf{A}, E)$,

$$L^s(M) \leq \mathrm{Time}_{\mathbf{A},E}(M) \leq L^s(M) + C.$$

HINT: For the second, non-trivial inequality, prove that for every pure, extended voc(E)-term $N(\vec{y})$, there is a number C_N such that for all $\vec{y} \in A^m$,

$$\overline{M} = \overline{N}(\vec{y})\downarrow \Longrightarrow L^s(M) \leq \mathrm{Time}(M) + C_N.$$

The next problem is a more detailed version of Theorem 2C.1, which relates the time complexity of an iterator in (2C-4) with the sequential calls-complexity of the associated tail recursive program.

---

[27]This leaves out so-called *call-by-need* implementations, a variety of *call-by-name* implementations which are outside the scope of this book.

x3A.3. **Problem.** Prove that for each iterator $\mathfrak{i}$ and the associated recursive program $E \equiv E_{\mathfrak{i}}$ on $\mathbf{A} = \mathbf{A}_{\mathfrak{i}}$ and for all $x \in X$,

$$\bar{\mathfrak{i}}(x) = \text{den}(\mathbf{A}_{\mathfrak{i}}, E_{\mathfrak{i}}(x)),$$
$$c^s(\mathbf{A}_{\mathfrak{i}}, E_{\mathfrak{i}}(x)) = 2\,\text{Time}_{\mathfrak{i}}(x) - 1.$$

HINT: Prove that if

$$\mathsf{p}(\mathsf{input}(\vec{\mathsf{x}})) \text{ where } \Big\{\mathsf{p}(\vec{\mathsf{u}}) = \text{if test}(\vec{\mathsf{u}}) \text{ then output}(\vec{\mathsf{u}}) \text{ else } \mathsf{p}(\sigma(\vec{\mathsf{u}}))\Big\}$$

is the tail recursive program which computes $\bar{\mathfrak{i}}(x)$ in $\mathbf{A}_{\mathfrak{i}}$, then for every state s of $\mathfrak{i}$ and $n \geq 1$,

$$\mathsf{p}(s) : \downarrow \ \& \ C^s(\mathsf{p}(s)) = 2n$$
$$\iff \text{there is a convergent computation } (s_1, \dots, s_n) \text{ of } \mathfrak{i} \text{ with } s = s_1.$$

This exact relation between the time complexity of $\mathfrak{i}$ and the number-of-calls complexity of the associated recursive program $E_{\mathfrak{i}}$ reflects some specific choices we made in how we count $\text{Time}_{\mathfrak{i}}$ and it is not of any particular interest; on the other hand, we would expect that any "natural" complexity measure $T_{\mathfrak{i}}(x)$ of an iterator $\mathfrak{i}$ is "coded" in $E_{\mathfrak{i}}$, in the sense that

$$T_{\mathfrak{i}}(x) = \Theta(C(\mathbf{A}_{\mathfrak{i}}, E_{\mathfrak{i}}(x)))$$

for some "natural" complexity measure on recursive programs. This is one version of the claim that algorithms can be faithfully expressed by recursive programs on which we have been harping.

x3A.4. **Problem.** Prove that for each extended Φ-program with empty body $E(\vec{x}) \equiv E(\vec{x})$ where $\{\ \}$ (a Φ-term), there is a number C_E such that for every Φ-structure $\mathbf{A}$ and $\vec{x} \in A^n$,

$$\mathbf{A} \models E(\vec{x})\downarrow \implies l^s_E(\vec{x}) \leq C_E.$$

HINT: Take C_E to be any reasonable definition of the *length* of E.

x3A.5. **Problem.** Prove that the following are equivalent for any structure $\mathbf{A}$ and $M \in \text{Conv}(\mathbf{A}, E)$:

(i) $C^p(M) = 0$.
(ii) $L^s(M) = 0$.
(iii) The value $\overline{M}$ is independent of the primitives of $\mathbf{A}$, i.e., for any Φ-structure $\mathbf{A}' = (A, \mathbf{\Phi}')$ with the same universe

$$\text{den}(\mathbf{A}, E, M) = \text{den}(\mathbf{A}', E, M).$$

(iv) There are no Φ-nodes in the computation tree $\mathcal{T}(M)$.

We will sometimes appeal silently to this simple observation to simplify formulas, for example by dividing by $c^p(\mathbf{A}, E(\vec{x}))$ or using it in the form $c^p(\mathbf{A}, E(\vec{x})) \geq 1$ when the value $\text{den}^{\mathbf{A}}_E(\vec{x})$ obviously depends on the primitives of $\mathbf{A}$.

x3A.6. **Problem.** Prove that for any term $M \in \text{Conv}(\mathbf{A}, E)$,

$$D(M) \leq L^p(M). \tag{3A-14}$$

x3A.7. **Problem.** Compute (up to a multiplicative constant) $c_E^p(x, y)$ for the program defined (informally) in Problem x1B.1.

x3A.8. **Problem.** Compute (up to a multiplicative constant) $c_E^p(x, y)$ for the program defined (informally) in Problem x1B.2.

x3A.9. **Problem.** Compute (up to a multiplicative constant) $c^p(x, y)$ for the program in Problem x1B.3.

x3A.10. **Problem.** Give an example of a structure $\mathbf{A}$, a recursive program of total arity 0 and, for any n, a term $M \in \text{Conv}(\mathbf{A}, E)$ such that $L^s(M) = n$, (so (3) or Theorem 3A.3 fails for $\ell = 0$).

x3A.11. **Problem.** Prove that if $\mathbf{A}$ is a Φ-structure and $E(\mathsf{x})$ is an extended Φ-program of total arity 1, then for every $M \in \text{Conv}(\mathbf{A}, E)$,

$$L^s(M) \leq 2^{L^p(M)} - 1 < 2^{L^p(M)}.$$

x3A.12. **Problem.** Consider the program E with the single equation

$$p(x) = \text{if } (x = 0) \text{ then } 0 \text{ else } p(\text{Pd}(x)) + p(\text{Pd}(x))$$

in the structure $(\mathbb{N}, 0, 1, \text{Pd}, +, \text{eq}_0)$. Determine the function $\text{den}_E(x)$ computed by this program, and verify that for some $r > 1$ and all sufficiently large x,

$$l_E^s(x) \geq r^{l_E^p(x)}, \quad c_E^s(x) \geq r^{c_E^p(x)}.$$

A node $(u_0, \dots, u_i)$ in a computation tree $\mathcal{T}(M)$ is a Φ_0-*node* if $u_i \equiv \phi(N_1, \dots, N_n)$ for some $\phi \in \Phi_0$.

x3A.13. **Problem.** Prove that $C^s(\Phi_0)(\mathbf{A}, E, M)$ is the number of Φ_0-*nodes* in $\mathcal{T}(M)$. HINT: Use induction on $D(M)$.

x3A.14. **Problem.** Prove that $C^s(\Phi_0)(\mathbf{A}, E, M)$ is the number of external calls $\vec{a}\ \phi : w_1 \cdots w_n\ \vec{b}$ with $\phi \in \Phi_0$ in the computation of the recursive machine for $\mathbf{A}$ which starts with M : .

Note that with $M \equiv E_0(\vec{x})$, this agrees with the definition of $c^s(\Phi_0)(\mathbf{A}, E(\vec{x}))$ in (2E-5) for the more general case, when E may be nondeterministic. HINT: Use induction on $D(M)$ (with frequent appeals to Lemma 2D.1).

x3A.15. **Open problem** (vague). Is there a conceptually simple and technically useful way to read $C^p(\Phi_0)(\mathbf{A}, E, M)$ from the tree $\mathcal{T}(M)$ or from the computation of the recursive machine for $\mathbf{A}$ which starts with M : , similar to the characterization of $C^s(\Phi_0)(\mathbf{A}, E, M)$ in the preceding two problems?

3B. Complexity inequalities

Next we derive the expected inequalities among these complexity measures.

3B.1. **Proposition.** *For each* Φ*-structure* $\mathbf{A}$, *each extended* Φ*-program* $E(\vec{x})$ *of total arity* $\ell \geq 1$ *and each* $M \in \mathrm{Conv}(\mathbf{A}, E)$:

$$\begin{array}{ccccccc}
 & & C^s(M) & & & & (\ell+1)^{L^p(M)} \\
 & \rotatebox{45}{$\leq$} & & \rotatebox{-45}{$\leq$} & & \rotatebox{45}{$\leq$} & \\
C^p(M) & & & & L^s(M) & & \\
 & \rotatebox{-45}{$\leq$} & & \rotatebox{45}{$\leq$} & & \rotatebox{-45}{$\leq$} & \\
 & & D(M) \leq L^p(M) & & & & (\ell+1)^{D(M)+1}
\end{array}$$

and, in particular, for all $\vec{x}$ *such that* $\mathrm{den}(\mathbf{A}, E(\vec{x}))\downarrow$,

$$\begin{array}{ccccccc}
 & & c^s(\vec{x}) & & & & (\ell+1)^{l^p(\vec{x})} \\
 & \rotatebox{45}{$\leq$} & & \rotatebox{-45}{$\leq$} & & \rotatebox{45}{$\leq$} & \\
c^p(\vec{x}) & & & & l^s(\vec{x}) & & \\
 & \rotatebox{-45}{$\leq$} & & \rotatebox{45}{$\leq$} & & \rotatebox{-45}{$\leq$} & \\
 & & d(\vec{x}) \leq l^p(\vec{x}) & & & & (\ell+1)^{d(\vec{x})+1}.
\end{array}$$

Proof. The five inequalities on the left are very easy to check by induction on $D(M)$ and the top one on the right is Theorem 3A.3. The bottom one on the right holds because

$$\begin{aligned}
L^s(M) &\leq |\mathcal{T}(M)| && \text{(by (1) of Theorem 3A.3)} \\
&< \mathrm{degree}(\mathcal{T}(M))^{\mathrm{depth}(\mathcal{T}(M))+1} && \text{(by (3) of Problem x1A.4)} \\
&\leq (\ell+1)^{\mathrm{depth}(\mathcal{T}(M))+1} && \text{(because } \mathrm{degree}(\mathcal{T}(M)) \leq \ell+1) \\
&\leq (\ell+1)^{D(M))+1},
\end{aligned}$$

the last by Proposition 3A.2. ⊣

Together with Problem x3A.12, Theorem 3A.3 gives the expected relation between the sequential and parallel logical complexities: $l^s_E(\vec{x})$ is bounded by an exponential function of $l^p_E(x)$ and in some cases it requires this rate of growth—although we have not established that $(\ell+1)$ is the smallest base for this.

Recursive vs. explicit definability. This last inequality bounds $l^s(\vec{x})$, the largest of the four basic complexity functions associated with a recursive program by an exponential of its tree-depth complexity $d(\vec{x})$. It implies a simple characterization of explicit definability in terms of tree-depth complexity, in "reasonable" structures:

3B.2. **Theorem.** *Suppose* Φ *has a relation symbol* R *of arity* $m > 0$ *and* $\mathbf{A}$ *is a* Φ*-structure such that* $\mathsf{R}^{\mathbf{A}} : A^m \to \mathbb{B}$ *is total.*

Then: for any $f : A^n \rightharpoonup A_s$ *and* $S \subseteq \{\vec{x} : f(\vec{x})\downarrow\}$ *the following are equivalent:*

(1) *There is an* **A**-*explicit* $f^* : A^n \rightharpoonup A_s$ *such that*

$$\vec{x} \in S \Longrightarrow f(\vec{x}) = f^*(\vec{x}).$$

(2) *There is an extended* Φ-*program* $E(\vec{x})$ *and a number* C *such that for every* $\vec{x} \in S$,

$$f(\vec{x}) = \mathrm{den}_E^{\mathbf{A}}(\vec{x}) \text{ and } d(\vec{x}) = d(\mathbf{A}, E(\vec{x})) \leq C.$$

PROOF. (1) $\Longrightarrow$ (2) follows immediately from (x3A.4), taking $E(\vec{x})$ to be any program (with empty body) that computes $f^*(\vec{x})$ on S.

For the converse, assume (2) for a program $E(\vec{x})$ with recursive variables $\mathsf{p}_1, \dots, \mathsf{p}_K$ and look back at the proof of Lemma 3A.1: the tree-depth complexity $D(M)$ of $M \in \mathrm{Conv}(\mathbf{A}, E)$ was defined in terms of

$$\overline{M}^k = \mathrm{den}((\mathbf{A}, \overline{p}_1^k, \dots, \overline{p}_K^k), M),$$

where $\overline{p}_1^k, \dots, \overline{p}_K^k$ are the iterates of the system that defines $\overline{M}$ as in the proof of the Fixed Point Lemma, Theorem 1B.1.

Lemma 1. For every $M \in \mathrm{Conv}(\mathbf{A}, E)$,

$$D(M) \geq \text{the least } k \text{ such that } \overline{M}^k \downarrow = \mathrm{stage}(M).$$

PROOF is by induction on the definition of $D(M)$, and it is very easy, except perhaps for Case (D4),

$$D(\mathsf{p}(M_1, M_2)) = \max\{D(M_1), D(M_2), D(E_{\mathsf{p}}(\overline{M}_1, \overline{M}_2))\} + 1,$$

taking $m = 2$ to keep the notation simple. By (3A-6)

$$\mathrm{stage}(\mathsf{p}(\overline{M}_1, \overline{M}_2)) = \mathrm{stage}(E_{\mathsf{p}}(\overline{M}_1, \overline{M}_2)) + 1;$$

and so, using the induction hypothesis,

$$\begin{aligned} D(\mathsf{p}(\overline{M}_1, \overline{M}_2)) &\geq \max\{\mathrm{stage}(M_1), \mathrm{stage}(M_2), \mathrm{stage}(\mathsf{p}(\overline{M}_1, \overline{M}_2)) - 1\} + 1 \\ &\geq \mathrm{stage}(\mathsf{p}(\overline{M}_1, \overline{M}_2)). \end{aligned}$$

⊣ (Lemma 1)

Lemma 2. For each k and each (**A**, E)-*term* $M \equiv N(\vec{x})$, *there is an* **A**-*explicit function* $f_{N,k}(\vec{x})$ (depending only on N and k) *such that*

$$\text{(3B-1)} \qquad \overline{M}^k \downarrow \ \& \ D(M) \leq k \Longrightarrow \overline{M}^k = f_{N,k}(\vec{x}).$$

PROOF. This is where we will use the hypothesis about **A**, which ensures that **Expl(A)** is closed under substitutions by Proposition 1E.4.

We define $f_{N,k}$ by recursion on $D(M)$, skipping the verifications of (3B-1) which are all routine.

(D1) $D(M) = 0$, so $M \equiv N(\vec{x})$ with $N(\vec{x})$ one of tt, ff, x_i; we set $f_{N,0}(\vec{x})$ to one of tt, ff, x_i accordingly.

(D2) $M \equiv \phi(M_1, \dots, M_m) \equiv N(\vec{x})$ with $N(\vec{x}) \equiv \phi(N_1(\vec{x}), \dots, N_m(\vec{x}))$. Now $D(M_i) < k = D(M)$, so we can set

$$f_{N,k}(\vec{x}) = \phi^{\mathbf{A}}(f_{N_1,k-1}(\vec{x}), \dots, f_{N_2,k-1}(\vec{x})).$$

(D3) where M is defined by branching is handled similarly.

(D4) Taking $m = 2$ again,

$$M \equiv \mathsf{p}(M_1, M_2) \equiv \mathsf{p}(N_1(\vec{x}), N_2(\vec{x}))$$

for a recursive variable p of E and

$$\overline{M} = \text{den}(\mathbf{A}, E_{\mathsf{p}}(\overline{M}_1, \overline{M}_2));$$

now $D(M_1), D(M_2), D(E_{\mathsf{p}}(\overline{M}_1, \overline{M}_2)) < D(M)$, so we can use the induction hypothesis and set

$$f_{N,k}(\vec{x}) = f_{E_{\mathsf{p}},k-1}(f_{N_1,k-1}(\vec{x}), f_{N_2,k-1}(\vec{x})). \qquad \dashv \text{(Lemma 2)}$$

To prove (1) with $f(\vec{x}) = \text{den}(\mathbf{A}, E(\vec{x}))$, we note that a uniform bound on $d(\vec{x})$ gives a uniform bound k for stage$(\vec{x})$ for $\vec{x} \in S$ by Lemma 1; and then Lemma 2 gives an explicit definition for $f(\vec{x})$ on S. $\dashv$

Tserunyan's inequalities. The constants in Proposition 3B.1 depend only on Φ and the total arity of a program which computes some $f : A^n \rightharpoonup A_s$ in any Φ-structure $\mathbf{A}$, and their proofs were quite simple. Less obvious is that for any program E, there are constants K_s, K_p such that for every Φ-structure $\mathbf{A}$, if $\text{den}(\mathbf{A}, \vec{x})\downarrow$, then

$$\text{(3B-2)} \qquad \text{(a)} \quad l^s_E(\vec{x}) \le K_s + K_s c^s_E(\vec{x}), \quad \text{(b)} \quad l^p_E(\vec{x}) \le K_p + K_p c^p_E(\vec{x}),$$

i.e., in both the sequential and the parallel measures, counting the logical steps in addition to the calls to the primitives produces at most a linear increase in complexity. From the point of view of deriving lower bounds, the significance of these inequalities becomes evident if we reverse them:

$$\text{(3B-3)} \quad \text{(a)} \quad c^s_E(\vec{x}) \ge \frac{1}{K_s}(l^s_E(\vec{x}) - K_s), \quad \text{(b)} \quad c^p_E(\vec{x}) \ge \frac{1}{K_p}(l^p_E(\vec{x}) - K_p).$$

Here (a) means that the high sequential complexity of a $f : A^n \rightharpoonup A_s$ from specified primitives is not caused by the large number of logical operations that we must execute to compute $f(\vec{x})$—i.e., (literally) by the "computational complexity" of f—but is due to the large number of calls to the primitives that are necessary to compute $f(\vec{x})$, at least up to a linear factor. Ditto for the parallel complexity $l^p_E(\vec{x})$ and its "calls-counting" counterpart $c^p_E(\vec{x})$. It is another manifestation of the fact that lower bound results are most often proved by counting calls to the primitives, which is well known and little understood; and, incidentally, it emphasizes the importance of identifying *all* the (non-logical) primitives that are used by an algorithm which computes a particular function.

The proofs of these inequalities require some new ideas and are due to Anush Tserunyan.[28]

We fix a Φ-structure $\mathbf{A}$ and a recursive program E with recursive variables $\mathsf{p}_1, \dots, \mathsf{p}_K$ and total arity $\ell = \ell(E) \geq 1$. We can insure that

$$\text{the number of free and bound variables in } E \leq \ell$$

by making innocuous alphabetic changes to the bound variables of E. It follows that if

$$t = t(E) = \text{the number of distinct subterms of the terms in } E,$$
$$H = H(E) = t\ell^{\ell},$$

then H is an overall upper bound to the number of terms that can be constructed by a single assignment of parameters to the variables in all the subterms of E.

We start with some preliminary estimates which are needed for the proofs of both inequalities in (3B-2).

3B.3. **Lemma.** *If $M \in \mathrm{Conv}(\mathbf{A}, E)$ and $C^p(M) = 0$, then either $\overline{M} \in \mathbb{B}$ or $\overline{M}$ occurs in M.*

PROOF is by an easy induction on $D(M)$. ⊣

3B.4. **Lemma.** *Suppose $M \in \mathrm{Conv}(\mathbf{A}, E)$.*

(a) *If $(M_1, \dots, M_k) \in \mathcal{T}(M)$, then $M_k \in \mathrm{Conv}(\mathbf{A}, E)$.*

(b) *If $(M_1, \dots, M_k) \in \mathcal{T}(M)$ and the parameters in every M_i occur in M, then $k \leq H$.*

PROOF. (a) is immediate by the construction of $\mathcal{T}(M)$.

(b) Suppose $x_1, \dots, x_m$ is a list of the parameters in $M_1 \equiv M$, so $m \leq \ell$. Each M_i is an $(\mathbf{A}, E)$-term whose parameters are among $x_1, \dots, x_m$, and there are at most H distinct such terms; so if $k > H$, then $M_i \equiv M_j$ for some $1 \leq i < j \leq k$, and then $\overline{M_1}\uparrow$. ⊣

It is clear from the construction of the computation tree that new parameters enter the tree only by Case ($\mathcal{T}4$), in the term $E_{\mathsf{p}}(\overline{M}_1, \dots, \overline{M}_m)$, and then only if some $\overline{M}_i$ does not occur in the parent node. Isolating and counting these critical nodes is the main new tool we need.

---

[28]The results in the remainder of this section are due to Anush Tserunyan, Part (3) of her Ph.D. Thesis, Tserunyan [2013].

Splitting. A term $M \in \text{Conv}(\mathbf{A}, E)$ is *splitting*, if $M \equiv \mathsf{p}(M_1, \dots, M_n)$ with a recursive variable p of E and

$$\max_{1 \le j \le n} C^p(M_j) > 0, \qquad C^p(E_i(\overline{M}_1, \dots, \overline{M}_n)) > 0.$$

By Problem x3A.5, these conditions are equivalent to their version with C^s rather than C^p.

3B.5. **Lemma.** *If* $(M_1, \dots, M_k) \in \mathcal{T}(M) = \mathcal{T}(M_1)$ *and no* M_i *is splitting, then* $k \le 2H$.

PROOF. If the parameters in every M_i occur in $M_1 \equiv M$, then we apply (b) of Lemma 3B.4. Otherwise, let i be least such that M_{i+1} has parameters that do not occur in M_1. Now $i \le H$ by Lemma 3B.4 again, and by the definition of $\mathcal{T}(M)$,

$$M_i \equiv \mathsf{p}(N_1, \dots, N_n), \text{ and } M_{i+1} \equiv E_{\mathsf{p}}(\overline{N}_1, \dots, \overline{N}_n).$$

Moreover, $\max\{C^p(N_j) : 1 \le j \le n\} > 0$ since otherwise, each $\overline{N}_j$ occurs in N_j and hence $\overline{N}_j \in M_i$ by Lemma 3B.3, contradicting the choice of i. But M_i is not splitting, so $C^p(M_{i+1}) = C^p(E_{\mathsf{p}}(\overline{N}_1, \dots, \overline{N}_n)) = 0$. Hence for all $j \ge i+1$, $C^p(M_j) = 0$ and then by Lemma 3B.3 again, all parameters in M_j occur in M_{i+1}, and the length of $(M_{i+1}, \dots, M_k)$ is $\le H$; which means that $k = i + (k - i) \le H + H = 2H$. ⊣

Let

$$v(M) = \Big\{(M_1, \dots, M_n) \in \mathcal{T}(M) : \forall i, M_i \text{ is not splitting}\Big\}.$$

This is the empty set if M is splitting and a subtree of $\mathcal{T}(M)$ if it is not. By Lemma 3B.5 and the fact that degree$(\mathcal{T}(M)) \le (\ell + 1)$,

$$|v(M)| \le V, \text{ where } V = (\ell + 1)^{2H}. \tag{3B-4}$$

3B.6. **Lemma.** *If* $C^p(M) = 0$, *then* $L^s(M) \le |v(M)|$.

PROOF. If $C^p(M) = 0$, then there are no splitting terms below M, and so $v(M) = \mathcal{T}(M)$ and the inequality follows from (a) of Theorem 3A.3. ⊣

For the proof of (a) in (3B-2), we need the *sequential splitting complexity* of a (convergent) $(\mathbf{A}, E)$-term:

$$F^s(M) = \text{the number of splitting nodes in } \mathcal{T}(M), \tag{3B-5}$$

where $(M_0, \dots, M_k) \in \mathcal{T}(M_0)$ is splitting if M_k is splitting. This satisfies some obvious recursive conditions which we will introduce and use in the proof of the next lemma. It is clear, however, that

$$C^p(M) = 0 \Longrightarrow F^s(M) = 0;$$

because if $C^p(M) = 0$, then $C^p(N) = 0$ for every N in $\mathcal{T}(M)$ and so no such term can be splitting.

3B.7. **Lemma.** *For every* $M \in \mathrm{Conv}(\mathbf{A}, E)$,

$$F^s(M) \leq C^s(M) \dot{-} 1.$$

PROOF is by induction on $D(M)$, as usual, and it is trivial in all cases except when M is splitting. If $M \equiv \mathsf{p}(M_1, \dots, M_m)$ is splitting, let $M_{m+1} \equiv E_{\mathsf{p}}(\overline{M}_1, \dots, \overline{M}_m)$ and compute:

$$\begin{aligned} F^s(M) &= F^s(M_1) + \cdots + F^s(M_m) + F^s(M_{m+1}) + 1 \\ &\leq (C^s(M_1) \dot{-} 1) + \cdots + (C^s(M_m) \dot{-} 1) + (C^s(M_{m+1}) \dot{-} 1) + 1 \\ &\leq C^s(M_1) + \cdots + C^s(M_m) - 1 + C^s(M_{m+1}) - 1 + 1 \\ &= C^s(M) - 1. \end{aligned}$$

The only thing we used here is that if M is splitting, then $C^s(M_i) \geq C^p(M_i) > 0$ for at least one i, and similarly $C^s(M_{m+1}) > 0$. ⊣

3B.8. **Lemma.** *If* $M \in \mathrm{Conv}(\mathbf{A}, E)$, *then there is a* (possibly empty) *sequence of splitting terms* $N_0, \dots, N_{k-1}$ *in* $\mathcal{T}(M)$ *such that*

$$F^s(M) = \textstyle\sum_{i<k} F^s(N_i), \qquad L^s(M) \leq \textstyle\sum_{i<k} L^s(N_i) + |v(M)|. \tag{3B-6}$$

PROOF. If M is splitting, we take just one $N_0 \equiv M$, and if there are no splitting terms in $\mathcal{T}(M)$ we set $k = 0$ and understand $\sum_{i<k} L^s(N_i) = 0$, so that $\mathcal{T}(M) = v(M)$ by definition and (3B-6) follows from (a) of Theorem 3A.3. The lemma is proved in the general case by induction on $D(M)$, and the argument is trivial in most of the cases. We consider only the case of a non-splitting recursive call

$$M \equiv \mathsf{p}(M_1, \dots, M_m).$$

Set $M_{m+1} \equiv E_{\mathsf{p}}(\overline{M}_1, \dots, \overline{M}_m)$. The induction hypothesis gives us (a possibly empty) sequence $N_0^i, \dots, N_{k_i}^i$ of splitting terms in each $\mathcal{T}(M_i)$ such that, to begin with,

$$\begin{aligned} L^s(M) &= L^s(M_1) + \cdots + L^s(M_m) + L^s(M_{m+1}) + 1 \\ &\leq \textstyle\sum_{j<k_1} L^s(N_j^1) + |v(M_1)| + \cdots + \sum_{j<k_m} L^s(N_j^m) + |v(M_m)| \\ &\qquad + \textstyle\sum_{j<k_{m+1}} L^s(N_j^{m+1}) + |v(M_{m+1})| + 1 \\ &\leq \textstyle\sum_{1\leq i\leq m+1, j<k_i} L^s(N_j^i) + |v(M_1)| + \cdots + |v(M_{m+1})| + 1. \end{aligned}$$

Now $v(M) = \bigcup_{i=1,\dots,m+1} v(M_i) \cup \{M\}$ because M is not splitting, and so

$$|v(M_1)| + \cdots + |v(M_{m+1})| + 1 = |v(M)|.$$

Moreover, $F^s(M_i) = \sum_{j<k_i} N_j^i$ by the induction hypothesis, and so

$$F^s(M) = F^s(M_1) + \cdots + F^s(M_{m+1}) = \textstyle\sum_{1\leq i\leq m+1, j<k_i} F^s(N_j^i)$$

as required, again because M is not splitting. ⊣

3B.9. **Theorem** (Tserunyan [2013]). *For every* Φ*-structure* $\mathbf{A}$*, every* Φ*-program* E *and every* $M \in \text{Conv}(\mathbf{A}, E)$*, if* V *is the constant defined in* (3B-4)*, then*:

(a) *If* M *is splitting, then* $L^s(M) \le ((\ell+1)V+1)F^s(M)$.

(b) *If* M *is not splitting, then* $L^s(M) \le ((\ell+1)V+1)F^s(M) + V$.

It follows that $L^s(M) \le V + ((\ell+1)V+1)C^s(M)$*, and so*

$$l^s_E(\vec{x}) \le K_s + K_s c^s_E(\vec{x}) \quad (\text{den}_E(\vec{x})\downarrow) \tag{3B-7}$$

with $K_s = (\ell+1)V+1$*, a constant which depends only on the program* E.

PROOF. The main result (3B-7) follows from (a) and (b) by taking $M \equiv E_0(\vec{x})$ and appealing to Lemma 3B.7.

We prove (a) and (b) together by induction on $D(M)$, noting that (b) is a weaker inequality than (a) and so we can use it whether M is splitting or not when we invoke the induction hypothesis.

Case 1, $M \equiv \mathsf{p}(M_1, \dots, M_m)$ is splitting. Set $M_{m+1} \equiv E_{\mathsf{p}}(\overline{M}_1, \dots, \overline{M}_m)$ as above and compute:

$$\begin{aligned} L^s(M) &= L^s(M_1) + \cdots + L^s(M_{m+1}) + 1 \\ &\le ((\ell+1)V+1)[F^s(M_1) + \cdots + F^s(M_{m+1})] + (\ell+1)V + 1 \\ &= ((\ell+1)V+1)[F^s(M_1) + \cdots + F^s(M_{m+1}) + 1] \\ &= ((\ell+1)V+1)F^s(M), \end{aligned}$$

because $F^s(M) = F^s(M_1) + \cdots + F^s(M_{m+1}) + 1$ for splitting M.

Case 2, M is not splitting. Choose splitting terms $N_0, \dots, N_{k-1}$ by Lemma 3B.8 so that

$$F^s(M) = \textstyle\sum_{i<k} F^s(N_i), \quad L^s(M) \le \sum_{i<k} L^s(N_i) + |v(M)|$$

and compute using the result from Case 1:

$$\begin{aligned} L^s(M) &\le \textstyle\sum_{i<k} L^s(N_i) + |v(M)| \\ &\le ((\ell+1)V+1)\textstyle\sum_{i<k} F^s(N_i) + V = ((\ell+1)V+1)F^s(M) + V \end{aligned}$$

as required. ⊣

We now turn to the proof of (b) in (3B-2), and for this we need a count $F^p(M)$ of the splitting terms which parallels the way in which $C^p(M)$ counts "the depth" of calls to the primitives. This is easiest to define by induction on $D(M)$:

(F^p1) $F^p(\mathrm{tt}) = F^p(\mathrm{ff}) = F^p(x) = 0 \quad (x \in A)$.

(F^p2) If $M \equiv \phi(M_1, \dots, M_m)$, then

$$F^p(M) = \max\{F^p(M_1), \dots, F^p(M_m)\}.$$

(F^p3) If $M \equiv \text{if } M_0 \text{ then } M_1 \text{ else } M_2$, then

$$F^p(M) = \begin{cases} \max\{F^p(M_0), F^p(M_1)\}, & \text{if } \overline{M}_0 = \text{tt}, \\ \max\{F^p(M_0), F^p(M_2)\}, & \text{if } \overline{M}_0 = \text{ff}. \end{cases}$$

(F^p4) If $M \equiv \mathsf{p}(M_1, \dots, M_m)$ of E, let $M_{m+1} \equiv E_{\mathsf{p}}(\overline{M}_1, \dots, \overline{M}_m)$ and set

$$F^p(M) = \begin{cases} \max\{F^p(M_1), \dots, F^p(M_m)\} + F^p(M_{m+1}), & \text{if } M \text{ is not splitting,} \\ \max\{F^p(M_1), \dots, F^p(M_m)\} + F^p(M_{m+1}) + 1, & \text{if } M \text{ is splitting.} \end{cases}$$

Notice that if M is not splitting, then

$$F^p(M) = \max\{F^p(M_1), \dots, F^p(M_m), F^p(M_{m+1})\},$$

since one of $\max\{F^p(M_1), \dots, F^p(M_m)\}$ and $F^p(M_{m+1})$ is 0, so that the sum of these two numbers is the same as their maximum.

With this splitting complexity, the proof of (b) in (3B-2) is only a minor modification (a parallel version) of the proof of Theorem 3B.9.

3B.10. **Lemma.** *For every* $M \in \text{Conv}(\mathbf{A}, E)$,

$$F^p(M) \le C^p(M) \dot{-} 1.$$

PROOF is by induction on $D(M)$ and it is again trivial in all cases except when M is splitting. In this case, if $M \equiv \mathsf{p}(M_1, \dots, M_m)$ and we set $M_{m+1} \equiv E_{\mathsf{p}}(\overline{M}_1, \dots, \overline{M}_m)$, then

$$\begin{aligned} F^p(M) &= \max\{F^p(M_1), \dots, F^p(M_m)\} + F^p(M_{m+1}) + 1 \\ &\le \max\{(C^p(M_1) \dot{-} 1), \dots, (C^p(M_m) \dot{-} 1)\} + C^p(M_{m+1}) \dot{-} 1 + 1 \\ &\le \max\{C^p(M_1), \dots, C^p(M_m)\} - 1 + C^p(M_{m+1}) - 1 + 1 \\ &= C^p(M) \dot{-} 1. \end{aligned}$$

As in the proof of Lemma 3B.7, the only thing we use here is that if M is splitting, then $C^p(M_i) > 0$ for at least one i, and similarly $C^p(M_{m+1}) > 0$. ⊣

3B.11. **Lemma.** *If* $M \in \text{Conv}(\mathbf{A}, E)$, *then there is a term* N *in* $\mathcal{T}(M)$ *which is either a leaf or splitting and such that*

$$L^p(M) \le L^p(N) + |v(M)|. \tag{3B-8}$$

PROOF is by induction on $D(M)$, as usual, and the result is trivial if M is a leaf or splitting, taking $N \equiv M$.

If $M \equiv \phi(M_1, \dots, M_m)$, then $L^p(M) = L^p(M_i) + 1$ for some i, and the induction hypothesis gives us a leaf or splitting term N in $\mathcal{T}(M_i)$ such that

$$L^p(M_i) \le L^p(N) + |v(M_i)|.$$

It follows that

$$L^p(M) = L^p(M_i) + 1 \le L^p(N) + |v(M_i)| + 1 \le L^p(N) + |v(M)|$$

since M is not splitting and so $v(M) = \bigcup_{j=1,\dots M} v(M_j) \cup \{M\}$.

The argument is similar for conditional terms.

If M is a non-splitting recursive call

$$M \equiv \mathsf{p}(M_1, \dots, M_m),$$

set again $M_{m+1} \equiv E_{\mathsf{p}}(\overline{M}_1, \dots, \overline{M}_m)$ and choose i such that $L^p(M_i) = \max\{L^p(M_1), \dots, L^p(M_m)\}$. If $C^p(M_i) = 0$, then then $L^p(M_i) \leq |v(M_i)|$ by Lemma 3B.6, and if we choose $N \in \mathcal{T}(M_{m+1})$ by the inductive hypothesis so that $L^p(M_{m+1}) \leq L^p(N) + |v(M_{m+1}|$, then

$$\begin{aligned} L^p(M) &= L^p(M_i) + L^p(M_{m+1}) + 1 \\ &\leq |v(M_i)| + L^p(N) + |v(M_{m+1})| + 1 = L^p(N) + |v(M)|. \end{aligned}$$

If $C^p(M_i) > 0$, then $C^p(M_{m+1}) = 0$, since M is not splitting, and we can repeat the argument with M_i and M_{m+1} interchanged. ⊣

3B.12. **Theorem** (Tserunyan [2013]). *For every Φ-structure **A**, every Φ-program E and every $M \in \text{Conv}(\mathbf{A}, E)$, if V is the constant defined in* (3B-4), *then*:

(a) *If M is splitting, then $L^p(M) \leq (2V + 1)F^p(M)$.*

(b) *If M is not splitting, then $L^p(M) \leq (2V + 1)F^p(M) + V$.*

It follows that $L^p(M) \leq V + (2V + 1)C^p(M)$ and

$$l_E^p(\vec{x}) \leq K_p + K_p c_E^p(\vec{x}) \qquad (\text{den}_E(\vec{x})\downarrow) \tag{3B-9}$$

with $K_p = 2V + 1$, which depends only on the program E.

PROOF is a minor adaptation of the proof of Theorem 3B.9, with (3B-9) following from (a) and (b) by appealing to Lemma 3B.10. We prove (a) and (b) together by induction on $D(M)$.

Case 1, $M \equiv \mathsf{p}(M_1, \dots, M_m)$ is splitting. Set $M_{m+1} \equiv E_{\mathsf{p}}(\overline{M}_1, \dots, \overline{M}_m)$ as above and compute:

$$\begin{aligned} L^p(M) &= \max_{1 \leq i \leq m} L^p(M_i) + L^p(M_{m+1}) + 1 \\ &\leq (2V + 1) \max_{1 \leq i \leq m} F^p(M_i) + V + (2V + 1)F^p(M_{m+1}) + V + 1 \\ &= (2V + 1)\Big(\max_{1 \leq i \leq m} F^p(M_i) + F^p(M_{m+1})\Big) + 2V + 1 \\ &= (2V + 1)\Big(\max_{1 \leq i \leq m} F^p(M_i) + F^p(M_{m+1}) + 1\Big) = (2V + 1)F^p(M), \end{aligned}$$

because $F^p(M) = \max_{1 \leq i \leq m} F^p(M_i) + F^p(M_{m+1}) + 1$ for splitting M.

Case 2, M is not splitting. There is nothing to prove if M is a leaf. If it is not, choose a leaf or splitting term N in $\mathcal{T}(M)$ by Lemma 3B.11 such that $L^p(M) \leq L^p(N) + |v(M)|$. If N is a leaf, then $L^p(N) = 0$ and so $L^p(M) \leq |v(M)| \leq V \leq (2V + 1)F(M) + V$ by (3B-4). If N is splitting, then the induction hypothesis applies to it since it is not M and hence $D(N) < D(M)$, and we have

$$L^p(M) \le L^p(N) + |v(M)| \le (2V+1)F^p(N) + V$$

by (3B-4) again, as required. ⊣

3B.13. **Corollary.** (1) *For every Φ-program E, there is are constants K_1, K_2 such that for every Φ-structure $\mathbf{A}$, if $(\mathrm{den}_E^{\mathbf{A}}(\vec{x})\downarrow)$, then*

$$c^s(\vec{x}) \le K_1^{c^p(\vec{x})} \text{ and } l^s(\vec{x}) \le K_2^{c^p(\vec{x})}. \tag{3B-10}$$

(2) *If an extended Φ-program $E(\vec{x})$ computes $f : A^n \rightharpoonup A_s$ in a structure $\mathbf{A}$ and one of the complexities $d(\vec{x})$, $c^p(\vec{x})$, $c^s(\vec{x})$, $l^p(\vec{x})$, $l^s(\vec{x})$ is bounded on the domain of convergence of f, then they are all bounded*—and Theorem 3B.2 applies, *so $f(\vec{x})$ is explicit if $\mathbf{A}$ is reasonably well behaved.*

PROOF. (1) Notice that these inequalities hold with any $K > 0$ if $c^p(\vec{x}) = 0$, because in that case $l^s(\vec{x}) = 0$ by Problem x3A.5, and then $c^s(\vec{x}) = 0$ because $c^s(\vec{x}) \le l^s(\vec{x})$. So we fix a structure $\mathbf{A}$ and we compute, assuming $\mathrm{den}_E^{\mathbf{A}}(\vec{x})\downarrow$ and $c^p(\vec{x}) > 0$:

For the second inequality,

$$\begin{aligned} l^s(\vec{x}) &\le (\ell+1)^{l^p(\vec{x})} && \text{(Theorem 3A.3)} \\ &\le (\ell+1)^{K_p + K_p c^p(\vec{x})} && \text{(Theorem 3B.12)} \\ &\le \left((\ell+1)^{2K_p}\right)^{c^p(\vec{x})}; \end{aligned}$$

and for the first, we start with $c^s(\vec{x}) \le l^s(\vec{x})$ and use the second one.

(2) If $d(\vec{x})$ is bounded, then all five complexities are bounded by Proposition 3B.1; and if $c^p(\vec{x})$ is bounded, then $l^s(\vec{x})$ is bounded by (1) and then $d(\vec{x})$ is bounded by Proposition 3B.1 again. This covers all possibilities. ⊣

Full vs. tail recursion (II). In Section 2G we described several examples of (pointed) structures in which not every recursive function is tail recursive. A related—perhaps more interesting—question is about the *relative complexities* of full and tail recursive definitions of the same function, and the best result I know about this is the following:

3B.14. **Theorem** (Bhaskar [2017]). *For every function $g : \mathbb{N} \to \mathbb{N}$, there is an increasing function $\gamma : \mathbb{N} \to \mathbb{N}$ with the following properties, where*

$$\mathbf{A} = (\mathbf{N}_{\mathrm{Pd}}, \gamma) = (\mathbb{N}, 0, 1, \mathrm{Pd}, \mathrm{eq}_0, \gamma) \text{ and } f(n, x) = \gamma^{2^n}(x).$$

(1) $\mathbf{Tailrec}^0(\mathbf{A}) = \mathbf{Rec}^0(\mathbf{A})$.

(2) *There is an extended recursive program E which computes f in $\mathbf{A}$ with sequential logical complexity* (time)

$$l^s(\mathbf{A}, E(n, x)) = O(2^n).$$

(3) *The function f is tail recursive in $\mathbf{A}$.*

(4) *For every extended tail recursive program F of $\mathbf{A}$ which computes f, there are increasing sequences $\{n_i\}, \{x_i\}$ such that for all i,*

$$l^s(\mathbf{A}, F(n_i, x_i)) \geq g(n_i).$$

Questions about *intensional*—especially *complexity*—*differences* between programs which compute the same function from the same primitives were an important part of schematology from the very beginning. My own interest in these problems was stimulated by the remarkable theorem in Colson [1991], whose call-by-value version (due to Fredholm [1995]) and with the natural understanding of the terminology has the following, dramatic consequence:

3B.15. **Colson's Corollary.** *If a primitive recursive definition of the minimum function on $\mathbb{N}$ is expressed faithfully in a programming language, then one of the two computations of* $\min(1, 1000)$ *and* $\min(1000, 1)$ *will take at least* 1000 *steps.*

The point is that the simple (tail) recursive program which computes $\min(x, y)$ in $\mathbf{N}_u$ in time $O(\min(x, y))$ cannot be matched in efficiency by a primitive recursive one, even though $\min(x, y)$ is a primitive recursive function; and so, as a practical and (especially) a foundational matter, we need to consider "recursive schemes" more general than primitive recursion, even if, ultimately, we are only interested in primitive recursive functions.

Moschovakis [2003] extends the results of Colson and Fredholm to primitive recursive programs from arbitrary Presburger primitives using $\gcd(x, y)$ rather than $\min(x, y)$ and then asks if the same gap in efficiency occurs between tail and full recursion in the Presburger structure (1E-10), cf. Open Problem x3B.2.[29] The corresponding question about $\mathbf{N}_u$ in Open Problem x3B.3 is (I think) the most important problem in this area.

Problems for Section 3B

x3B.1. **Open problem.** Can a complexity gap like that in Theorem 3B.14 be realized by a relation R rather than a function $f : \mathbb{N}^2 \to \mathbb{N}$? For a precise version: is there a total, pointed structure $\mathbf{A}$ and an n-ary relation R on A such that the following hold:

(1) $\mathbf{Tailrec}^0(\mathbf{A}) = \mathbf{Rec}^0(\mathbf{A})$.

(2) R is tail recursive in $\mathbf{A}$.

(3) Some extended program E decides R in $\mathbf{A}$ so that for every tail recursive F which also decides R, there is a sequence $\{\vec{x}\}_n$ of inputs such that

$$\lim_{n\to\infty} l^s(\mathbf{A}, F(\vec{x}_n)) = \infty \;\&\; (\forall n)[l^s(\mathbf{A}, F(\vec{x}_n)) \geq (l^s(\mathbf{A}, E(\vec{x}_n)))^2].$$

---

[29] van den Dries [2003] answered another question in this paper, and this started the collaboration which led to van den Dries and Moschovakis [2004], [2009] and eventually to many of the results in Chapters 5 – 8 of this book.

x3B.2. **Open problem.** Can a complexity gap between tail and full recursion like that in Theorem 3B.14 or Open Problem x3B.1 be realized in the Presburger structure?

x3B.3. **Open problem.** Can a complexity gap between tail and full recursion like that in Theorem 3B.14 or Open Problem x3B.2 be realized in unary arithmetic $\mathbf{N}_u$ or in the List structure $\mathbf{L}^*$?

Part II. Intrinsic complexity

CHAPTER 4

THE HOMOMORPHISM METHOD

Most of the known lower bound results in the literature are established for specific computation models and the natural complexity measures associated with them, and so any claim that they are absolute—or even that they hold for a great variety of models—must be inferred from the proofs. The results about recursive programs in van den Dries and Moschovakis [2004], [2009] are somewhat more robust: they are proved directly from the abstract definitions of complexities in Chapter 3 using basically nothing more than the *homomorphism* and *finiteness properties* in Problems x2E.2, x2E.4, without reference to the recursive machine or any other implementations of recursive programs. They imply lower bounds for most computation models, because of the representation of iterators by tail recursive programs in Theorem 2C.1, the complexity refinement of it in Problem x3A.3 (and the discussion after it) and what comes next.

Our main aim in this chapter is to extract from the homomorphism and finiteness properties of nondeterministic recursive programs in Problems x2E.2 and x2E.4 a general, algebraic method for deriving robust lower bounds for algorithms from specified primitives assuming very little about what these objects are.[30]

The key notions are those of a *uniform process* and *certification* in Sections 4B and 4D and the main result is the *Homomorphism Test*, Lemma 4E.3. We will start, however, with a brief discussion of "algorithms from primitives" which motivates our choice of notions without assuming any specific, rigorous definition of "what algorithms are".

4A. Axioms which capture the uniformity of algorithms

The basic intuition is that *an n-ary algorithm of sort s from the primitives of a structure* $\mathbf{A} = (A, \mathbf{\Phi})$ computes (in some one or in many different ways) an

[30]It is sometimes doubted whether it makes sense to talk of *nondeterministic algorithms from specified primitives*, and with good reason. Here, it is useful to assume that it does: the precise results will be stated about *nondeterministic recursive programs* and are not affected by surrounding loose talk.

n-ary partial function

$$\overline{\alpha} = \overline{\alpha}^{\mathbf{A}} : A^n \rightharpoonup A_s \quad (\text{with } A_{\texttt{boole}} = \{\text{tt}, \text{ff}\}, A_{\texttt{ind}} = A)$$

using the primitives in $\mathbf{\Phi}$ as *oracles*. We understand this to mean that in the course of a computation of $\overline{\alpha}(\vec{x})$, the algorithm may request from the oracle for any $\phi^{\mathbf{A}}$ any particular value $\phi^{\mathbf{A}}(u_1, \ldots, u_{n_\phi})$, where each u_i is either given by the input or has already been computed; and that if the oracles cooperate and respond to all requests, then this computation of $\overline{\alpha}(\vec{x})$ is completed in a finite number of steps.

The three axioms we formulate in this section capture part of this minimal understanding of how algorithms from primitives operate in the style of *abstract model theory*.

The crucial first axiom expresses the possibility that the oracles may choose not to respond to a request for $\phi^{\mathbf{A}}(u_1, \ldots, u_{n_\phi})$ unless

$$u_1, \ldots, u_n \in U \;\&\; \phi^{\mathbf{U}}(u_1, \ldots, u_n) \in U_s$$

for some fixed, arbitrary substructure $\mathbf{U} \subseteq_p \mathbf{A}$: the algorithm will still compute a partial function, which simply diverges on those inputs $\vec{x}$ for which no computation of $\overline{\alpha}(\vec{x})$ by α can be executed "inside" $\mathbf{U}$ (as far as calls to the oracles are involved).

A *process* of a structure $\mathbf{A}$ of arity n and sort s is an operation

$$\alpha : \mathbf{U} \to \overline{\alpha}^{\mathbf{U}} \quad (\mathbf{U} \subseteq_p \mathbf{A},\ \overline{\alpha}^{\mathbf{U}} : U^n \rightharpoonup U_s)$$

which assigns to each $\mathbf{U} \subseteq_p \mathbf{A}$ a partial function $\overline{\alpha}^{\mathbf{U}} : U^n \rightharpoonup U_s$. We call $\overline{\alpha}^{\mathbf{U}}$ the partial function *defined by* α *on* $\mathbf{U}$, and we write

$$\mathbf{U} \models \alpha(\vec{x}) = w \iff \vec{x} \in U^n \;\&\; \overline{\alpha}^{\mathbf{U}}(\vec{x}) = w,$$
$$\mathbf{U} \models \alpha(\vec{x})\downarrow \iff (\exists w)[\mathbf{U} \models \alpha(\vec{x}) = w];$$

in particular, α *defines* (on $\mathbf{A}$) the partial function $\overline{\alpha}^{\mathbf{A}} : A^n \rightharpoonup A_s$.

I. Locality Axiom. *Every algorithm of a structure* $\mathbf{A}$ *induces a process* α *of* $\mathbf{A}$ *of the same arity and sort which defines the partial function computed by the algorithm.*[31]

For example, if $E(\vec{x})$ is a nondeterministic recursive program which computes a partial function in $\mathbf{A}$, then the process α_E induced in $\mathbf{A}$ by $E(\vec{x})$ is defined by setting for each $\mathbf{U} \subseteq_p \mathbf{A}$,

$$\overline{\alpha}_E^{\mathbf{U}}(\vec{x}) = w \iff_{\text{df}} \mathbf{U} \models E(\vec{x}) = w$$

in the notation of (2E-4). It is important for this definition, of course, that if $E(\vec{x})$ computes a partial function in $\mathbf{A}$, then it computes a partial function in every $\mathbf{U} \subseteq_p \mathbf{A}$, Problem x2E.1.

[31]The Locality Axiom is much stronger than it might appear, because of the very liberal definition of substructures on page 32: it is possible that $w \in U, \phi(w) \in U$ but still $\phi^{\mathbf{U}}(w)\uparrow$.

The second axiom tries to capture the *uniformity* of algorithms, that they follow "the same" procedure on any two inputs $\vec{x}, \vec{y} \in A^n$. We would expect—at the least—that an isomorphism $\pi : \mathbf{U} \rightarrowtail\!\!\!\!\rightarrow \mathbf{V}$ between two substructures of $\mathbf{A}$ induces an isomorphism between the computations of $\overline{\alpha}^{\mathbf{U}}$ and those of $\overline{\alpha}^{\mathbf{V}}$; we postulate something which is weaker, in that it does not refer to "computations" but stronger in that it requires uniformity for arbitrary homomorphisms, which need not be injective or surjective:

II. Homomorphism Axiom. *If α is the process induced by an algorithm of $\mathbf{A}$ and $\pi : \mathbf{U} \to \mathbf{V}$ is a homomorphism of one substructure of $\mathbf{A}$ into another, then*

$$\text{(4A-1)} \qquad \mathbf{U} \models \alpha(\vec{x}) = w \Longrightarrow \mathbf{V} \models \alpha(\pi(\vec{x})) = \pi(w) \quad (\vec{x} \in U^n).$$

In particular, by applying this to the identity embedding $\mathrm{id}_U : \mathbf{U} \rightarrowtail \mathbf{A}$,

$$\mathbf{U} \subseteq_p \mathbf{A} \Longrightarrow \overline{\alpha}^{\mathbf{U}} \sqsubseteq \overline{\alpha}^{\mathbf{A}} = \overline{\alpha}.$$

The idea here is that the oracle for each $\phi^{\mathbf{A}}$ may consistently respond to each request for $\phi^{\mathbf{U}}(\vec{u})$ by delivering $\phi^{\mathbf{V}}(\pi(\vec{u}))$. This transforms any computation of $\overline{\alpha}^{\mathbf{U}}(\vec{x})$ into one of $\overline{\alpha}^{\mathbf{V}}(\pi(\vec{x}))$, which in the end delivers the value $\pi(w) = \pi(\overline{\alpha}^{\mathbf{U}}(\vec{x}))$.

This argument is convincing for the identity embedding $\mathrm{id}_U : \mathbf{U} \rightarrowtail \mathbf{V}$ and works for nondeterministic recursive programs, Problem x2E.2. It is not quite that simple in full generality, because algorithms may employ in their computations complex data structures and rich primitives, e.g., stacks, queues, trees, the introduction of higher type objects by λ-abstraction and subsequent application of these objects to suitable arguments, etc. The claim is that any homomorphism $\pi : \mathbf{U} \to \mathbf{V}$ lifts naturally to these data structures, and so the image of a convergent computation of $\overline{\alpha}^{\mathbf{U}}(\vec{x})$ is a convergent computation of $\overline{\alpha}^{\mathbf{V}}(\pi(\vec{x}))$. Put another way: if some $\pi : \mathbf{U} \to \mathbf{V}$ does not lift naturally to a mapping of the relevant computations, then the algorithm *is using essentially some hidden primitives not included in* $\mathbf{A}$ and so *it is not an algorithm from* $\{\phi^{\mathbf{A}}\}_{\phi \in \Phi}$. It is clear, however, that the Homomorphism Axiom demands something more of algorithms (and how they use oracles) than the Locality Axiom, and we will discuss it again in Section 4G.

The Homomorphism Axiom is at the heart of this approach to the derivation of lower bounds.

III. Finiteness Axiom. *If α is the process induced by an algorithm of $\mathbf{A}$, then*

$$\text{(4A-2)} \quad \mathbf{A} \models \alpha(\vec{x}) = w \Longrightarrow \Big(\mathbf{U} \models \alpha(\vec{x}) = w \\ \text{for some finite } \mathbf{U} \subseteq_p \mathbf{A} \text{ which is generated by } \vec{x}\Big);$$

or, equivalently, using the *certificates* defined on page 34,

$$\text{(4A-3)} \quad \mathbf{A} \models \alpha(\vec{x}) = w \Longrightarrow \Big(\mathbf{U} \models \alpha(\vec{x}) = w \quad \text{for some certificate } (\mathbf{U}, \vec{x}) \text{ of } \mathbf{A}\Big).$$

This combines two ingredients of the basic intuition: first, that in the course of a computation the algorithm may only request of the oracles values $\phi^{\mathbf{A}}(\vec{u})$ for arguments in the input $\vec{x}$ or that it has already computed from $\vec{x}$, and second, that computations are finite. A suitable $\mathbf{U}$ is then determined by putting in eqdiag($\mathbf{U}$) all the *calls made by the algorithm in the computations of* $\overline{\alpha}(\vec{x})$.

The Finiteness Axiom implies, in particular, that partial functions computed by an $\mathbf{A}$-algorithm take values in the substructure generated by the input, since for any process of $\mathbf{A}$ which satisfies Axiom **III**,

$$\overline{\alpha}^{\mathbf{A}}(\vec{x}) = w \Longrightarrow w \in G_\infty(\mathbf{A}, \vec{x}) \cup \mathbb{B}.$$

By Problems x2E.2 and x2E.4, axioms **I** – **III** are satisfied by nondeterministic recursive algorithms. They also hold for all *concrete algorithms* expressed by *computation models*, e.g., *Turing machines*, *finite register programs*, *Random Access machines*, *decision trees* ..., and their *nondeterministic* versions; this can be proved by applying Theorem 2C.1, Proposition 2C.3 and their natural extensions to the nondeterministic versions of these models, cf. Problem x4B.2. We can express succinctly the claim that all algorithms satisfy these axioms in terms of the following basic notion:

A process α of $\mathbf{A}$ is *uniform* if it has the homomorphism and finiteness properties (4A-1), (4A-2); a partial function $f : A^n \rightharpoonup A_s$ is *uniform in* $\mathbf{A}$ if it is defined by a uniform process of $\mathbf{A}$; and we set

$$\mathbf{Unif}(\mathbf{A}) = \Big\{f : A^n \rightharpoonup A_s : f \text{ is uniform in } \mathbf{A}\Big\}.$$

Uniformity Thesis. *Every algorithm which computes a partial function* $f : A^n \rightharpoonup A_s$ *from the primitives of a* Φ*-structure* $\mathbf{A}$ *induces a uniform process* α *of* $\mathbf{A}$ which defines f.

This is a weak Church-Turing-type assumption about algorithms which, of course, cannot be established rigorously absent a precise definition of algorithms. It limits somewhat the notion of "algorithm", but not in any novel way which yields new undecidability results. We will show, however, that it can be used to derive *absolute lower bounds* for many, natural complexity measures on decidable relations, very much like the Church-Turing Thesis is used to establish *absolute undecidability*.

We record for easy reference the part of the Uniformity Thesis which is an immediate consequence of Problems x2E.2 and x2E.4:

4A.1. **Proposition.** *Every nondeterministic* Φ*-program* $E(\vec{x})$ *induces on each* Φ*-structure* $\mathbf{A}$ *the uniform process* $\alpha_E = \alpha_E^{\mathbf{A}}$ *defined by*

$$\text{(4A-4)} \qquad \overline{\alpha}_E^{\mathbf{U}}(\vec{x}) = w \iff_{\text{df}} \mathbf{U} \models E(\vec{x}) = w \quad (\mathbf{U} \subseteq_p \mathbf{A}).$$

4B. Examples and counterexamples

We will be using consistently for processes the notations

$$\begin{aligned} \mathbf{U} \models \alpha(\vec{x}) = w &\iff \vec{x} \in U^n \;\&\; \overline{\alpha}^{\mathbf{U}}(\vec{x}) = w, \\ \mathbf{U} \models \alpha(\vec{x})\downarrow &\iff (\exists w)[\mathbf{U} \vdash \alpha(\vec{x}) = w] \end{aligned}$$

introduced above, and it is also useful to set

$$\begin{aligned} \mathbf{U} \models_c \alpha(\vec{x}) = w &\iff (\mathbf{U}, \vec{x}) \text{ is a certificate and } \mathbf{U} \models \alpha(\vec{x}) = w, \\ \mathbf{U} \models_c \alpha(\vec{x})\downarrow &\iff (\exists w)[\mathbf{U} \models_c \alpha(\vec{x}) = w]. \end{aligned}$$

In this notation, the finiteness property takes the simple form

$$\text{(4B-1)} \qquad \overline{\alpha}(\vec{x}) = w \Longrightarrow (\exists \mathbf{U} \subseteq_p \mathbf{A})[\mathbf{U} \models_c \alpha(\vec{x}) = w].$$

If we read "$\models_c$" as *computes*, then this form of the axiom suggests that the triples $(\mathbf{U}, \vec{x}, w)$ such that $\mathbf{U} \models_c \alpha(\vec{x}) = w$ play the role of (very abstract) *computations* for uniform processes—which, however, is very misleading.

We have shown in Proposition 4A.1 that $\mathbf{Rec}^0_{\text{nd}}(\mathbf{A}) \subseteq \mathbf{Unif}(\mathbf{A})$ and we have argued that every algorithm from specified primitives induces a uniform process. The converse, however, is far from true: nothing in the homomorphism and finiteness properties suggests that $\mathbf{A}$-uniform functions are "computable" from the primitives of $\mathbf{A}$ in any intuitive sense, and in general, they are not:

4B.1. **Proposition.** *If a* Φ*-structure* $\mathbf{A}$ *is generated by the empty tuple, then every* $f : A^n \rightharpoonup A_s$ *is defined by some uniform process of* $\mathbf{A}$.

In particular, every $f : \mathbb{N}^n \rightharpoonup \mathbb{N}_s$ *is uniform in* $\mathbf{A} = (\mathbb{N}, 0, \Phi^{\mathbf{A}})$ *if* $\Phi^{\mathbf{A}}$ *includes either the successor function* S *or the primitives of binary arithmetic* $\text{em}_2(x) = 2x$ *and* $\text{om}_2(x) = 2x + 1$.

Proof. Let $G_m = G_m(\mathbf{A}, \emptyset)$ be the set generated in $\leq m$ steps by the empty set, so that $G_0 = \emptyset$, G_1 comprises the distinguished elements of $\mathbf{A}$, etc. Let

$$d(\vec{x}, w) = \min\{m : x_1, \dots, x_n, w \in G_m \cup \mathbb{B}\},$$

and define α by setting for each $\mathbf{U} \subseteq_p \mathbf{A}$,

$$\overline{\alpha}^{\mathbf{U}}(\vec{x}) = w \iff f(\vec{x}) = w \;\&\; \mathbf{G}_{d(\vec{x},w)} \subseteq_p \mathbf{U}.$$

The finiteness property is immediate taking $\mathbf{U} = \mathbf{G}_{d(\vec{x},w)}$, and α has the homomorphism property because if $\mathbf{G}_m \subseteq_p \mathbf{U}$, then every homomorphism $\pi : \mathbf{U} \to \mathbf{V}$ fixes every $u \in G_m$. ⊣

The axioms aim to capture the *uniformity* of algorithms—that they compute all their values following "the same procedure"— but surely do not capture their *effectiveness*.

An example of a non-uniform process. Consider the standard structure of Peano arithmetic $\mathbf{N} = (\mathbb{N}, 0, 1, +, \cdot, =)$, and let α be the "procedure" which "computes" the function

$$f(x) = (x+1)^2 = x^2 + 2x + 1 \tag{4B-2}$$

using the first or the second of these two expressions for $f(x)$ accordingly as x is odd or even. We can view this as a unary process α as follows: first we define for each $x \in \mathbb{N}$ a finite substructure $\mathbf{U}_x \subseteq_p \mathbf{N}$ by giving its equational diagram as on page 32,

$$\begin{aligned}&\text{eqdiag}(\mathbf{U}_x)\\ &= \begin{cases} \{(1,1), (+, x, 1, x+1), (\cdot, x+1, x+1, (x+1)^2)\}, & \text{if } x \text{ is odd},\\ \{(1,1), (\cdot, x, x, x^2), (+, x, x, 2x), \\ \qquad (+, 2x, 1, 2x+1), (+, x^2, 2x+1, (x+1)^2)\}, & \text{if } x \text{ is even;}\end{cases}\end{aligned}$$

and then we define $\overline{\alpha}^{\mathbf{U}} : U \rightharpoonup U$ for each $\mathbf{U} \subseteq_p \mathbf{N}$ by setting

$$\overline{\alpha}^{\mathbf{U}}(x) = w \iff w = (x+1)^2 \;\&\; \mathbf{U}_x \subseteq_p \mathbf{U}.$$

This operation α is a process, it has the nice property

$$\mathbf{U} \subseteq_p \mathbf{V} \Longrightarrow \overline{\alpha}^{\mathbf{U}} \subseteq_p \overline{\alpha}^{\mathbf{V}} \quad \text{(monotonicity)},$$

it satisfies the Finiteness Axiom **III** directly from its definition, and it defines the map $x \mapsto (x+1)^2$ in $\mathbf{N}$ since, obviously, $\overline{\alpha}^{\mathbf{N}}(x) = (x+1)^2$. It is not uniform, however, because it fails to satisfy the Homomorphism Axiom **II**, as follows: if

$$\begin{aligned}\text{eqdiag}(\mathbf{U}) &= \text{eqdiag}(\mathbf{U}_3) = \{(1,1), (+, 3, 1, 4), (\cdot, 4, 4, 16)\},\\ \pi(3) &= 4, \pi(1) = 1, \pi(4) = \pi(3) + \pi(1) = 5, \pi(16) = \pi(4)\cdot\pi(4) = 25,\\ &\text{and } \mathbf{V} = \pi[\mathbf{U}] \text{ so } \text{eqdiag}(\mathbf{V}) = \{(1,1), (+, 4, 1, 5), (\cdot, 5, 5, 25)\},\end{aligned}$$

then $\pi : \mathbf{U} \to \mathbf{V}$ is a homomorphism, $\pi(3) = 4$, but $\overline{\alpha}^{\mathbf{V}}(4)\uparrow$, because $\mathbf{U}_4 \not\subseteq_p \mathbf{V}$. What spoils the homomorphism property is that α uses exactly one of two different algorithms to compute $\overline{\alpha}(x)$ depending on whether x is odd or even *without checking first which of the cases applies*, and, in fact, we would not consider it an algorithm for computing $x \mapsto (x+1)^2$ in $\mathbf{N}$ for exactly this reason.

One might argue that the remarks following (4B-2) describe intuitively a nondeterministic algorithm which computes the function $x \mapsto (x+1)^2$, and this is sort-of true: we can put down a nondeterministic recursive program of $\mathbf{N}$ which chooses whether to compute $(x+1)^2$ or $x^2 + 2x + 1$ but *makes this choice autonomously, independently of whether x is even or odd.* The process β induced by this program is

$$\text{(4B-3)}\quad \overline{\beta}^{\mathbf{U}}(x) = w \iff w = (x+1)^2$$
$$\& \Big(\text{either } \{(1,1), (+, x, 1, x{+}1), (\cdot, x{+}1, x{+}1, (x{+}1)^2)\} \subseteq \text{eqdiag}(\mathbf{U})$$
$$\text{or } \{(1,1), (+, x, x, 2x), (+, 2x, 1, 2x{+}1), (+, x^2, 2x{+}1, x{+}1)^2)\} \subseteq \text{eqdiag}(\mathbf{U})\Big),$$

and it is uniform, cf. Problem x4B.3.

Problems for Section 4B

x4B.1. **Problem.** (1) Suppose $\mathbf{F} = (F, 0, 1, +, -, \cdot, \div, =)$ is a field,

$$N_F(a_0, a_1, a_2, x) \iff a_0 + a_1x + a_2x^2 = 0$$

is the nullity relation of degree 2 on F defined by (1C-12) and α is the process (of sort `boole` and four arguments) defined by

$$\overline{\alpha}^{\mathbf{U}}(a_0, a_1, a_2, x) = w \iff [w = \mathrm{tt} \iff a_0 + a_1x + a_2x^2 = 0]$$
$$\& \{(\cdot, a_2, x, a_2x), (+, a_1, a_2x, a_1 + a_2x)$$
$$(\cdot, x, a_1 + a_2x, a_1x + a_2x^2),$$
$$(+, a_0, a_1x + a_2x^2, a_0 + a_1x + a_2x^2),$$
$$(0, 0), (=, a_0 + a_1x + a_2x^2, 0)\} \subseteq \text{eqdiag}(\mathbf{U}).$$

Prove that α is uniform and defines the nullity relation of degree 2.

(2) Define the uniform process on $\mathbf{F}$ (of sort `boole` and arity $n + 2$) which is induced by the Horner Rule described on page 25.

x4B.2. **Problem.** For each (possibly nondeterministic) iterator $\mathfrak{i}$ which computes a partial function $f : X \rightharpoonup W$, define the uniform process $\alpha_{\mathfrak{i}}$ which is induced by $\mathfrak{i}$ in the associated structure $\mathbf{A}_{\mathfrak{i}}$ and prove that it defines f.

x4B.3. **Problem.** Prove that the process β defined by (4B-3) is uniform and defines the map $x \mapsto (x + 1)^2$ in $\mathbf{N}$.

x4B.4*. **Problem.** Prove that if $f : A^k \rightharpoonup A_s$ is uniform in $\mathbf{A}$, then so is every $g \in \mathbf{Unif}(\mathbf{A}, f)$. HINT: If

$$f(\vec{u}) = \overline{\alpha}^{\mathbf{A}}(\vec{u}) \text{ and } g(\vec{x}) = \overline{\beta}^{(\mathbf{A},f)}(\vec{x})$$

with uniform processes α of $\mathbf{A}$ and β of $(\mathbf{A}, f)$, then the process

$$\overline{\gamma}^{\mathbf{U}}(\vec{x}) = \overline{\beta}^{(\mathbf{U}, f \upharpoonright U^k)}(\vec{x}) \quad (\mathbf{U} \subseteq_p \mathbf{A}, \vec{x} \in U^n)$$

of $\mathbf{A}$ defines g. The proof that it is uniform requires some computation.

x4B.5. **Problem.** Prove that if $f \in \mathbf{Unif}(\mathbf{A})$ and $\rho : \mathbf{A} \rightarrowtail\!\!\!\!\twoheadrightarrow \mathbf{A}$ is an automorphism of $\mathbf{A}$, then $f(\rho(\vec{x})) = \rho(f(\vec{x})) \quad (f(\vec{x})\downarrow)$.

x4B.6. **Problem.** Give an example of a finite structure $\mathbf{A}$ and a unary relation $P \subseteq A$ which is respected by all automorphisms of $\mathbf{A}$ but is not uniform in $\mathbf{A}$. HINT: Use a structure (A, E) where E is an equivalence relation on A with suitable properties.

x4B.7. **Problem.** Characterize the total functions $f^n : A \to A_s$ which are uniform in $\mathbf{A} = (A)$ (with no primitives), where A is infinite.

x4B.8*. **Problem.** Suppose $\mathbf{A} = (A, R_1, \dots, R_K)$ is a structure whose primitives are total relations on A. What are the (total) relations $P \subseteq A^n$ which are uniform in $\mathbf{A}$?

4C. Complexity measures on uniform processes

A *substructure norm* on a Φ-structure $\mathbf{A}$ is an operation μ which assigns to every certificate $(\mathbf{U}, \vec{x})$ in $\mathbf{A}$ a number $\mu(\mathbf{U}, \vec{x})$ and respects isomorphisms, i.e.,

$$(4C\text{-}1) \quad \mathbf{U} = \mathbf{G}_m(\mathbf{U}, \vec{x}) \subseteq_p \mathbf{A} \;\&\; \pi : \mathbf{U} \rightarrowtail\!\!\!\!\twoheadrightarrow \mathbf{V} \subseteq_p \mathbf{A} \Longrightarrow \mu(\mathbf{U}, \vec{x}) = \mu(\mathbf{V}, \pi(\vec{x})).$$

Typical examples are[32]

$$\begin{aligned} \text{depth}(\mathbf{U}, \vec{x}) &= \min\{m : \mathbf{U} = \mathbf{G}_m(\mathbf{U}, \vec{x})\}, \\ \text{values}(\mathbf{U}, \vec{x}) &= |\{w \in U : (\phi, \vec{u}, w) \in \text{eqdiag}(\mathbf{U}) \text{ for some } \phi, \vec{u}\}|, \\ \text{calls}(\Phi_0)(\mathbf{U}, \vec{x}) &= |\text{eqdiag}(\mathbf{U} \restriction \Phi_0)| \quad (\Phi_0 \subseteq \Phi) \end{aligned}$$

which we introduced on page 34 (with $\Phi_0 = \Phi$ for the last one). By Proposition 1D.1,

$$(4C\text{-}2) \qquad \text{depth}(\mathbf{U}, \vec{x}) \le \text{values}(\mathbf{U}, \vec{x}) \le \text{calls}(\mathbf{U}, \vec{x}).$$

If μ is a substructure norm on $\mathbf{A}$ and α is a uniform process of $\mathbf{A}$, then the *μ-complexity measure* of α is the partial function

$$(4C\text{-}3) \qquad \mu(\alpha, \vec{x}) =_{\text{df}} \min\{\mu(\mathbf{U}, \vec{x}) : \mathbf{U} \models_c \alpha(\vec{x})\downarrow\},$$

defined on the domain of convergence of $\overline{\alpha}^{\mathbf{A}}$. By using the norms above, we get three natural complexity measures on uniform processes,

$$(4C\text{-}4) \qquad \text{depth}(\alpha, \vec{x}), \ \text{values}(\alpha, \vec{x}), \ \text{calls}(\Phi_0)(\alpha, \vec{x}).$$

---

[32]The depth and values norms can also be relativized to arbitrary $\Phi_0 \subseteq \Phi$, but it is tedious, we have no interesting results about them, and we leave the relativization for depth for Problem x4C.8*.

The first and last of these three correspond to familiar complexity measures with roughly similar names for concrete algorithms but not exactly:[33]

– The "sequential" measure $\text{calls}(\Phi_0)(\alpha, \vec{x})$ intuitively counts *the least number of distinct calls to primitives in* Φ_0 *required to define* $\overline{\alpha}(\vec{x})$ *by the process* α, where $(\phi, \vec{u}, w)$ and $(\phi', \vec{u}', w')$ are distinct if either $\phi \not\equiv \phi'$ or $\vec{u} \neq \vec{u}'$;

– the "parallel" measure $\text{depth}(\alpha, \vec{x})$ counts *the least number of distinct calls to the primitives of* **A** *which must be executed in sequence to define* $\overline{\alpha}(\vec{x})$; and

– the less familiar middle measure $\text{values}(\alpha, \vec{x})$ counts *the least number of points in* A *that* α *must compute from* $\vec{x}$ *to define* $\overline{\alpha}(\vec{x})$.

These measures are typically lower than their versions for concrete algorithms, because they count *distinct* calls and points, while an algorithm may (stupidly or by design, e.g., to simplify the code) make the same call many times, cf. Problem x4C.6.

4C.1. **Lemma.** *For every uniform process* α *of a* Φ*-structure* **A** *and all* $\vec{x}, w$ *such that* $\overline{\alpha}(\vec{x}) = w$,

$$\text{depth}(\overline{\alpha}(\vec{x}); \mathbf{A}, \vec{x}) \leq \text{depth}(\alpha, \vec{x}) \leq \text{values}(\alpha, \vec{x}) \leq \text{calls}(\alpha, \vec{x}), \tag{4C-5}$$

where, by our convention, $\text{depth}(\text{tt}, \mathbf{A}) = \text{depth}(\text{ff}, \mathbf{A}) = 0$.

PROOF. The first inequality is trivial if $w \in \mathbb{B}$ and immediate for $w \in A$, because if $\mathbf{U} \models_c \alpha(\vec{x}) = w$, then $w \in U$ and so

$$\text{depth}(w; \mathbf{A}, \vec{x}) \leq \text{depth}(w; \mathbf{U}, \vec{x}) \leq \text{depth}(\mathbf{U}, \vec{x}).$$

For the third claimed inequality, suppose $\overline{\alpha}(\vec{x}) = w$ and choose a certificate $(\mathbf{U}, \vec{x})$ of **A** with least $|\text{eqdiag}(\mathbf{U})|$ such that $\mathbf{U} \models_c \alpha(\vec{x}) = w$, so that $\text{calls}(\alpha, \vec{x}) = |\text{eqdiag}(\mathbf{U})|$. Now $\text{values}(\mathbf{U}, \vec{x}) \leq |\text{eqdiag}(\mathbf{U})|$ by (1D-16) in Proposition 1D.1, and since $(\mathbf{U}, \vec{x})$ is among the certificates in **A** considered in the definition of $\text{values}(\alpha, \vec{x})$, we have

$$\text{values}(\alpha, \vec{x}) \leq |\text{eqdiag}(\mathbf{U})| = \text{calls}(\alpha, \vec{x}).$$

The second inequality is proved by a similar argument. ⊣

We record for easy reference the relation between these complexity measures and the complexity measures for programs we studied in Chapter 3, in the two cases that are most important for the applications:

4C.2. **Proposition.** *Let* α_E *be the uniform process induced on a* Φ*-structure* **A** *by a nondeterministic extended program* $E(\vec{x})$.

(1) *If* $\Phi_0 \subseteq \Phi$ *and* $\text{den}(\mathbf{A}, E(\vec{x}))\downarrow$, *then*

$$\text{calls}(\Phi_0)(\alpha_E, \vec{x}) \leq c^s(\Phi_0)(\mathbf{A}, E(\vec{x})), \tag{4C-6}$$

[33]There are, of course, many other substructure norms which induce useful complexity measures, including those which come by combining the three basic ones: for example

$$\mu(\mathbf{U}, \vec{x}) = \text{values}(\mathbf{U}, \vec{x}) \cdot 6^{\text{depth}(\mathbf{U}, \vec{x})}$$

comes up naturally in the proof of Theorem 8B.2!

where $c^s(\Phi_0)(\mathbf{A}, E(\vec{x}))$ is defined for nondeterministic programs by (2E-5) and agrees with the definition (3A-12) when E is deterministic by Problem x3A.14.

(2) *If E is deterministic and* $\text{den}(\mathbf{A}, E(\vec{x}))\downarrow$, *then*

$$\text{(4C-7)} \qquad \text{depth}(\alpha_E, \vec{x}) \le c^p(\mathbf{A}, E(\vec{x})) \textit{ and } \text{depth}(\alpha_E, \vec{x}) \le d(\mathbf{A}, E(\vec{x})).$$

PROOF is easy and we leave it for Problems x4C.4, x4C.6 and x4C.7. ⊣

The time complexity on RAMs. The requirement that substructure norms respect isomorphisms is weak but not trivial, cf. Problems x4C.11 and x4C.12*; the first of these implies that the (most natural) time complexity $T(n)$ on Random Access Machines in Cook and Reckhow [1973] is not induced by a substructure norm. This can be handled in many ways, of course, but we will not go into it here.

Problems for Section 4C

x4C.1. **Problem.** Prove that $\text{depth}(\mathbf{U}, \vec{x})$, $\text{values}(\mathbf{U}, \vec{x})$ and $\text{calls}(\Phi_0)(\mathbf{U}, \vec{x})$ are preserved by isomorphisms, as in (4C-1).

x4C.2. **Problem.** Prove that if $\mathbf{U} \subseteq_p \mathbf{V}$ and $(\mathbf{U}, \vec{x})$, $(\mathbf{V}, \vec{x})$ are both certificates of a structure $\mathbf{A}$, then

$$\text{calls}(\mathbf{U}, \vec{x}) \le \text{calls}(\mathbf{V}, \vec{x}), \quad \text{values}(\mathbf{U}, \vec{x}) \le \text{values}(\mathbf{V}, \vec{x}).$$

The point of the next problem is that there is no similar, useful relation between the depth complexity and the substructure relation:

x4C.3. **Problem.** (1) Give an example of two certificates $(\mathbf{U}, \vec{x})$ and $(\mathbf{V}, \vec{x})$ of the same structure $\mathbf{A}$, such that $\mathbf{U} \subseteq_p \mathbf{V}$ and

$$\text{depth}(\mathbf{U}, x) < \text{depth}(\mathbf{V}, x).$$

(2) Give an example of two certificates $(\mathbf{U}, \vec{x})$ and $(\mathbf{V}, \vec{x})$ of the same structure $\mathbf{A}$, such that $\mathbf{U} \subseteq_p \mathbf{V}$ but

$$\text{depth}(\mathbf{U}, x) > \text{depth}(\mathbf{V}, x).$$

HINT: Use the vocabulary $\{\phi, \psi\}$ with both ϕ, ψ unary of sort `ind`.

x4C.4. **Problem.** Prove that if α_E is the uniform process induced by a nondeterministic recursive program in $\mathbf{A}$ by (4A-4) and $c^s(\Phi_0)(\mathbf{A}, E(\vec{x}))$ is the calls-complexity for E in $\mathbf{A}$ as defined in (2E-5), then

$$\text{(4C-8)} \qquad \text{calls}(\Phi_0)(\alpha_E, \vec{x}) \le c^s(\Phi_0)(\mathbf{A}, E(\vec{x})) \quad (\overline{\alpha}(\vec{x})\downarrow).$$

There is no useful bound of $c^s(\Phi_0)(\vec{x})$ in terms of $\text{calls}(\Phi_0)(\alpha_E, \vec{x})$ which holds generally, even for deterministic programs:

x4C.5. **Problem.** Consider the following deterministic, extended program $E(\mathsf{x})$ in unary arithmetic $\mathbf{N}_u$:

$$p(\mathsf{x}) \text{ where } \Big\{\mathsf{p}(x) = \text{if } \mathsf{x} = 0 \text{ then } 0 \text{ else } \mathsf{q}(\mathsf{p}(\mathrm{Pd}(\mathsf{x})), \mathsf{p}(\mathrm{Pd}(\mathsf{x}))),$$

$$\mathsf{q}(\mathsf{u}, \mathsf{v}) = \mathsf{u}\Big\}.$$

Prove that if α_E is the uniform process induced by $E(\mathsf{x})$ in $\mathbf{N}_u$, then

$$\text{calls}(\mathrm{Pd})(\alpha_E, x) = x, \quad c^s(\mathrm{Pd})(\mathbf{N}_u, E(x)) = 2^x \quad (x \geq 1).$$

x4C.6. **Problem.** Let α_E be the uniform process induced by a deterministic extended program $E(\vec{x})$ in a Φ-structure $\mathbf{A}$ by (4A-4). Prove that for all $\vec{x} \in A^n$ such that $\text{den}_E(\vec{x})\downarrow$,

$$\text{(4C-9)} \qquad \text{depth}(\alpha_E, \vec{x}) \leq c^p(\mathbf{A}, E(\vec{x}))$$

as $c^p(\mathbf{A}, E(\vec{x}))$ is defined by (3A-13), and give an example where the inequality is strict.

HINT: Prove the following refinement of Problems x2E.2 and x2E.4 for deterministic programs: *If $M \in \text{Conv}(\mathbf{A}, E)$, $X \subseteq A$ contains all the parameters which occur in M, $m = C^p(\mathbf{A}, M)$ and $\mathbf{U} = \mathbf{G}_m(\mathbf{A}, X)$, then*

$$\mathbf{U} = \mathbf{G}_m(\mathbf{U}, X) \text{ and } \mathbf{U} \models M = w.$$

This implies (4C-9) directly from the definitions.

x4C.7. **Problem.** Let α_E be the uniform process induced by a deterministic extended program $E(\vec{x})$ in a Φ-structure $\mathbf{A}$ by (4A-4). Prove that for all $\vec{x} \in A^n$ such that $\text{den}_E(\vec{x})\downarrow$,

$$\text{(4C-10)} \qquad \text{depth}(\alpha_E, \vec{x}) \leq d(\mathbf{A}, E(\vec{x}))$$

as $d(\mathbf{A}, E(\vec{x}))$ is defined by (3A-3). HINT: Appeal to Proposition 3A.2.

x4C.8*. **Problem** $(\text{depth}(\Phi_0)(\mathbf{U}, \vec{x}))$. For a Φ-structure $\mathbf{A}$, any $X \subseteq A$ and any $\Phi_0 \subseteq \Phi$, define $\text{depth}(\Phi_0)(\mathbf{G}_m(\mathbf{A}, X))$ by the following recursion on m (skipping $\mathbf{A}$):

$$\text{depth}(\Phi_0)(\mathbf{G}_0(X)) = 0,$$

$$\text{depth}(\Phi_0)(\mathbf{G}_{m+1}(X) = \begin{cases} \text{depth}(\Phi_0)(\mathbf{G}_m(X)) + 1, \\ \quad \text{if for some } \phi \in \Phi_0, \vec{u} \text{ and } w, \\ \quad (\phi, \vec{u}, w) \in \text{eqdiag}(\mathbf{G}_{m+1}(X)) \setminus \text{eqdiag}(\mathbf{G}_m(X)), \\ \text{depth}(\Phi_0)(\mathbf{G}_m(X)), \quad \text{otherwise.} \end{cases}$$

(1) Prove that $\text{depth}(\Phi_0)(\mathbf{U}, \vec{x})$ is invariant under isomorphisms.

(2) Prove that if α_E is the uniform process induced by a deterministic program $E(\vec{x})$, then

$$\text{depth}(\Phi_0)(\alpha_E, \vec{x}) \leq c^p_{\Phi_0}(\mathbf{A}, E, \vec{x}) \quad (\text{den}_E(\vec{x})\downarrow).$$

HINT: Formulate and prove a Φ_0-version of the hint in Problem x4C.6.

x4C.9. **Problem.** Prove that if the successor S is a primitive of a structure $\mathbf{A} = (\mathbb{N}, 0, \mathbf{\Phi})$, then every $f : \mathbb{N}^n \rightharpoonup \mathbb{N}_s$ is defined by some uniform process α of $\mathbf{A}$ such that

$$\text{calls}(\alpha, \vec{x}) \leq \max\{\vec{x}, f(\vec{x})\} \quad (f(\vec{x})\downarrow)$$

where $\max\{\vec{x}, w\} = \max\{\vec{x}\}$ if $w \in \mathbb{B}$. HINT: Look up the proof of Lemma 4B.1.

x4C.10. **Problem.** Prove that if 0, 1 and the binary primitives $\text{em}_2(x) = 2x$, $\text{om}_2(x) = 2x + 1$ are among the primitives of $\mathbf{A} = (\mathbb{N}, \mathbf{\Phi})$, then every $f : \mathbb{N}^n \rightharpoonup \mathbb{N}_s$ is defined by some uniform process α of $\mathbf{A}$ with

$$\text{calls}(\alpha, \vec{x}) \leq 2\max\{\lfloor\log(x_1)\rfloor, \dots, \lfloor\log(x_n)\rfloor, \lfloor\log(f(\vec{x}))\rfloor\} \quad (f(\vec{x})\downarrow).$$

x4C.11. **Problem.** Fix a Φ-structure $\mathbf{A}$.

(1) Prove that the function

$$\mu_1(\mathbf{U}, \vec{x}) = \text{the cardinality of the set } \{x_1, \dots, x_n\}$$

is a substructure norm on $\mathbf{A}$.

(2) Prove that if weight : $\Phi \to \mathbb{N}$ assigns a "weight" to every function symbol in Φ, then

$$\mu_2(\mathbf{U}, \vec{x}) = \sum\{\text{weight}(\phi) : (\phi, \vec{u}, w) \in \text{eqdiag}(\mathbf{U})\};$$

is a substructure norm on $\mathbf{A}$.

(3) Prove that if $f : A \to \mathbb{N}$ and

$$\nu(\mathbf{U}, \vec{x}) = f(x_1) \quad (\mathbf{U} \subseteq_p \mathbf{A}),$$

then ν is not a substructure norm on $\mathbf{A}$ unless f is constant.

x4C.12*. **Problem.** Prove that for every Φ-structure $\mathbf{A}$ and every substructure norm μ on $\mathbf{A}$, there is some $K_{n,m} \in \mathbb{N}$, such that for every certificate $(\mathbf{U}, \vec{x})$ of $\mathbf{A}$ with $\vec{x} \in A^n$,

$$\text{calls}(\mathbf{U}, \vec{x}) \leq m \Longrightarrow \mu(\mathbf{U}, \vec{x}) \leq K_{n,m}.$$

Briefly, without specifying the constant involved,

$$\mu(\mathbf{U}, \vec{x}) = O(\text{calls}(\mathbf{U}, \vec{x})).$$

HINT: For any two certificates $(\mathbf{U}, \vec{x})$ and $(\mathbf{V}, \vec{y})$ of $\mathbf{A}$ with $\vec{x}, \vec{y} \in A^n$, set

$$(\mathbf{U}, \vec{x}) \approx (\mathbf{V}, \vec{y}) \iff |\text{eqdiag}(\mathbf{U})| = |\text{eqdiag}(\mathbf{V})| \\ \& \ (\exists \text{ an isomorphism } \pi : \mathbf{U} \rightarrowtail\!\!\!\!\!\rightarrow \mathbf{V})[\pi(\vec{x}) = \vec{y}],$$

and prove that $\approx$ is an equivalence relation with finitely many equivalence classes.

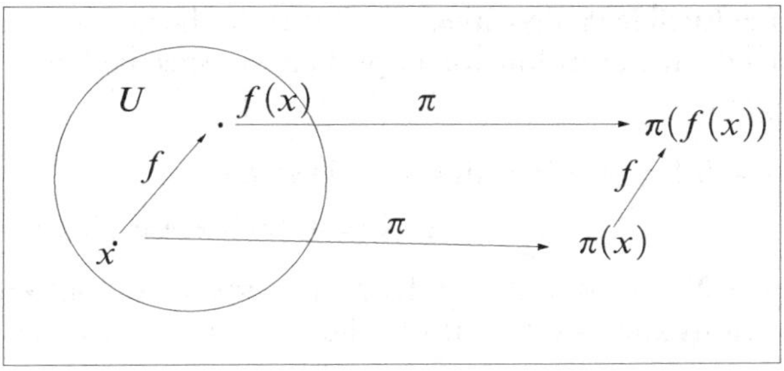

FIGURE 4. $\pi : \mathbf{U} \to \mathbf{A}$ respects $f : A \rightharpoonup A$ at x.

4D. Forcing $\Vdash^{\mathbf{A}}$ and certification $\Vdash_c^{\mathbf{A}}$

Suppose $\mathbf{A}$ is a Φ-structure, $f : A^n \rightharpoonup A_s$ (with $s \in \{\texttt{ind}, \texttt{boole}\}$), $\mathbf{U} \subseteq_p \mathbf{A}$, and $f(\vec{x})\downarrow$. A homomorphism $\pi : \mathbf{U} \to \mathbf{A}$ *respects f at $\vec{x}$* if

(4D-1) $$\vec{x} \in U^n \;\&\; f(\vec{x}) \in U_s \;\&\; \pi(f(\vec{x})) = f(\pi(\vec{x})).$$

The definition takes a particularly simple form when $R : A^n \to \mathbb{B}$ is a total relation, Problem x4D.1: $\pi : \mathbf{U} \to \mathbf{A}$ *respects R at $\vec{x}$ if*

(4D-2) $$\vec{x} \in U^n \;\&\; \Big(R(\vec{x}) \iff R(\pi(\vec{x}))\Big).$$

Next come *forcing* and *certification*, the two basic notions of this chapter:

$$\begin{aligned}
\mathbf{U} \Vdash^{\mathbf{A}} f(\vec{x}) = w &\iff f(\vec{x}) = w \\
&\qquad \& \text{ every homomorphism } \pi : \mathbf{U} \to \mathbf{A} \text{ respects } f \text{ at } \vec{x}, \\
\mathbf{U} \Vdash_c^{\mathbf{A}} f(\vec{x}) = w &\iff \mathbf{U} \text{ is finite, generated by } \vec{x} \;\&\; \mathbf{U} \Vdash^{\mathbf{A}} f(\vec{x}) = w, \\
\mathbf{U} \Vdash_c^{\mathbf{A}} f(\vec{x})\downarrow &\iff (\exists w)[\mathbf{U} \Vdash_c^{\mathbf{A}} f(\vec{x}) = w].
\end{aligned}$$

If $\mathbf{U} \Vdash_c^{\mathbf{A}} f(\vec{x})\downarrow$, we call $(\mathbf{U}, \vec{x})$ a *certificate* for f at $\vec{x}$ in $\mathbf{A}$,

$$\mathbf{U} \text{ certifies } f \text{ at } \vec{x} \text{ in } \mathbf{A} \iff \mathbf{U} \Vdash_c^{\mathbf{A}} f(\vec{x})\downarrow .$$

We will sometimes write $\Vdash$ and $\Vdash_c$ for $\Vdash^{\mathbf{A}}$ and $\Vdash_c^{\mathbf{A}}$ when the relevant Φ-structure is clear from the context and also

$$\mathbf{U} \Vdash R(\vec{x}) \text{ for } \mathbf{U} \Vdash R(\vec{x}) = \mathrm{tt}, \quad U \Vdash \neg R(\vec{x}) \text{ for } \mathbf{U} \Vdash R(\vec{x}) = \mathrm{ff}$$

when $R : U^n \rightharpoonup \mathbb{B}$ is a partial relation.

Example: the Euclidean algorithm. To illustrate the notions, consider once more the Euclidean algorithm for coprimeness, specified by the extended recursive program

$$\varepsilon(\mathsf{x}, \mathsf{y}) \equiv (\mathsf{x} \neq 0 \;\&\; \mathsf{y} \neq 0 \;\&\; \mathrm{gcd}(\mathsf{x}, \mathsf{y}) = 1) \text{ where}$$
$$\{\mathrm{gcd}(\mathsf{x}, \mathsf{y}) = \text{if } (\mathsf{y} = 0) \text{ then } \mathsf{x} \text{ else } \mathrm{gcd}(\mathsf{y}, \mathrm{rem}(\mathsf{x}, \mathsf{y}))\}$$

on the structure $\mathbf{N}_\varepsilon$. Given $x, y \geq 1$, the Euclidean computes $\gcd(x, y)$ by successive divisions and 0-tests (calls to the rem- and eq_0-oracles)

$$\mathrm{rem}(x, y) = r_1, r_1 \neq 0, \mathrm{rem}(y, r_1) = r_2, r_2 \neq 0,$$
$$\dots, r_{n+1} \neq 0, \mathrm{rem}(r_n, r_{n+1}) = r_{n+2}, r_{n+2} = 0$$

until the remainder 0 is obtained, at which time it is known that $\gcd(x, y) = r_{n+1}$; and to decide if $x \perp\!\!\!\perp y$, it must then do one last check to test whether $r_{n+1} = 1$. Suppose $x \perp\!\!\!\perp y$ and collect all these calls into a substructure $\mathbf{U}_0$, writing $u \neq 0, u = 1$ for $\mathrm{eq}_0(u) = \mathrm{ff}, \mathrm{eq}_1(u) = \mathrm{tt}$ as above:

$$\mathrm{eqdiag}(\mathbf{U}_0) = \{\mathrm{rem}(x, y) = r_1, r_1 \neq 0, \mathrm{rem}(y, r_1) = r_2, r_2 \neq 0,$$
$$\dots, r_{n+1} \neq 0, \mathrm{rem}(r_n, r_{n+1}) = r_{n+2}, r_{n+2} = 0, r_{n+1} = 1\};$$

it is now easy to check that

$$\mathbf{U}_0 \Vdash_c^{\mathbf{N}_\varepsilon} x \perp\!\!\!\perp y,$$

because if $\pi : \mathbf{U}_0 \to \mathbf{N}_\varepsilon$ is a homomorphism, then

$$\text{(4D-3)} \quad \mathrm{rem}(\pi(x), \pi(y)) = \pi(r_1), \pi(r_1) \neq 0,$$
$$\mathrm{rem}(\pi(y), \pi(r_1)) = \pi(r_2), \pi(r_2) \neq 0,$$
$$\dots, \pi(r_{n+1}) \neq 0, \mathrm{rem}(\pi(r_n), \pi(r_{n+1})) = \pi(r_{n+2}), \pi(r_{n+2}) = 0,$$
$$\pi(r_{n+1}) = 1,$$

and this in turn guarantees that $\pi(x) \perp\!\!\!\perp \pi(y)$, so that π respects the coprimeness relation at x, y. This is how certificates for functions and relations can be constructed from computations, and it is the basic method of applying uniform process theory to the derivation of lower bounds for concrete algorithms, cf. Problem 2E.1.

On the other hand, $\mathbf{U}_0$ is not a *minimal* substructure of $\mathbf{N}_\varepsilon$ which certifies that $x \perp\!\!\!\perp y$. Let

$$\text{(4D-4)} \quad \mathbf{U}_1 = \{\mathrm{rem}(x, y) = r_1, \mathrm{rem}(y, r_1) = r_2,$$
$$\dots, \mathrm{rem}(r_n, r_{n+1}) = r_{n+2}, r_{n+1} = 1\},$$

be the substructure of $\mathbf{U}_0$ with all the 0-tests deleted. We claim that $\mathbf{U}_1$ is also a certificate for $x \perp\!\!\!\perp y$, and to see this suppose that $\pi : \mathbf{U}_1 \to \mathbf{N}_\varepsilon$ is a homomorphism. To verify that π respects $x \perp\!\!\!\perp y$, check first that for

$i = 1, \dots, n+1$, $\pi(r_i) \neq 0$; otherwise $\mathrm{rem}(\pi(r_{i-1}), \pi(r_i))$ would not be defined (with $r_0 = y$), since rem requires its second argument to be non-zero, and so π would not be totally defined in $\mathbf{U}_1$. So the homomorphism property for π guarantees that

$$\mathrm{rem}(\pi(x), \pi(y)) = \pi(r_1), \pi(r_1) \neq 0, \mathrm{rem}(\pi(y), \pi(r_1)) = \pi(r_2), \pi(r_2) \neq 0, \\ \dots, \pi(r_{n+1}) \neq 0, \mathrm{rem}(\pi(r_n), \pi(r_{n+1})) = \pi(r_{n+2}), \pi(r_{n+1}) = 1.$$

The last two of these equations mean that for some q,

$$\pi(r_n) = q \cdot 1 + \pi(r_{n+2}), \quad 0 \leq \pi(r_{n+2}) < 1$$

so that we must have $\pi(r_{n+2}) = 0$; and then all the equations in (4D-3) hold and we have the required $\pi(x) \perp\!\!\!\perp \pi(y)$.

This is typical: although computations by concrete algorithms define certificates, they generally do not give minimal certificates.

The connection with Pratt certificates for primality. To the best of my knowledge, *certificates* were first introduced in Pratt [1975] in his proof that primality is NP and they have been used extensively since then, primarily to design nondeterministic algorithms for primality. We are using the term in a related but different way.

With our terminology, a *Pratt certificate* for primality in a structure $\mathbf{A} = (\mathbb{N}, \mathbf{\Phi})$ with universe $\mathbb{N}$ is a nondeterministic program $E(\mathsf{p})$ such that

$$\text{(4D-5)} \qquad \mathrm{den}(\mathbf{A}, E(p))\downarrow \iff \mathrm{Prime}(p) \quad (p \in \mathbb{N});$$

and then each convergent computation $\mathfrak{c}$ of $E(\mathsf{p})$ defines by Proposition 2E.1 a certificate $(\mathbf{U}_\mathfrak{c}, p)$ in our sense. **However**: $(\mathbf{U}_\mathfrak{c}, p)$ depends only on $\mathfrak{c}$ and p, and it is possible for two different Pratt certificates $E_1(\mathsf{p})$ and $E_2(\mathsf{p})$ to determine the same certificate $(\mathbf{U}_\mathfrak{c}, p)$; and there is no requirement that (4D-5) should hold for any p' other than the one used to define $(\mathbf{U}_\mathfrak{c}, p)$, unless $p' = \pi(p)$ for a homomorphism $\pi : \mathbf{U} \to \mathbf{A}$.

What may be the most substantial difference between these two notions is how they are used: a Pratt certificate is as good as the Turing complexity of verifying (4D-5), which should be polynomial in $\log p$ and *as small as possible*, while the specific choice of the primitives $\mathbf{\Phi}$ does not matter; our certificates depend essentially on the choice of primitives in $\mathbf{\Phi}$ and they should be *as large as possible* under various complexity measures, since they will be used to derive intrinsic (absolute) lower bounds for primality (in this case).

Problems for Section 4D

x4D.1. **Problem.** Prove that if $R : A^n \to \mathbb{B}$ is a total relation on A, $\mathbf{U} \subseteq_p \mathbf{A}$ and $\pi : \mathbf{U} \to \mathbf{A}$ is a homomorphism, then

$$\pi \text{ respects } R \text{ at } \vec{x} \iff \vec{x} \in U^n \ \& \ \Big(R(\vec{x}) \iff R(\pi(\vec{x}))\Big).$$

x4D.2. **Problem.** Prove that if $P(\vec{x})$ and $Q(\vec{x})$ are n-ary partial relations on A, then

$$\mathbf{U} \Vdash_c^{\mathbf{A}} P(\vec{x}) \ \& \ Q(\vec{x}) \iff \mathbf{U} \Vdash_c^{\mathbf{A}} P(\vec{x}) \ \& \ \mathbf{U} \Vdash_c Q(\vec{x}),$$

and similarly for $\neg P(\vec{x}), \neg Q(\vec{x})$.

x4D.3. **Problem.** Prove that if $\mathbf{U} \subseteq_p \mathbf{V} \subseteq_p \mathbf{A}$, then

$$\mathbf{U} \Vdash^{\mathbf{A}} f(\vec{x}) = w \Longrightarrow \mathbf{V} \Vdash^{\mathbf{A}} f(\vec{x}) = w.$$

x4D.4. **Problem.** Prove that for any coprime $x \geq y \geq 1$, the structure $\mathbf{U}_1$ defined in (4D-4) is a minimal certificate of $x \perp\!\!\!\perp y$ in $\mathbf{N}_\varepsilon$, i.e., no proper substructure of $\mathbf{U}_1$ certifies $x \perp\!\!\!\perp y$. HINT: For example, if we delete the last equation $r_{n+1} = 1$ from eqdiag($\mathbf{U}_1$), then the function $\pi(u) = 2u$ defines a homomorphism on the resulting substructure such that $\gcd(\pi(x), \pi(y)) = 2$.

4E. Intrinsic complexities of functions and relations

If μ is a substructure norm on a Φ-structure $\mathbf{A}$ and $f : A^n \rightharpoonup A_s$, set

$$\mu(\mathbf{A}, f, \vec{x}) = \min\{\mu(\mathbf{U}, \vec{x}) : \mathbf{U} \Vdash_c^{\mathbf{A}} f(\vec{x})\!\downarrow\} \qquad (f(\vec{x})\!\downarrow), \tag{4E-1}$$

where, as usual, $\min(\emptyset) = \infty$. This is the *intrinsic μ-complexity* (of f, in $\mathbf{A}$, at $\vec{x}$). It records the μ-smallest size of a substructure of $\mathbf{A}$ that is needed to determine the value $f(\vec{x})$, and its significance derives from the following, trivial

4E.1. **Proposition.** *If a uniform process α defines $f : A^n \rightharpoonup A_s$ in a Φ-structure $\mathbf{A}$, then for any substructure norm μ on $\mathbf{A}$,*

$$\mu(\mathbf{A}, f, \vec{x}) \leq \mu(\alpha, \vec{x}) \qquad (f(\vec{x})\!\downarrow). \tag{4E-2}$$

PROOF is immediate, because

$$\mathbf{U} \models_c \alpha(\vec{x}) = w \Longrightarrow \mathbf{U} \Vdash_c^{\mathbf{A}} f(\vec{x}) = w, \tag{4E-3}$$

directly from Axioms **II** and **III** for uniform processes and the definition of certification. ⊣

4E.2. **Corollary.** *If $f : A^n \rightharpoonup A_s$ is computed by a nondeterministic extended program $E(\vec{x})$ in a Φ-structure* **A**, *then*

$$\text{depth}(\mathbf{A}, f, \vec{x}) \le c^p(\mathbf{A}, E(\vec{x})) \quad (f(\vec{x})\downarrow),$$

and for every $\Phi_0 \subseteq \Phi$,

$$\text{calls}(\Phi_0)(\mathbf{A}, f, \vec{x}) \le c^s(\Phi_0)(\mathbf{A}, E(\vec{x})) \quad (f(\vec{x})\downarrow).$$

PROOF is immediate from Problems x4C.6, x4C.4 and the Proposition. ⊣

The key point here is that $\mu(\mathbf{A}, f, \vec{x})$ is defined directly from **A**, μ and f, but it provides a lower bound for the μ-complexity of any (deterministic or nondeterministic) algorithm which might compute f in **A**; provably for algorithms specified by nondeterministic recursive programs by Proposition 4A.1 and Problem 4A.1 (and hence for all the computation models we discussed in Section 2F), and plausibly for all algorithms from the primitives of **A** by the Uniformity Thesis on page 132. The situation is most interesting, of course, when $\mu(\mathbf{A}, f, \vec{x})$ matches the corresponding complexity of some known algorithm which computes f in **A**, at least up to a multiplicative constant.

Moreover, the definitions yield a purely algebraic method for deriving these intrinsic lower bounds:

4E.3. **Lemma** (Homomorphism Test). ***If*** *μ is a substructure norm on a Φ-structure* **A**, *$f : A^n \rightharpoonup A_s$, $f(\vec{x})\downarrow$, and*

(4E-4) *for every certificate* $(\mathbf{U}, \vec{x})$ *of* **A**,

$$\Big(f(\vec{x}) \in U_s \;\&\; \mu(\mathbf{U}, \vec{x}) < m\Big) \Longrightarrow (\exists \pi : \mathbf{U} \to \mathbf{A})[f(\pi(\vec{x})) \ne \pi(f(\vec{x}))],$$

then $\mu(\mathbf{A}, f, \vec{x}) \ge m$.

Other than the definitions, this is the main—in fact the only—tool we will use in the sequel to derive intrinsic lower bounds from specified primitives. The results will be (primarily) about the most important intrinsic complexity measures, when $\mu(\mathbf{U}, \vec{x})$ is $\text{depth}(\mathbf{U}, \vec{x})$, $\text{values}(\mathbf{U})$ or $\text{calls}(\Phi_0)(\mathbf{U}, \vec{x})$. We will use two notations for them:

$$\begin{aligned} \text{depth}_f(\mathbf{A}, \vec{x}) &= \text{depth}(\mathbf{A}, f, \vec{x}) &&= \min\{\text{depth}(\mathbf{U}, \vec{x}) : \mathbf{U} \Vdash_c^{\mathbf{A}} f(\vec{x})\downarrow\}, \\ \text{values}_f(\mathbf{A}, \vec{x}) &= \text{values}(\mathbf{A}, f, \vec{x}) &&= \min\{\text{values}(\mathbf{U}) : \mathbf{U} \Vdash_c^{\mathbf{A}} f(\vec{x})\downarrow\}, \\ \text{calls}_f(\Phi_0)(\mathbf{A}, \vec{x}) &= \text{calls}(\Phi_0)(\mathbf{A}, f, \vec{x}) &&= \min\{\text{calls}(\mathbf{U} \restriction \Phi_0) : \mathbf{U} \Vdash_c^{\mathbf{A}} f(\vec{x})\downarrow\}. \end{aligned}$$

As usually, $\text{calls}_f(\mathbf{A}, \vec{x}) = \text{calls}_f(\Phi)(\mathbf{A}, \vec{x})$, so that by(1D-16), as in the proof of Lemma 4C.1,

(4E-5) $$\text{depth}_f(\mathbf{A}, \vec{x}) \le \text{values}_f(\mathbf{A}, \vec{x}) \le \text{calls}_f(\mathbf{A}, \vec{x}).$$

The output complexities. Note that for any $f : A^n \rightharpoonup A_s$ and $\vec{x} \in A^n$,

$$(4E\text{-}6) \quad \text{depth}(f(\vec{x}); \mathbf{A}, \vec{x}) = \min\{m : f(\vec{x}) \in \mathbf{G}_m(\mathbf{A}, \vec{x})\} \le \text{depth}_f(\mathbf{A}, \vec{x}),$$

because if $\mathbf{U} \subseteq_p \mathbf{A}$ is chosen so that $\text{depth}(\mathbf{A}, f, \vec{x}) = \text{depth}(\mathbf{U}, \vec{x})$, then $\mathbf{U} \Vdash_c^{\mathbf{A}} f(\vec{x})\downarrow$, so $f(\vec{x}) \in U$ and

$$\text{depth}(f(\vec{x}); \mathbf{A}, \vec{x}) \le \text{depth}(f(\vec{x}); \mathbf{U}, \vec{x}) \le \text{depth}(\mathbf{U}, \vec{x}) = \text{depth}_f(\mathbf{A}, \vec{x}).$$

Similarly for the number of calls to the primitives needed to construct $f(\vec{x})$,

$$(4E\text{-}7) \quad \text{calls}(w; \mathbf{A}, \vec{x}) = \min\{|\text{eqdiag}(\mathbf{U}, \vec{x})| : \mathbf{U} \subseteq_p \mathbf{A}, w \in U\} \\ \le \text{calls}(\mathbf{A}, f, \vec{x}),$$

cf. Problem x4E.11.

These *output complexities* provide lower bounds for any reasonable notion of algorithmic complexity measure which counts (among other things) the applications of primitives that must be executed (in sequence or altogether), simply because an algorithm must (at least) construct from the input the value $f(\vec{x})$. This is well understood and used extensively to derive lower bounds in arithmetic and algebra which are clearly absolute.[34] We will consider some results of this type in Section 5A and also in Chapter 9. The more sophisticated complexity measure $\text{depth}_f(\mathbf{A}, \vec{x})$ is especially useful when f takes simple values, e.g., when f is a relation: in this case $\text{calls}(f(\vec{x}); \mathbf{A}, \vec{x}) = 0$ and (4E-6), (4E-7) do not give any information.

Explicit (term) reduction and equivalence. Intrinsic complexities are very fine measures of information. Sometimes, especially in algebra, we want to know the exact value $\mu(\mathbf{A}, f, \vec{x})$, which might be the degree of a polynomial or the dimension of a space. In other cases, especially in arithmetic, we may try to computes $\mu(\mathbf{A}, f, \vec{x})$ up to a factor (a multiplicative constant), either because the exact value is too difficult to compute or because we only care for asymptotic estimates of computational complexity. The next, simple proposition gives a trivial way to relate the standard complexity measures of two structures when the primitives of one are explicitly definable in the other.

A structure $\mathbf{A} = (A, \mathbf{\Phi})$ is *explicitly reducible* to a structure $\mathbf{A}' = (A, \mathbf{\Psi})$ on the same universe if $\mathbf{\Phi} \subseteq \mathbf{Expl}(\mathbf{A}')$, and *explicitly equivalent* to $\mathbf{A}'$ if in addition $\mathbf{\Psi} \subseteq \mathbf{Expl}(\mathbf{A})$.

[34]The most interesting result of this type that I know is Theorem 4.1 in van den Dries and Moschovakis [2009], an $O\left(\sqrt{\log\log a}\right)$-lower bound on $\text{depth}(\text{gcd}(a+1, b), \mathbf{A}, a, b)$ with $\mathbf{A} = (\mathbb{N}, 0, 1, +, -, \cdot, \div)$ and (a, b) a *Pell pair*, cf. page 177. This is due to van den Dries, and it is the largest lower bound known for the gcd on infinitely many inputs from primitives that include multiplication. It is not known whether it holds for coprimeness, for which the best known result is a log log log-lower bound for algebraic decision trees in Mansour, Schieber, and Tiwari [1991a].

4E.4. **Proposition** (Explicit reduction). *If* $\mathbf{A} = (A, \mathbf{\Phi})$ *is explicitly reducible to* $\mathbf{A}' = (A, \mathbf{\Psi})$, *then there is constant* $K \in \mathbb{N}$ *such that for every* $f : A^n \rightharpoonup A_s$ *and each of the standard complexities* $\mu =$ depth, values, calls,

$$\mu(\mathbf{A}', f, \vec{x}) \leq K\, \mu(\mathbf{A}, f, \vec{x}) \quad (f(\vec{x})\downarrow).$$

It follows that if $\mathbf{A}$ *and* $\mathbf{A}'$ *are explicitly equivalent, then for suitable rational constants* $K, r > 0$, *every* $f : A^n \rightharpoonup A_s$ *and* $\mu =$ depth, values, calls,

$$r\, \mu(\mathbf{A}, f, \vec{x}) \leq \mu(\mathbf{A}', f, \vec{x}) \leq K\, \mu(\mathbf{A}, f, \vec{x}) \quad (f(\vec{x})\downarrow).$$

PROOF is fairly simple and we will leave it for Problem x4E.10*. ⊣

Problems for Section 4E

The first problem says, in effect, that for any substructure norm, $\mu(\mathbf{A}, f, \vec{x})$ is determined by the value $f(\vec{x})$ and the isomorphism type of the substructure $\mathbf{G}_\infty(\mathbf{A}, \vec{x})$ of $\mathbf{A}$ generated by $\vec{x}$.

x4E.1*. **Problem.** Suppose that $\mathbf{A}$ is a Φ-structure, $\vec{x}, \vec{y} \in A^n$, and there is an isomorphism

$$\pi : \mathbf{G}_\infty(\mathbf{A}, \vec{x}) \rightarrowtail\!\!\!\!\twoheadrightarrow \mathbf{G}_\infty(\mathbf{A}, \vec{y})$$

such that $\pi(\vec{x}) = \vec{y}$. Prove that for every substructure norm μ on $\mathbf{A}$ and every $f : A^n \rightharpoonup A_s$,

$$\mu(\mathbf{A}, f, \vec{x}) = \mu(\mathbf{A}, f, \vec{y}).$$

HINT: Use the fact that substructure norms are respected by isomorphisms and that the hypothesis is symmetric in $\vec{x}$ and $\vec{y}$.

x4E.2. **Problem.** Prove that every Φ-structure $\mathbf{A}$ and $f : A^n \rightharpoonup A_s$,

$$\text{depth}(\mathbf{A}, f, \vec{x}) \leq \text{calls}(\mathbf{A}, f, \vec{x}) \quad (f(\vec{x})\downarrow).$$

x4E.3. **Problem** (Obstruction to calls$(\mathbf{A}, R, \vec{x}) = 0$). Suppose $R(\vec{x})$ is a (total) n-ary relation on the universe of a Φ-structure $\mathbf{A}$ and for some tuple $\vec{x} \in A^n$ of distinct elements of A, calls$(\mathbf{A}, R, \vec{x}) = 0$. Prove that R is constant, i.e.,

$$\text{either } (\forall \vec{y})R(\vec{y}) \text{ or } (\forall \vec{y})\neg R(\vec{y}).$$

This allows us in most cases to see immediately that calls$_f(\mathbf{A}, \vec{x}) > 0$: for example, calls$(\mathbf{A}, \text{Prime}, x) > 0$ for every $\mathbf{A}$ with $A = \mathbb{N}$ and for every x, simply because there are primes and composites. There is no equally simple, general test for depth$(\mathbf{A}, f, \vec{x}) > 0$, but it is worth putting down some sufficient conditions which are easy to verify in special cases—we will need them in Chapters 5 – 8 when we want to divide by depth$(\mathbf{A}, f, \vec{x})$.

x4E.4. **Problem** (Obstruction to depth$(\mathbf{A}, R, \vec{x}) = 0$). Suppose $R(\vec{x})$ is a total n-ary relation on the universe of a Φ-structure $\mathbf{A}$, $\vec{x} = (x_1, \dots, x_n)$ and $\vec{y} = (y_1, \dots, y_n \in A^n$ are tuples of distinct elements of A, and the following conditions hold:

(1) The map $x_i \mapsto y_i$ respects all the primitives of $\mathbf{A}$ of sort `boole`.
(2) No x_i is the value of some primitive of $\mathbf{A}$ on members of $\{x_1, \dots, x_n\}$.
(3) $R(\vec{x})$ is true, but $\neg R(\vec{y})$.

Prove that depth$(\mathbf{A}, R, \vec{x}) > 0$. Infer that

$$\text{depth}(\mathbf{A}, \text{Prime}, p) > 0 \text{ and depth}(\mathbf{A}, \perp\!\!\!\perp, x, y) > 0 \quad (\text{Prime}(p), x \perp\!\!\!\perp y),$$

when $\mathbf{A}$ is one of $\mathbf{N}_u, \mathbf{N}_b, \mathbf{N}$ or their expansions by $=$ and/or $<$.

x4E.5. **Problem.** Prove that for the coprimeness relation,

$$\text{depth}(\mathbf{N}_\varepsilon, \perp\!\!\!\perp, x, y) \le 2\log(\min(x, y)) + 1 \quad (x, y \ge 1).$$

x4E.6*. **Problem** (Upper bound for poly 0-testing). Let

$$N_F(a_0, \dots, a_n, x) \iff a_0 + a_1 x + \cdots + a_n x^n = 0,$$

be the nullity relation (of degree $n \ge 1$) on a field $\mathbf{F} = (F, 0, 1, +, -, \cdot, \div, =)$ and prove the following inequalities:

$$\begin{aligned} \text{calls}(\cdot, \div)(\mathbf{F}, N_F, \vec{a}, x) &\le n, \\ \text{calls}(\cdot, \div, =)(\mathbf{F}, N_F, \vec{a}, x) &\le n + 1, \\ \text{calls}(+, -)(\mathbf{F}, N_F, \vec{a}, x) &\le n - 1, \\ \text{calls}(+, -, =)(\mathbf{F}, N_F, \vec{a}, x) &\le n + 1. \end{aligned}$$

HINT: Use Horner's Rule and Problem x1C.17* (for the last two claims).

Recall from page 9 that for $f : A^n \rightharpoonup A$

$$\text{Graph}_f(\vec{x}, w) \iff f(\vec{x}) = w.$$

x4E.7*. **Problem.** Suppose that $\mathbf{A}$ is a Φ-structure and $f : A^n \rightharpoonup A$.

(1) Prove that for any $\mathbf{U} \subseteq_p \mathbf{A}$ and all $\vec{x}, w$,

$$\mathbf{U} \Vdash f(\vec{x}) = w \iff \mathbf{U} \Vdash \text{Graph}_f(\vec{x}, w) = \mathrm{tt}.$$

(2) Prove that for any $\mathbf{U} \subseteq_p \mathbf{A}$,

$$\mathbf{U} \Vdash_c f(\vec{x}) = w \Longrightarrow \mathbf{U} \Vdash_c \text{Graph}_f(\vec{x}, w) = \mathrm{tt},$$

but the converse is not always true.

(3) Infer that for any substructure norm μ,

$$\mu(\mathbf{A}, \text{Graph}_f, \vec{x}, f(\vec{x})) \le \mu(\mathbf{A}, f, \vec{x}).$$

(4) Give an example where

$$\text{depth}(\mathbf{A}, \text{Graph}_f, x, w) < \text{depth}(\mathbf{A}, f, x) < \infty.$$

HINT: For the counterexamples in (2) use a structure where Graph_f is among the primitives; and for (4), you need a specific such example which keeps $\mu(\mathbf{A}, f, \vec{x})$ finite.

x4E.8. **Problem.** Prove that for the coprimeness relation, some K and all $t \geq 3$,

$$\text{depth}(\mathbf{N}_\varepsilon, \perp\!\!\!\perp, F_{t+1}, F_t) \leq K(\log t) = O(\log\log F_t),$$

where $F_0, F_1, \dots$ is the Fibonacci sequence. HINT: Use Pratt's algorithm.

x4E.9*. **Problem.** Prove that if $f : A^n \rightharpoonup A_s$ is explicit in a Φ-structure $\mathbf{A}$ and μ is any substructure norm on $\mathbf{A}$, then there is a number K such that

$$\mu(\mathbf{A}, f, \vec{x}) \leq K \quad (f(\vec{x})\downarrow).$$

HINT: Use Proposition x1E.14 to prove that for explicit f, there is an m such that

$$f(\vec{x}) = w \Longrightarrow (\exists \mathbf{U})[\mathbf{U} \models_c f(\vec{x}) = w \;\&\; |\text{eqdiag}(\mathbf{U})| \leq m],$$

and then appeal to Problem x4C.12*.

x4E.10*. **Problem.** Prove Proposition 4E.4. HINT: Start by appealing to Problem x4E.9* to set

$$K_\phi = \max\{\mu(\mathbf{A}', \phi^{\mathbf{A}}, \vec{x}) : \phi \in \Phi\}.$$

x4E.11. **Problem.** Prove (4E-7).

x4E.12. **Open problem** (vague). For a (total) structure $\mathbf{A}$ and a function $f : A^n \to A_s$, do any of the complexity functions $\mu(\mathbf{A}, f, \vec{x})$ encode interesting model theoretic properties of $\mathbf{A}$? Perhaps when μ is one of depth, values or calls and $f(\vec{x})$ codes an invariant of some substructure of $\mathbf{A}$ determined by $\vec{x}$—a degree, dimension, etc.

4F. The best uniform process

Is there a "best algorithm" which computes a given $f : A^n \rightharpoonup A_s$ from specified primitives on A? The question is vague, of course—and the answer is almost certainly negative in the general case, no matter how you make it precise. The corresponding question about uniform processes has a positive (and very simple) answer.

For given $f : A^n \rightharpoonup A_s$, set

$$\text{(4F-1)} \qquad \overline{\beta}^{\mathbf{U}}_{f,\mathbf{A}}(\vec{x}) = w \iff \mathbf{U} \Vdash^{\mathbf{A}} f(\vec{x}) = w \quad (\mathbf{U} \subseteq_p \mathbf{A}).$$

4F.1. **Theorem.** *The following are equivalent for any Φ-structure $\mathbf{A}$ and any partial function $f : A^n \rightharpoonup A_s$, $s \in \{\texttt{ind}, \texttt{boole}\}$.*

(i) *Some uniform process α of $\mathbf{A}$ defines f.*

(ii) $(\forall \vec{x})\Big(f(\vec{x})\downarrow \implies (\exists \mathbf{U} \subseteq_p \mathbf{A})[\mathbf{U} \Vdash_c^{\mathbf{A}} f(\vec{x})\downarrow]\Big)$.

(iii) $\boldsymbol{\beta}_{f,\mathbf{A}}$ *is a uniform process of* **A** *which defines* f.

Moreover, if these conditions hold, then

$$\mu(\boldsymbol{\beta}_{f,\mathbf{A}}, \vec{x}) = \mu(\mathbf{A}, f, \vec{x}) \leq \mu(\alpha, \vec{x}) \qquad (f(\vec{x})\downarrow),$$

for any substructure norm μ *and uniform process* α *of* **A** *which defines* f.

PROOF. (iii) $\implies$ (i) is immediate and (i) $\implies$ (ii) follows from (4E-3).

(ii) $\implies$ (iii). The operation $(\mathbf{U} \mapsto \overline{\boldsymbol{\beta}}^{\mathbf{U}}_{f,\mathbf{A}})$ satisfies the Finiteness Axiom **III** by (ii). To verify the Homomorphism Axiom **II**, suppose

$$\mathbf{U} \Vdash^{\mathbf{A}} f(\vec{x}) = w \;\&\; \pi : \mathbf{U} \to \mathbf{V}$$

so that $\pi(\vec{x}) \in V^n$, $\pi(w) \in V_s$ and (since $\pi : \mathbf{U} \to \mathbf{V}$ is a homomorphism), $f(\pi(\vec{x})) = \pi(w)$. Let $\rho : \mathbf{V} \to \mathbf{A}$ be a homomorphism. The composition $\rho \circ \pi : \mathbf{U} \rightarrowtail \mathbf{A}$ is also a homomorphism, and so by (ii) it respects f at $\vec{x}$, i.e.,

$$f(\rho(\pi(\vec{x})) = \rho(\pi(f(\vec{x})) = \rho(\pi(w)) = \rho(f(\pi(\vec{x}))).$$

So ρ respects f at $\pi(\vec{x})$, and since it is arbitrary, we have the required

$$\mathbf{V} \Vdash^{\mathbf{A}} f(\pi(\vec{x})) = \pi(w).$$

The second claim follows from the definition of $\boldsymbol{\beta}_{f,\mathbf{A}}$ and (4E-3). ⊣

Optimality and weak optimality. The best uniform process $\overline{\boldsymbol{\beta}}_{f,\mathbf{A}}$ is clearly optimal (for f, from the primitives of **A**) by any natural measure of efficiency, but it is not reasonable to expect that it is induced by a nondeterministic program for any specific, interesting f—much less a deterministic one. We formulate here three, natural notions of (partial, worst-case) optimality which will help understand and discuss the intrinsic lower bounds we will derive in the sequel.

Suppose **A** is an infinite Φ-structure, $f : A^n \rightharpoonup A_s$ and μ is a substructure norm on **A**.

(1) A uniform process α on **A** is *μ-optimal for f on $D \subseteq A^n$*, if

(4F-2) $$\vec{x} \in D \Longrightarrow \Big(f(\vec{x}) = \overline{\alpha}(\vec{x})\downarrow \;\&\; \mu(\alpha, \vec{x}) = \mu(\mathbf{A}, f, \vec{x})\Big).$$

The optimality results about Horner's rule in Chapter 9 are of this sort, typically with D the set of *generic* (algebraically independent) inputs.

The next weaker notion allows a multiplicative factor:

(2) A uniform process α on **A** is *weakly μ-optimal for f on $D \subseteq A^n$*, if for some $K > 0$,

(4F-3) $$\vec{x} \in D \Longrightarrow \Big(f(\vec{x}) = \overline{\alpha}(\vec{x})\downarrow \;\&\; \mu(\alpha, \vec{x}) \leq K\mu(\mathbf{A}, f, \vec{x})\Big).$$

Almost all the results for arithmetical problems in Chapters 5 – 8 establish the weak optimality of simple programs for various norms—mostly depth.

For some of these results, the set D is not of any particular interest and we abstract it away:

(3) A uniform process α on $\mathbf{A}$ is *weakly μ-optimal for f* if it is weakly-μ-optimal on some $D \subseteq A^n$ such that

$$\text{(4F-4)} \qquad \sup\{\mu(\mathbf{A}, f, \vec{x}) : \vec{x} \in D\} = \infty.$$

One precise version of the Main Conjecture on page 2 claims something stronger than the weak calls(rem)-optimality of (the process) ε (induced by) the Euclidean in $\mathbf{N}_\varepsilon$: it asks for a set $D \subseteq \mathbb{N}^2$ and a $K > 0$ such that

$$\text{(4F-5)} \quad (x, y) \in D \Longrightarrow \Big(\text{calls(rem)}(\varepsilon, x, y) \leq L(x, y) \leq K\text{calls(rem)}(\mathbf{A}, \text{gcd}, x, y)\Big)$$

where $L(x, y) = 2\log(\max(x, y))$; which then satisfies (4F-3) and (4F-4) by $(*)$ on page 2. Most of the results in Chapters 5 – 8 establish the weak depth-optimality of some recursive algorithms which compute various arithmetical functions and relations from various primitives in this enhanced form, proving that

$$\vec{x} \in \text{den} \Longrightarrow \Big(\mu(\alpha, \vec{x}) \leq KL(\vec{x}) \leq \mu(\mathbf{A}, f, \vec{x})\Big) \quad (K > 0)$$

with some interesting $L(\vec{x})$— $\log(x)$, $\log\log(x)$, $\sqrt{\log\log(x)}$, etc.

Other than that, weak optimality looks like a very weak condition on processes, but it is not, cf. Problem x4F.3*.

Problems for Section 4F

x4F.1. **Problem.** Prove that if α is a uniform process of $\mathbf{A}$ which defines $f : A^n \rightharpoonup A_s$, $D \subseteq A^n$ is contained in the domain of convergence of f and for some substructure norm on $\mathbf{A}$

$$\sup\{\mu(\alpha, \vec{x}) : \vec{x} \in D\} < \infty \text{ and } \min\{\mu(\mathbf{A}, f, \vec{x}) : \vec{x} \in D\} > 0,$$

then α is weakly μ-optimal for f on D.

Infer that

(1) Every α which defines f in $\mathbf{A}$ is weakly μ-optimal on every finite subset D of the domain of convergence of f which contains some $\vec{x}$ such that $\mu(\mathbf{A}, f, \vec{x}) > 0$.

(2) If f is $\mathbf{A}$-explicit, then f is defined by some uniform process which is μ-optimal on the entire domain of convergence of f, for every substructure norm μ.

Hint: For (2), appeal to Problem x4E.9*.

x4F.2. **Problem.** Prove that if α is (the process induced by) the Euclidean algorithm in $\mathbf{N}_\varepsilon$, then

$$\text{(for infinitely many } x, y)[x > y \;\&\; \text{calls(rem)}(x, y) \leq 1].$$

(This trivial fact is why the definition of weak optimality demands (4F-4) of D.)

x4F.3*. **Problem.** Suppose α is a uniform process of $\mathbf{A}$ which defines a total function $f : A^n \rightharpoonup A_s$. Prove that if α is not weakly μ-optimal for f, then for every rational $r > 0$, there exists a set $E \subsetneq A^n$ and a number m such that

$$\text{(4F-6)} \qquad \sup\{\mu(\alpha, \vec{x}) : \vec{x} \in E\} \leq m \;\&\; (\forall \vec{x} \notin E)[\mu(\mathbf{A}, f, \vec{x}) \leq r\mu(\alpha, \vec{x})].$$

4G. Logical extensions

We introduce here a relation between structures which helps express and prove the robustness of intrinsic complexities and the complexities of recursive programs that we studied in Chapter 3.

A $(\Phi \cup \Psi)$-structure $\mathbf{B} = (B, \{\phi^{\mathbf{B}}\}_{\phi\in\Phi}, \{\psi^{\mathbf{B}}\}_{\psi\in\Psi})$ is a *logical extension*[35] of the Φ-structure $\mathbf{A} = (A, \{\phi^{\mathbf{A}}\}_{\phi\in\Phi})$ (or *logical over* $\mathbf{A}$) if

(LE1) $A \subseteq B$,

(LE2) each $\phi^{\mathbf{B}} : B^n \rightharpoonup B_s$ has the same domain of convergence as $\phi^{\mathbf{A}}$ and agrees with $\phi^{\mathbf{A}}$ on it, and

(LE3) every permutation $\pi^A : A \rightarrowtail\!\!\!\twoheadrightarrow A$ has an extension $\pi^B : B \rightarrowtail\!\!\!\twoheadrightarrow B$ such that for every "fresh" primitive $\psi^{\mathbf{B}} : B^n \rightharpoonup B_s$ and all $\vec{x} \in B^n$,

$$\text{(4G-1)} \qquad \pi^B(\psi^{\mathbf{B}}(x_1, \dots, x_n)) = \psi^{\mathbf{B}}(\pi^B(x_1), \dots \pi^B(x_n)).$$

If $B = A$, we call $\mathbf{B} = (A, \{\phi^{\mathbf{A}}\}_{\phi\in\Phi}, \{\psi^{\mathbf{B}}\}_{\psi\in\Psi})$ a *logical expansion* of $\mathbf{A}$.

For example, the expansion $(\mathbf{L}^*, \leq)$ of the Lisp structure on a set L by an ordering of L and the structure RAM($\mathbf{A}$) over a Φ-structure $\mathbf{A}$ are logical extensions of $(L, \leq)$ and $\mathbf{A}$ respectively, cf. Problems x4G.5 and x4G.9*.

For every $\mathbf{A}$, the expansion $(\mathbf{A}, =_A)$ by the identity relation on A is logical, trivially, taking $\pi^B = \pi^A$, so logical expansions may compute functions on A which are not $\mathbf{A}$-recursive; on the other hand, we will show that on the level of uniformity (rather than recursiveness), adding the identity relation on A is the only way to enlarge **Unif**($\mathbf{A}$) by adding new, logical primitives: *if* $=_A$ *is uniform in* $\mathbf{A}$, *then every* $f : A^n \rightharpoonup A_s$ *which is uniform in some logical extension of* $\mathbf{A}$ *is already uniform in* $\mathbf{A}$. The central result of this section is a complexity

[35]These were called *inessential extensions* in van den Dries and Moschovakis [2004], although the connection with the *logical notions* in Tarski [1986] was pointed out in a footnote. The definition and (a version of) Lemma 4G.2 below are due to Itay Neeman and they were used to simplify considerably both the statements and the proofs of the lower bound results about RAMs in that paper, but the notion deserves the fancier name—it has legs.

version of this fact: we will show (in part) that if $=_A$ is among the primitives of **A**, then *lower bounds for intrinsic complexities which hold in* **A** *persist in all its logical extensions*—in a precise and very general way.

First, however, we prove a very simple application of logical extensions which uses nothing but their definition:

The lower bound for comparison sorting. One of the few lower bound results that is regularly taught in undergraduate courses on algorithms is that "$\Omega(n \log n)$ *comparisons are* [on some input] *necessary for sorting n elements*" in the eloquent formulation of Dasgupta, Papadimitriou, and Vazirani [2011, 2.3], which also notes that [their] "neat argument applies only to algorithms that use comparisons". We give here a version of the usual proof which is based on a precise definition of the "comparison algorithms" for which it works in terms of logical extensions.

4G.1. **Theorem** (Lower bound for sorting). ***Suppose***

- $\leq$ *is a linear ordering of a set L which has at least* $n \geq 2$ *elements*;
- $\mathbf{B} = (B, \leq, \{\psi^{\mathbf{B}}\}_{\psi \in \Psi})$ *is a logical extension of* $(L, \leq)$ *which contains the set of strings from L,* $L^* \subseteq B$;
- $E(\mathsf{u})$ *is an extended* (deterministic) *recursive program in the vocabulary of* **B** *which computes* $\mathrm{sort}_{\leq}(u)$ *for every* $u \in L^*$; *and*
- $c^s(\leq)(u) = c^s(\leq)(\mathbf{B}, E(u))$ *is the number of* $\leq$*-calls complexity of* $E(\mathsf{u})$ *in* **B**;

then *there is some* $u \in L^*$ *of length n such that* $c^s(\leq)(u) \geq \log(n!)$.

Proof is by way of two Lemmas:

Lemma 1. For every ordering $\leq'$ *of L,* $E(\mathsf{v})$ *computes*

$$\mathrm{sort}_{\leq'} : L^n \to L^n$$

in every structure $\mathbf{B}' = (B, \leq', \{\psi^{\mathbf{B}}\}_{\psi \in \Psi})$ *derived from* **B** *by replacing* $\leq$ *by* $\leq'$.

Proof. Fix a set $X \subseteq L$ with exactly n members and let $\pi : L \rightarrowtail\!\!\!\!\twoheadrightarrow L$ be the unique permutation of L such that

$$s, t \in X \Longrightarrow s \leq t \iff \pi(s) \leq' \pi(t) \text{ and } s \notin X \Longrightarrow \pi(s) = s.$$

Since **B** is a logical extension of $(L, \leq)$, π can be extended to a permutation of B which respects the fresh primitives, and so it is an isomorphism

$$\pi : \mathbf{B} = (B, \leq \restriction X, \{\psi^{\mathbf{B}}\}_{\psi \in \Psi}) \rightarrowtail\!\!\!\!\twoheadrightarrow (B, \leq' \restriction X, \{\psi^{\mathbf{B}}\}_{\psi \in \Psi}) = \mathbf{B}'.$$

The hypothesis on the extended program $E(\mathsf{u})$ is that

$$\mathbf{B} \models E(u) = \mathrm{sort}_{\leq}(u);$$

so Problem x3A.1 implies that if $u \in X^n$, then

$$\mathbf{B}' \models E(\pi(u)) = \pi(\mathrm{sort}_{\leq}(u)) = \mathrm{sort}_{\leq'}(\pi(u));$$

and if we set in this $u = \pi^{-1}(v)$ for any $v \in X^n$, we get the required

$$\text{(4G-2)} \qquad \mathbf{B}' \models E(v) = \pi(\mathrm{sort}_{\leq}(\pi^{-1}(v)) = \mathrm{sort}_{\leq'}(v). \qquad \dashv \text{(Lemma 1)}$$

Problem x3A.1 also gives the following equation on the complexity measures,

$$\text{(4G-3)} \qquad c^s(\leq')(\mathbf{B}', E(v)) = c^s(\leq)(\mathbf{B}, E(\pi^{-1}(v))),$$

which we now use:

Lemma 2. Fix a tuple $v = (v_1, \dots, v_n) \in X^n$ *of distinct members of* X *and consider the computations of the recursive machine for* $E(\mathsf{v})$ *with respect to all orderings* $\leq$ *of* X*; there is at least one such* $\leq$ *for which*

$$c^s(\leq)(\mathbf{B}, E(v)) \geq \log(n!).$$

Proof. For each ordering $\leq$ of X, let $s_0^{\leq}, \dots, s_m^{\leq}$ be the computation of the recursive machine for the structure $\mathbf{B}$ which computes sort(v), The set of all these computations is a tree (on the states) with degree $= 2$ and $n!$ leaves, one for each permutation of X which returns $\mathrm{sort}_{\leq}(v)$. By (1A-18), $n! \leq 2^{\mathrm{spdepth}(\mathcal{T})}$, so $\mathrm{spdepth}(\mathcal{T}) \geq \log(n!)$; and so there must be one of these computations, corresponding to some $\leq$, and the splitting nodes are precisely the calls to $\leq$ which we wanted to count. $\dashv$ (Lemma 2)

If we now set $u = \pi^{-1}(v)$, then (4G-3) gives

$$c^s(\leq)(\mathbf{B}, E(u)) = c^s(\leq')(\mathbf{B}', E(v)) \geq \log(n!),$$

which completes the proof of the theorem. $\dashv$

The $\log(n!)$ lower bound does not hold for non-deterministic programs, cf. Problem x4G.6.

Embedding Test. The most important applications of logical extensions to complexity theory derive from the following basic fact:

4G.2. **Lemma** (Neeman, cf. Footnote 35). ***If*** $\mathbf{B}$ *is logical over an infinite structure* $\mathbf{A}$, $\mathbf{V} \subseteq_p \mathbf{B}$, *and*

$$\text{(4G-4)} \qquad \mathbf{U} = \mathbf{V} \restriction \Phi \restriction A \subseteq_p \mathbf{A},$$

then *for every finite* $X \subseteq U$ *and every* k,

(1) $G_k(\mathbf{V}, X) \cap A = G_k(\mathbf{U}, X)$, *and*

(2) *if* $\pi^A : A \rightarrowtail\!\!\!\!\twoheadrightarrow A$ *and* $\pi^B \rightarrowtail\!\!\!\!\twoheadrightarrow B$ *are as in* (LE3) *above and* π^A *fixes every* $s \in G_k(\mathbf{U}, X)$, *then* π^B *fixes every* $t \in G_k(\mathbf{V}, X)$.

It follows that for all $\vec{x} \in A^n$,

$$\text{(4G-5)} \qquad (\mathbf{V}, \vec{x}) \text{ is a certificate in } \mathbf{B} \Longrightarrow (\mathbf{U}, \vec{x}) \text{ is a certificate in } \mathbf{A}.$$

PROOF. We prove (1) and (2) together by induction on k, the basis being trivial since

$$G_0(\mathbf{U}, X) = G_0(\mathbf{V}, X) = X.$$

At the induction step, we assume (1) and (2) for k.

To prove (2) first at $k + 1$, suppose π^A fixes every $s \in G_{k+1}(\mathbf{U}, X)$ and $t \in G_{k+1}(\mathbf{V}, X)$. If

$$t = \phi^{\mathbf{V}}(z_1, \dots, z_n) = \phi^{\mathbf{B}}(\vec{z}) \text{ with } \phi \in \Phi \text{ and } z_1, \dots, z_n \in G_k(\mathbf{V}, X),$$

then $z_1, \dots, z_n \in A$, since $\phi^{\mathbf{B}}(\vec{z}) = \phi^{\mathbf{A}}(\vec{z})\downarrow$; so $z_1, \dots, z_n \in G_k(\mathbf{U}, X)$ by (1) of the induction hypothesis; so $t \in G_{k+1}(\mathbf{U}, X)$ and π^B fixes it by the hypothesis. If

$$t = \psi^{\mathbf{V}}(\vec{z}) = \psi^{\mathbf{B}}(\vec{z}) \text{ with } z_1, \dots, z_n \in \mathbf{G}_k(\mathbf{V}, X),$$

then using the properties of π^B and (2) for k,

$$\pi^B(t) = \pi^B(\psi^{\mathbf{B}}(\vec{z})) = \psi^{\mathbf{B}}(\pi^B(\vec{z})) = \psi^{\mathbf{B}}(\vec{z}) = t.$$

To prove (1) at $k + 1$, suppose there is some $t \in G_{k+1}(\mathbf{V}, X) \cap A$ such that $t \notin G_{k+1}(\mathbf{U}, X)$, so it must be that $t = \psi^{\mathbf{V}}(z_1, \dots, z_n)$ with a fresh primitive and $z_1, \dots, z_n \in G_k(\mathbf{V}, X)$. Choose a permutation $\pi^A : A \rightarrowtail\!\!\twoheadrightarrow A$ which fixes every member of $G_k(\mathbf{A}, X)$ but moves t—which exists since $G_k(\mathbf{A}, X)$ is finite while A is infinite; now $\pi^B(t) = \pi^A(t) \neq t$, which contradicts (2) at $k + 1$.

Proof of (4G-5). If $\mathbf{V} = \mathbf{G}_k(\mathbf{V}, \vec{x})$ and $X = \{x_1, \dots, x_n\}$, then

$$V = G_k(\mathbf{V}, \vec{x}) = G_k(\mathbf{V}, X), \quad U = G_k(\mathbf{U}, \vec{x}) = G_k(\mathbf{U}, X),$$

and the two structures $\mathbf{U} = \mathbf{G}_k(\mathbf{V}, \vec{x}) \restriction \Phi \restriction A$ and $\mathbf{G}_k(\mathbf{U}, \vec{x})$ have the same universe $G_k(\mathbf{V}, \vec{x}) \cap A = G_k(\mathbf{U}, \vec{x})$ by (1). They are both Φ-structures; and for every $\phi \in \Phi$,

$$\mathbf{V} \restriction \Phi \restriction A \models \phi(\vec{z}) = w \iff \mathbf{U} \models \phi(\vec{z}) = w,$$

because $\phi^{\mathbf{B}} = \phi^{\mathbf{A}}$, and so the restrictions of these two partial functions to the same set are identical, i.e., $\mathbf{U} = \mathbf{G}_k(\mathbf{U}, \vec{x})$ and so $(\mathbf{U}, \vec{x})$ is a certificate. ⊣

Substructure norms on logical extensions. If $\mathbf{B}$ is logical over an infinite structure $\mathbf{A}$ and μ is a substructure norm on $\mathbf{A}$, set

(4G-6) $\quad \mu^{\mathbf{B}}(\mathbf{V}, \vec{x}) = \mu(\mathbf{V} \restriction \Phi \restriction A, \vec{x}) \quad (\vec{x} \in A^n, (\mathbf{V}, \vec{x}) \text{ a certificate in } \mathbf{B}).$

This makes sense for $\vec{x} \in A^n$ by (4G-5) and specifies the only values of $\mu^{\mathbf{B}}$ that we care about, but it is quite trivial to extend it so it is a full substructure norm on $\mathbf{B}$, cf. Problem x4G.8).

For example, if $\mathbf{U} = \mathbf{V} \restriction \Phi \restriction A$, then

$$\text{depth}^{\mathbf{B}}(\mathbf{V}, \vec{x}) = \min\{k : \mathbf{U} = \mathbf{G}_k(\mathbf{U}, k)\} \leq \min\{k : \mathbf{V} = \mathbf{G}_k(\mathbf{V}, k)\}$$

by Lemma 4G.2. If

$$\boldsymbol{c} = (E_0(\vec{x}) :, \dots, : w)$$

is a (recursive machine) computation of a non-deterministic program $E(\vec{x})$ of $\mathbf{B}$ which computes $f(\vec{x}) \in A_s$ on some input $\vec{x} \in A^n$ and $(\mathbf{V}_c, \vec{x})$ is the certificate in $\mathbf{B}$ with

$$\text{eqdiag}(\mathbf{V}_c) = \{(\chi, \vec{u}, v) : \text{a transition } \vec{a}\ \chi : \vec{u}\ \vec{b} \to \vec{a} : v\ \vec{b} \text{ occurs in } \boldsymbol{c}\},$$

then $\text{calls}(\Phi_0)^{\mathbf{B}}(\mathbf{V}_c, \vec{x})$ is the number of calls in $\boldsymbol{c}$ to primitives in Φ_0.

4G.3. **Lemma** (Embedding Test). *Suppose* $\mathbf{A}$ *is an infinite* Φ*-structure,* μ *is a substructure norm on* $\mathbf{A}$*,* $f : A^n \rightharpoonup A_s$ *and* $f(\vec{x})\downarrow$*. If*

(4G-7) for every certificate $(\mathbf{U}, \vec{x})$ of $\mathbf{A}$,

$$\Big(f(\vec{x}) \in U_s \ \&\ \mu(\mathbf{U}, \vec{x}) < m\Big) \Longrightarrow (\exists \pi : \mathbf{U} \rightarrowtail \mathbf{A})[f(\pi(\vec{x})) \neq \pi(f(\vec{x}))],$$

then *for every logical extension* $\mathbf{B}$ *of* $\mathbf{A}$*,* $\mu^{\mathbf{B}}(\mathbf{B}, f, \vec{x}) \geq m$*.*

The difference between the Embedding and the Homomorphism tests is that the hypothesis of the first supplies an *embedding* $\pi : \mathbf{U} \rightarrowtail \mathbf{A}$ which does not respect f at $\vec{x}$ when $\mu(\mathbf{U}, \vec{x}) < m$, and the injectivity of π suffices to guarantee that m is a lower bound for $\mu^{\mathbf{B}}(\mathbf{B}, f, \vec{x})$ in every logical extension of $\mathbf{A}$.

PROOF. By the Homomorphism Test, to prove $\mu^{\mathbf{B}}(\mathbf{B}, f, \vec{x}) \geq m$ it is enough to show that for every certificate $(\mathbf{V}, \vec{x})$ of $\mathbf{B}$,

$$\mu^{\mathbf{B}}(\mathbf{V}, \vec{x}) < m \Longrightarrow (\exists \rho : \mathbf{V} \to \mathbf{B})[f(\rho(\vec{x})) \neq \rho(f(\vec{x}))],$$

so assume the hypothesis of this implication for $(\mathbf{V}, \vec{x})$ and let $\mathbf{U} = \mathbf{V} \restriction \Phi \restriction \mathbf{A}$. By (4G-5), $(\mathbf{U}, \vec{x})$ is a certificate of $\mathbf{A}$,

$$\mu(\mathbf{U}, \vec{x}) = \mu^{\mathbf{B}}(\mathbf{V}, \vec{x}) < m,$$

and so the hypothesis of the lemma supplies an embedding $\pi : \mathbf{U} \rightarrowtail \mathbf{A}$ which does not respect f at $\vec{x}$. Since U and $\pi[U]$ have the same (finite) cardinality, there is a permutation $\pi^A : A \rightarrowtail\!\!\!\!\twoheadrightarrow A$ which extends π; and since $\mathbf{B}$ is a logical extension of $\mathbf{A}$, there is an extension $\pi^B : B \rightarrowtail\!\!\!\!\twoheadrightarrow B$ of π^A which respects the fresh primitives. It is now easy to check that

$$\rho = \pi^B \restriction V : \mathbf{V} \rightarrowtail \mathbf{B}$$

and ρ does not respect f at $\vec{x}$ (since it agrees with π on U), which completes the argument. ⊣

The version for depth is important for complexity in arithmetic, because in the proofs in Chapters 5 – 8 we will actually construct *embeddings*, and so the lower bounds we will derive apply to all logical extensions of the relevant structures. The version for calls yields the expected lower bound for RAMS in Problem x4G.9*, and also has robustness implications for the lower bounds in algebra that we will derive in Chapter 9.

4G.4. **Corollary.** *If* $=_A$ *is uniform in* $\mathbf{A}$ *and* $f : A^n \rightharpoonup A_s$ *is uniform in some logical extension* $\mathbf{B}$ *of* $\mathbf{A}$*, then* f *is already uniform in* $\mathbf{A}$*.*

PROOF. We may assume by Problem x4B.4* that $=$ is a primitive of **A**.

Suppose **B** is a logical extension of **A**, $f : A^n \rightharpoonup A_s$ is uniform in **B** and $f(\vec{x})\downarrow$, so there is a certificate $(\mathbf{V}, \vec{x})$ of **B** such that every $\pi : \mathbf{V} \to \mathbf{B}$ respects f at $\vec{x}$. If we set $\mathbf{U} = \mathbf{V} \upharpoonright \Phi \upharpoonright A$, then $(\mathbf{U}, \vec{x})$ is a certificate in **A**, by (4G-5); and then—easily—so is $(\mathbf{U}^*, \vec{x})$ where the universe U of $\mathbf{U}^*$ is the same as the universe of **U** and

$$\mathrm{eqdiag}(\mathbf{U}^*) = \mathrm{eqdiag}(\mathbf{U}) \cup \{x \neq y : x, y \in U\}.$$

Now every homomorphism $\pi^* : \mathbf{U}^* \to \mathbf{A}$ is an embedding and it can be exended to some $\pi : \mathbf{V} \rightarrowtail \mathbf{B}$, as in the proof of the Embedding Test, so it respects f at $\vec{x}$. ⊣

Problems for Section 4G

x4G.1. **Problem.** Prove that if **B** is logical over **A** and **C** is logical over **B**, then **C** is logical over **A**.

x4G.2. **Problem.** Prove that $(\mathbf{A}, P_i^n)$ is logical over **A**, if $P_i^n(\vec{x}) = x_i$.

x4G.3. **Problem.** Prove that if

$$\mathrm{ap}_n(x_1, \dots, x_n, p) = p(x_1, \dots, x_n)$$

is the n-ary application functional, then

$$\mathbf{B} = (A, (A^n \rightharpoonup A_s), \mathbf{\Phi}, \mathrm{ap}_n) = (A \uplus (A^n \rightharpoonup A_s), \mathbf{\Phi}, \mathrm{ap}_n)$$

is logical over $(A, \mathbf{\Phi})$.

x4G.4. **Problem.** True or false:

(1) If $\psi(x) = w_0$ is a constant function on A, then $(\mathbf{A}, \psi)$ is a logical expansion of **A**.

(2) If $\psi(x) = \phi^{\mathbf{A}}(x)$ is a primitive of **A**, then $(\mathbf{A}, \psi)$ is logical over **A**.

x4G.5. **Problem.** Prove that if $\leq$ is a linear ordering of a set L, then

$$(\mathbf{L}^*, \leq) = (L^*, \mathrm{nil}, \mathrm{eq}_{\mathrm{nil}}, \mathrm{head}, \mathrm{tail}, \mathrm{cons}, \leq)$$

is a logical extension of $(L, \leq)$.

x4G.6. **Problem.** Construct an extended nondeterministic recursive program which sorts in $(\mathbf{L}^*, \leq)$ every string of length $n > 1$ using $n - 1$ comparisons. Infer that

$$\mathrm{calls}(\leq)((\mathbf{L}^*, \leq), \mathrm{sort}, s) \leq |s| \dot{-} 1 \quad (s \in S^*).$$

HINT: Use your favorite sorting algorithm, guessing whether $a \leq b$ without calling the comparison function, and then output the one string you have constructed which is, in fact sorted. For the second claim, use Problems x4C.4 and 4E.1.

x4G.7*. **Problem.** Prove that if $\leq$ is a linear ordering of a set L with at least two members and $\mathbf{A} = (A, \leq, \mathbf{\Phi})$ is a logical extension of $(L, \leq)$ such that $L^* \subseteq A$, then for every $s = (s_1, \dots, s_n) \in L^*$ with length $|n| > 1$

$$\bigwedge\!\!\!\bigwedge_{1 \leq i < j \leq n}[s_i \neq s_j] \Longrightarrow \mathrm{calls}(\leq)(\mathbf{A}, \mathrm{sort}, s) = n - 1.$$

HINT: Problem x4G.6 for the upper bound.

x4G.8. **Problem.** Prove that every substructure norm μ on a Φ-structure $\mathbf{A}$ can be extended to a substructure norm $\mu^{\mathbf{B}}$ on any logical extension of $\mathbf{A}$ so that (4G-6) holds when $\vec{x} \in A^n$.

x4G.9*. **Problem.** Prove that the structure $\mathsf{RAM}(\mathbf{A})$ associated by (2F-3) with an infinite, pointed structure $\mathbf{A}$ is a logical extension of $\mathbf{A}$. Infer that if a RAM over a structure $\mathbf{A}$ computes $f : A^n \rightharpoonup A_s$, then

$$\mathrm{calls}(\mathbf{A}, f, \vec{x}) \leq 2\,\mathrm{Time}(\vec{x}) - 1 \quad (f(\vec{x})\downarrow),$$

where $\mathrm{Time}(\vec{x})$ is the time complexity of the RAM when we view it as an iterator, cf. (2C-4). (This is not the most natural complexity measure for RAMs introduced in Cook and Reckhow [1973], cf. the comment on page 138.) HINT: For the second claim use Problem x3A.3.

4H. Deterministic uniform processes

An n-ary uniform process of a structure $\mathbf{A}$ is *deterministic* if it satisfies the following, stronger form of the Finiteness Axiom as expressed in (4B-1):

$$\text{(4H-1)}\quad \overline{\alpha}(\vec{x}) = w \Longrightarrow (\exists \mathbf{U} \subseteq_p \mathbf{A})\Big(\mathbf{U} \models_c \alpha(\vec{x}) = w]$$
$$\&\ (\text{for all } \mathbf{V} \subseteq_p \mathbf{A})[\mathbf{V} \models_c \alpha(\vec{x}) = w \Longrightarrow \mathbf{U} \subseteq_p \mathbf{V}]\Big),$$

i.e., if whenever $\overline{\alpha}(\vec{x})\downarrow$, then there is a unique, $\subseteq_p$-least "abstract computation" of $\overline{\alpha}(\vec{x})$ by α. The notion is natural and interesting. I record it here for completeness, but I have no real understanding of deterministic uniform processes and no methods for deriving lower bounds for them which are greater than the lower bounds for all uniform processes which define the same function.

Problems for Section 4H

x4H.1. **Problem.** Prove that the uniform process α_E induced by a deterministic recursive $\mathbf{A}$-program is deterministic.

x4H.2*. **Problem.** Give an example of a total, finite structure $\mathbf{A}$ and a unary relation R on A such that for some a, $\mathrm{depth}(\mathbf{A}, R, a) = 1$, but for every deterministic uniform α which defines R in $\mathbf{A}$, $\mathrm{depth}(\alpha, a) > 1$.

CHAPTER 5

LOWER BOUNDS FROM PRESBURGER PRIMITIVES

We establish here log-lower bounds for $\mathrm{depth}_f(\mathbf{N}_d, \vec{x})$ of various functions on the natural numbers, where

$$\mathbf{N}_d = (\mathbb{N}, \boldsymbol{Lin}_d) \text{ with } \boldsymbol{Lin}_d = \{0, 1, \ldots, d, +, \dot{-}, \mathrm{iq}_d, =, <\}.$$

The structure $\mathbf{N}_d$ is clearly explicitly equivalent to its reduct without $=$ and the constants $2, \ldots, d$, but including these among the primitives simplifies some of the formulas below. Lower bounds for $\mathbf{N}_d$ have wide applicability: binary arithmetic $\mathbf{N}_b$ and the Stein structure $\mathbf{N}_{\mathrm{st}}$ are explicitly reducible to $\mathbf{N}_2$, and every expansion of $(\mathbb{N}, 0, 1, +, \dot{-}, <, =)$ by finitely many Presburger primitives is explicitly equivalent with some $\mathbf{N}_d$, cf. Problems x5A.1 and x5A.3*.

The results in this chapter are interesting on their own, but they also illustrate the use of the Homomorphism Test 4E.3 in a very simple context, where the required arithmetic is trivial. They are mostly from van den Dries and Moschovakis [2004], [2009].

5A. Representing the numbers in $G_m(\mathbf{N}_d, \vec{a})$

To illustrate the use of the primitives of $\boldsymbol{Lin}_d$, consider the following.

5A.1. **Lemma.** *There is a recursive program E which computes the product $x \cdot y$ from $\boldsymbol{Lin}_2$ with parallel complexity*

$$c^p_E(x, y) \le 3 \log(\min(x, y)) \quad (x, y \ge 2).$$

Proof. The idea (from Problem x1B.2) is to reduce multiplication by x to multiplication by 2 and $\mathrm{iq}_2(x)$ (and addition), using the identity

$$(2x_1 + r) \cdot y = 2(x_1 \cdot y) + r \cdot y,$$

which means that the multiplication function satisfies and is determined by the recursive equation

$$\begin{aligned}f(x,y) = \text{if } (x=0) &\text{ then } 0\\ &\text{else if } (x=1) \text{ then } y\\ &\text{else if } (\text{parity}(x)=0) \text{ then } 2(f(\text{iq}_2(x),y))\\ &\text{else } 2(f(\text{iq}_2(x),y))+y.\end{aligned}$$

Now, obviously,

$$c_f^p(0,y)=1, \quad c_f^p(1,y)=\max\{1,1\}=1,$$

and with a careful reading of the equation, for $x \geq 2$,

$$c_f^p(x,y) \leq c_f^p(\text{iq}_2(x),y)+2.$$

To get an explicit form for an upper bound to $c_f^p(x,y)$, we prove by (complete) induction the inequality

$$c_f^p(x,y) \leq 3\log(x) \quad (x \geq 2),$$

the basis being trivial, since $c_f^p(2,y) = c_f^p(1,y)+2 = 3 = 3\log 2$, directly from the definition. In the inductive step,

$$c_f^p(x,y) \leq c_f^p(\text{iq}_2(x),y)+2 \leq 3\log(\frac{x}{2})+3 = 3(\log(\frac{x}{2})+1) = 3\log x.$$

Finally, to complete the proof, we add a head equation which insures that the first argument of f is the minimum of x and y:

$$g(x,y) = \text{if } (y<x) \text{ then } f(y,x) \text{ else } f(x,y);$$

the resulting program E has the claimed complexity bound. ⊣

The basic tool for the derivation of lower bounds in $\mathbf{N}_d$ is a canonical representation of numbers in $G_m(\mathbf{N}_d,\vec{a})$.

For a fixed d and each tuple of natural numbers $\vec{a}=(a_1,\dots,a_n)$, let

(5A-1) $$B_m(\vec{a}) = B_m^d(\vec{a}) = \Big\{\frac{x_0+x_1a_1+\cdots+x_na_n}{d^m} \in \mathbb{N} \\ : x_0,\dots,x_n \in \mathbb{Z} \text{ and } |x_i| \leq d^{2m}, i \leq n\Big\}.$$

The members of $B_m(\vec{a})$ are *natural numbers*. In full detail:

$$\begin{aligned}x \in B_m(\vec{a}) \iff\ & x \in \mathbb{N} \text{ and there exist } x_0,\dots,x_n \in \mathbb{Z}\\ & \text{such that } x = \frac{x_0+x_1a_1+\cdots+x_na_n}{d^m},\\ & \text{and for } i=0,\dots,n, |x_i| \leq d^{2m}.\end{aligned}$$

5A.2. **Lemma** (***Lin***$_d$-inclusion). *For all $\vec{a} \in \mathbb{N}^n$ and all m:*

(1) $a_1,\dots,a_n \in B_m(\vec{a}) \subseteq B_{m+1}(\vec{a})$.

(2) *For every primitive* $\phi : \mathbb{N}^k \to \mathbb{N}$ *in* $\boldsymbol{Lin}_d$,

$$x_1, \dots, x_k \in B_m(\vec{a}) \Longrightarrow \phi(x_1, \dots, x_k) \in B_{m+1}(\vec{a}).$$

(3) $G_m(\vec{a}) = G_m(\mathbf{N}_d, \vec{a}) \subseteq B_m(\vec{a})$.

PROOF. We take $n = 2$ to simplify the formulas, the general argument being only a notational variant.

(1) The first inclusion holds because $a_i = \dfrac{d^m a_i}{d^m}$ and the second because

$$\frac{x_0 + x_1 a_1 + x_2 a_2}{d^m} = \frac{dx_0 + dx_1 a_1 + dx_2 a_2}{d^{m+1}}$$

and $|dx_i| \leq d \cdot d^{2m} < d^{2(m+1)}$.

(2) Clearly $0, \dots, d \in B_m(\vec{a})$ for every $m \geq 1$, and so the constants stay in $B_m(\vec{a})$ once they get in.

For addition, let $x, y \in B_m(\vec{a})$, so

$$\begin{aligned} x + y &= \frac{x_0 + x_1 a_1 + x_2 a_2}{d^m} + \frac{y_0 + y_1 a_1 + y_2 a_2}{d^m} \\ &= \frac{d(x_0 + y_0) + d(x_1 + y_1)a_1 +_1 + d(x_2 + y_2)a_n}{d^{m+1}} \end{aligned}$$

and the coefficients in the numerator satisfy

$$|d(x_i + y_i)| \leq d(d^{2m} + d^{2m}) \leq dd^{2m+1} = d^{2m+2}.$$

The same works for arithmetic subtraction. Finally, for integer division by d, if $i = \mathrm{rem}_d(x) < d$, then

$$\mathrm{iq}_d(x) = \frac{1}{d}(x - i) = \frac{(x_0 - id^m) + x_1 a_1 + x_2 a_2}{d^{m+1}} \text{ for some } 1 \leq i < d$$

and this number is in $B_{m+1}(\vec{a})$ as above.

(3) follows immediately from (2), by induction on m. ⊣

5A.3. **Proposition** (Multiplication from $\boldsymbol{Lin}_d$). *For every number* $a \geq 2$,

$$\mathrm{depth}(a^2; \mathbf{N}_d, a) \geq \frac{1}{\log d} \log\left(\frac{a^2}{a+1}\right).$$

PROOF. It is enough to show that for $a \geq 2$,

$$a^2 \in G_m(\mathbf{N}_d, a) \Longrightarrow m \geq \frac{1}{\log d} \log\left(\frac{a^2}{a+1}\right),$$

so assume that $a^2 \in G_m(a)$. By Lemma 5A.2, there exist $x_0, x_1 \in \mathbb{Z}$ such that $|x_0|, |x_1| \leq d^{2m}$ and

$$a^2 = \frac{x_0 + x_1 a}{d^m},$$

from which we get $d^m a^2 = |x_0 + x_1 a| \le d^{2m} + d^{2m} a$; thus $a^2 \le d^m + d^m a$, which yields the required

$$d^m \ge \frac{a^2}{a+1}. \qquad \dashv$$

Similar arguments can be used to establish output-depth lower bounds from $\boldsymbol{Lin}_d$ for all functions which grow faster than x.

Problems for Section 5A

x5A.1. **Problem.** Prove that the structures

$$\mathbf{N}_b = (\mathbb{N}, 0, \text{parity}, \text{iq}_2, \text{em}_2, \text{om}_2, \text{eq}_0),$$

$$\mathbf{N}_{\text{st}} = (\mathbb{N}, \text{parity}, \text{em}_2, \text{iq}_2, \dot{-}, =, <)$$

of binary arithmetic and the Stein algorithm are explicitly reducible to $\mathbf{N}_2$.

x5A.2. **Problem.** Prove that $\text{rem}_d(x)$ is explicit in $\mathbf{N}_d$.

x5A.3*. **Problem.** Prove that for all $m, n \ge 2$:
(i) $\text{iq}_m \in \mathbf{Expl}(\mathbf{N}_{mn})$;
(ii) $\text{iq}_{mn} \in \mathbf{Expl}(\mathbb{N}, 0, 1, \dots, d, +, \dot{-}, \text{iq}_m, \text{iq}_n, =, <)$.
Infer that if $\mathbf{\Phi}$ is any finite set of Presburger primitives, then the structure

$$(\mathbb{N}, 0, 1, +, \dot{-}, <, =, \mathbf{\Phi})$$

is explicitly equivalent with some $\mathbf{N}_d$. (For the definition of Presburger functions see (1E-10).)

x5A.4. **Problem.** Specify a system of two recursive equations

$$q(x, y) = E_q(x, y, q, r)$$
$$r(x, y) = E_r(x, y, q, r),$$

in the vocabulary $\boldsymbol{Lin}_2 \cup \{q, r\}$, such that in $\mathbf{N}_2$,

$$\overline{q}(x, y) = \text{iq}(x, y), \quad \overline{r}(x, y) = \text{rem}(x, y),$$

and the corresponding complexities are $O(\log(x))$, i.e., for some B and all sufficiently large x,

$$c_q^p(x, y) \le B \log x, \quad c_r^p(x, y) \le B \log x.$$

(With the appropriate head equations, this system defines two programs from $\boldsymbol{Lin}_2$, one for $\text{iq}(x, y)$ and the other for $\text{rem}(x, y)$.)

x5A.5. **Problem.** Prove that the recursive program for the integer quotient function in Problem x5A.4 is weakly depth-optimal from $\boldsymbol{Lin}_d$, for any $d \ge 2$.

x5A.6*. **Problem.** Prove that for every $d \geq 2$, there is an $r > 0$ and infinitely pairs of numbers (a, b) such that for every $d \geq 2$,

$$\text{depth}(\text{rem}(a, b); \boldsymbol{Lin}_d, a, b) > r \log(\max(a, b)).$$

Infer that the recursive program for $\text{rem}(x, y)$ in Problem x5A.4 is weakly depth-optimal for $\text{rem}(x, y)$ in every $\mathbf{N}_d$. HINT: Use the pairs $(a^3 + a, a^2)$. (Note: It is possible to give an easier proof of an $O(\log \max(a, b))$ lower bound for $\text{depth}_{\text{rem}}(\mathbf{N}_d, a, b)$ using the Homomorphism Test, see Problem x5B.1. The proof suggested here uses output-depth complexity and requires a simple divisibility argument. It is due to Tim Hu.)

x5A.7. **Problem.** Define a weakly depth-optimal program from $\boldsymbol{Lin}_d$ which computes the exponential function $f(x, y) = x^y$ (with $0^0 = x^0 = 1$).

5B. Primality from $\boldsymbol{Lin}_d$

To establish lower bounds from $\boldsymbol{Lin}_d$ for decision problems, we need to complement Lemma 5A.2 with a uniqueness result.

5B.1. **Lemma** ($\boldsymbol{Lin}_d$-Uniqueness). *If* $x_i, y_i \in \mathbb{Z}$, $|x_i|, |y_i| < \frac{a}{2}$ *and* $\lambda \geq 1$, *then*

$$\begin{aligned} x_0 + x_1\lambda a = y_0 + y_1\lambda a &\iff [x_0 = y_0 \;\&\; x_1 = y_1], \\ x_0 + x_1\lambda a > y_0 + y_1\lambda a &\iff [x_1 > y_1 \vee (x_1 = y_1 \;\&\; x_0 > y_0)]. \end{aligned}$$

PROOF. It is enough to prove the result for $\lambda = 1$, since $a \leq \lambda a$, and so the general result follows from the special case applied to λa.

The second equivalence implies the first, and it follows from the following fact applied to $(x_0 - y_0) + (x_1 - y_1)a$:

If $x, y \in \mathbb{Z}$ *and* $|x|, |y| < a$, *then*

$$x + ya > 0 \iff [y > 0] \vee [y = 0 \;\&\; x > 0].$$

Proof. This is obvious if $y = 0$; and if $y \neq 0$, then $|x| < a \leq |ya|$, so that $x + ya$ has the same sign as ya, which has the same sign as y. ⊣

5B.2. **Lemma** ($\boldsymbol{Lin}_d$-embedding). *Suppose* $d^{2m+2} < a$, *and let* $\lambda > 1$ *be any number such that* $d^{m+1} \mid \lambda - 1$; *then there exists an embedding*

$$\pi : \mathbf{G}_m(\mathbf{N}_d, a) \rightarrowtail \mathbf{N}_d,$$

such that $\pi a = \lambda a$.

PROOF. By Lemma 5B.1 and part (3) of Lemma 5A.2, the equation

$$\pi\Big(\frac{x_0 + x_1 a}{d^m}\Big) = \frac{x_0 + x_1\lambda a}{d^m} \qquad \Big(|x_0|, |x_1| \leq d^{2m}, \frac{x_0 + x_1 a}{d^m} \in G_m(a)\Big)$$

defines a map $\pi : \mathbf{G}_m(\mathbf{N}_d, a) \to \mathbb{Q}$, since

$$d^{2m} < d^{2m+1} < \frac{a}{2}$$

by the hypothesis. This map takes values in $\mathbb{N}$, because

$$x_0 + \lambda x_1 a = x_0 + x_1 a + (\lambda - 1)x_1 a, \tag{5B-1}$$

so that if $d^m \mid (x_0 + x_1 a)$, then also $d^m \mid (x_0 + \lambda x_1 a)$ since $d^m \mid (\lambda - 1)$ by the hypothesis. It is injective and order-preserving, by Lemma 5B.1 again, applied to both a and λa.

To check that it preserves addition when the sum is in $G_m(a) = G_m(\mathbf{N}_d, a)$, suppose that $X, Y, X + Y \in G_m(a)$, and write

$$X = \frac{x_0 + x_1 a}{d^m}, \quad Y = \frac{y_0 + y_1 a}{d^m}, \quad X + Y = Z = \frac{z_0 + z_1 a}{d^m}$$

with all $|x_i|, |y_i|, |z_i| \le d^{2m}$. Now

$$Z = \frac{(x_0 + y_0) + (x_1 + y_1)a}{d^m},$$

and $|x_0 + y_0|, |x_1 + y_1| \le 2 \cdot d^{2m} \le d^{2m+1} < \frac{a}{2}$, and so by the Uniqueness Lemma 5B.1,

$$x_0 + y_0 = z_0, \quad x_1 + y_1 = z_1,$$

which gives $\pi X + \pi Y = \pi Z$.

The same argument works for arithmetic subtraction.

Finally, for division by d, suppose

$$X = \frac{x_0 + x_1 a}{d} = d\,\mathrm{iq}_d(X) + i \quad (i < d)$$

where $|x_0|, |x_1| \le d^{2m}$ as above, so that

$$\mathrm{iq}_d(X) = \frac{1}{d}\Big(\frac{x_0 + x_1 a}{d^m} - i\Big) = \frac{x_0 - id^m + x_1 a}{d^{m+1}} = Z = \frac{z_0 + z_1 a}{d^m}$$

for suitable z_0, z_1 with $|z_0|, |z_1| \le d^{2m}$, if $Z \in G_m$. These two representations of Z must now be identical since $|dz_i| \le dd^{2m} = d^{2m+1} < \frac{a}{2}$, and

$$|x_0 - id^m| \le d^{2m} + id^m < d^{2m} + d^{m+1}$$
$$= d^m(d^m + d) \le d^m d^{m+1} = d^{2m+1} < \frac{a}{2}.$$

So $x_0 - id^m = dz_0$ and $x_1 = dz_1$. These two equations imply that

$$\frac{x_0 + x_1 \lambda a}{d^m} = d\frac{z_0 + z_1 \lambda a}{d^m} + i,$$

which means that

$$\mathrm{iq}_2(\pi X) = \frac{z_0 + z_1 \lambda a}{d^m} = \pi(Z) = \pi(\mathrm{iq}_d(X))$$

as required. ⊣

5B.3. **Theorem.** *For every prime number p,*

$$\mathrm{depth}_{\mathrm{Prime}}(\mathbf{N}_d, p) \geq \frac{1}{4 \log d} \log p.$$

PROOF. Let $m = \mathrm{depth}_{\mathrm{Prime}}(\mathbf{N}_d, p)$ and suppose that

$$d^{2m+2} < p. \tag{5B-2}$$

Lemma 5B.2 guarantees an embedding

$$\pi : \mathbf{G}_m(\mathbf{N}_d, p) \rightarrowtail \mathbf{N}_d$$

with $\lambda = 1 + d^{m+1}$ such that $\pi(p) = \lambda p$, and this π does not respect the primality relation at p, which is absurd. So (5B-2) fails, and so (taking logarithms and using the fact that $m \geq 1$ by Problem x4E.4),

$$4m \log d \geq (2m + 2) \log d \geq \log p.$$

⊣

Using non-trivial number theory. The proof of Theorem 5B.3 used nothing but the definition of prime numbers, and none of the results about intrinsic lower bound of number-theoretic relations in the sequel will require much more. One might expect that truly significant results about primality depend on deep properties of primes, and it is worth stating here one example of this.

5B.4. **Theorem.** *There are infinitely many primes p, such that every number whose binary expansion differs from that of p in any one digit is composite.*

This was proved by Cohen and Selfridge [1975] and Sun [2000], Tao [2011] extended it to expansions of numbers relative to any $k > 1$, and all three of these papers established much stronger results about the form of primes with this property, various transformations (other than flipping a digit) which spoils their primeness, their density relative to the density of primes, etc. Sun and Tao also observed the obvious complexity implications of the theorem: to decide correctly whether x is prime, an algorithm which is given its input x in binary form (and basically nothing else) must read all the digits of x, on infinitely many inputs. This is usually made precise in terms of *boolean circuit complexity* which we have not defined, but a strong version of it is a fact about the intrinsic calls-complexity of primality in the appropriate structure.

If $x = \sum_j x_j 2^j$ (with $x_i < 2$) is the binary expansion of $x > 0$, let

$$\mathrm{bit}_i(x) = x_i, \quad \mathrm{length}(x) = \max\{i : \mathrm{bit}_i(x) > 0\} + 1,$$
$$\mathrm{bitbound}_i(x) \iff i \geq \mathrm{length}(x) - 1,$$

and consider the structure:[36]

$$\mathbf{Bits} = (\mathbb{N}, 0, 1, \{\text{bit}_i\}_{i\in\mathbb{N}}, \{\text{bitbound}_i\}_{i\in\mathbb{N}}, =).$$

Let $\mathbf{\Phi}_{\text{bits}} = \{\{\text{bit}_i\}_{i\in\mathbb{N}}\}$.

5B.5. **Proposition.** (1) *For every* $x > 0$,

$$\text{calls}(\mathbf{\Phi}_{\text{bits}})_{\text{Prime}}(\mathbf{Bits}, x) \le \text{length}(x).$$

(2) *For infinitely many primes p and every logical extension* **B** *of* **Bits**,

$$\text{calls}(\mathbf{\Phi}_{\text{bits}})_{\text{Prime}}(\mathbf{B}, p) \ge \text{length}(p).$$

Unwinding the technical terms, (1) says that *there is a uniform process of* **Bits** *which defines* Prime(x) *using no more than* length(x) *calls to the bit primitives*; and (2) says that *every uniform process of a logical extension* **B** *of* **Bits** *which defines* Prime(p)*—and so every nondeterministic recursive program of* **B** *which decides* Prime(p)*—makes at least* length(p) *calls to the bit primitives, on infinitely many primes.*

PROOF. (1) For any $x > 0$ with $\text{length}(x) = n + 1$, let $\mathbf{U} \subseteq_p \mathbf{Bits}$ be the structure with universe $\{x, 0, 1\}$ and

$$\text{eqdiag}(\mathbf{U}) = \{\text{bit}_0(x) = x_0, \dots, \text{bit}_n(x) = x_n, \text{bitbound}_n(x) = \text{tt}\}.$$

Every homomorphism $\pi : \mathbf{U} \to \mathbf{Bits}$ fixes all the bits and the length of x, so $\pi(x) = x$ and (trivially) π respects the relation Prime(x); so $(\mathbf{U}, x)$ is a certificate of Prime(x) or $\neg$Prime(x), whichever of these is true, and hence

$$\text{calls}(\mathbf{\Phi}_{\text{bits}})_{\text{Prime}}(\mathbf{Bits}, x) \le \text{calls}(\mathbf{\Phi}_{\text{bits}})(\mathbf{U}, x) = n + 1.$$

(2) By the Embedding Test 4G.3, it is enough to check that if p is any of the primes guaranteed by Theorem 5B.4, $\mathbf{U} \subseteq_p \mathbf{Bits}$ is finite, generated by p and $\text{calls}(\mathbf{\Phi}_{\text{bits}})(\mathbf{U}) < n+1 = \text{length}(p)$, then there is an embedding $\pi : \mathbf{U} \rightarrowtail \mathbf{Bits}$ such that $\pi(p)$ is composite. The hypothesis on **U** means that there is an $i < n + 1$ such the condition $(\text{bit}_i, p, \text{bit}_i(p))$ is not in eqdiag(**U**); and if x is the composite number produced by flipping $\text{bit}_i(p)$ and we set

$$\pi(p) = x, \quad \pi(0) = 0, \quad \pi(1) = 1,$$

then $\pi : \mathbf{U} \rightarrowtail \mathbf{B}$ is an embedding which does not respect Prime(p). ⊣

Problems for Section 5B

x5B.1. **Problem.** Suppose $e \ge 2$, set

$$|_e\,(x) \iff e \mid x \iff \text{rem}_e(x) = 0$$

and assume that $e \perp\!\!\!\perp d$.

[36] The vocabulary of **Bits** is infinite, see Footnote 8 on page 31.

(a) Prove that for all a which are not divisible by e,

$$\text{depth}_{|_e}(\mathbf{N}_d, a) \geq \frac{1}{4\log d} \log a.$$

(b) For some $r > 0$ and all a which are not divisible by e,

$$\text{depth}(\mathbf{N}_d, \text{iq}_e, a) > r \log a, \quad \text{depth}(\mathbf{N}_d, \text{rem}_e, a) > r \log a.$$

In particular, if e is coprime with d, then the relation $|_e\ (x)$ is not explicit in $\mathbf{N}_d$; the divisibility relation $x \mid y$ is not explicit in any $\mathbf{N}_d$; and the recursive programs for $\text{iq}(x, y)$ and $\text{rem}(x, y)$ in Problem x5A.4 are weakly depth-optimal in $\mathbf{N}_2$ and in every $\mathbf{N}_d$ (such that $2 \mid d$, so they can be expressed).

HINT: Use the fact that if $x \perp\!\!\!\perp y$, then there are constants $A \in \mathbb{Z}$ and $B \in \mathbb{N}$ such that $1 = Ax - By$.

x5B.2. **Problem.** If you know what boolean circuits are, check that they can be viewed as recursive programs of **Bits** and so (2) of Proposition 5B.5 applies—albeit the complexity measure $\text{calls}(\mathbf{\Phi}_{\text{bits}})_{\text{Prime}}(\mathbf{B}, p)$ on logical extensions of **Bits** is not close to the measures usually studied in boolean circuit theory.

5C. Good examples: perfect square, square-free, etc.

The method in the preceding section can be easily adapted to derive lower bound results for many unary relations on $\mathbb{N}$. Some of these are covered by the next, fairly general notion.

A unary relation $R(x)$ is a *good example* if for some polynomial

(5C-1) $$\lambda(\mu) = 1 + l_1\mu + l_2\mu^2 + \cdots + l_s\mu^s$$

with coefficients in $\mathbb{N}$, constant term 1, and degree > 0 and for all $\mu \geq 1$,

(5C-2) $$R(x) \Longrightarrow \neg R(\lambda(\mu)x).$$

For example, primality is good, taking $\lambda(\mu) = 1 + \mu$, and being a power of 2 is good with $\lambda(\mu) = 1 + 2\mu$. We leave for the problems several interesting results of this type.

Problems for Section 5C

x5C.1. **Problem.** Design a weakly depth-optimal recursive program from $\boldsymbol{Lin}_d$ for the relation

$$P(x) \iff (\exists y)[x = 2^y].$$

x5C.2. **Problem** (van den Dries and Moschovakis [2004]). Prove that if $R(x)$ is a good example, then for all $a \geq 2$,

$$R(a) \Longrightarrow \text{depth}_R(\mathbf{N}_d, a) \geq \frac{1}{4\log d} \log a.$$

x5C.3. **Problem.** Prove that if $m > 0$, then $(1+m^2)n^2$ is not a perfect square. HINT: Prove first that $1+m^2$ is not a perfect square, and then reduce the result to the case where $m \perp\!\!\!\perp n$.

x5C.4. **Problem.** Prove that the following two relations are good examples:

$$R_1(a) \iff a \text{ is a perfect square}$$
$$R_2(a) \iff a \text{ is square-free.}$$

x5C.5. **Problem.** Prove that if $\lambda(\mu)$ is as in (5C-1), then there is a constant C such that

$$\text{(5C-3)} \qquad \log \lambda(\mu) \leq C \log \mu \quad (\mu \geq 2).$$

The next problem gives a logarithmic lower bound for $\text{depth}_R(\mathbf{N}_d, a)$ with good R at many points where $R(a)$ fails to hold.

x5C.6. **Problem.** Suppose $R(x)$ is a good example with associated polynomial $\lambda(\mu)$. Prove that there is a rational constant $r > 0$, such that for all $a \geq 2$ and $m \geq 1$,

$$R(a) \Longrightarrow \text{depth}_R(\mathbf{N}_d, \lambda(d^{m+1})a) \geq r \log(\lambda(d^{m+1})a).$$

5D. Stein's algorithm is weakly depth-optimal from Lin_d

We extend here (mildly) the methods in the preceding section so they apply to binary functions, and we show the result in the heading.

For the remainder of this section, a, b, c range over $\mathbb{N}$ and x, y, z, x_i, y_i, z_i range over $\mathbb{Z}$.

5D.1. **Lemma.** *Suppose $a > 2$ and set $b = a^2 - 1$.*

(1) *$a \perp\!\!\!\perp b$, and if $|x_i|, |y_i| < \frac{a}{4}$ for $i = 0, 1, 2$ and $\lambda \geq 1$, then*

$$x_0 + x_1\lambda a + x_2\lambda b = y_0 + y_1\lambda a + y_2\lambda b \iff x_0 = y_0 \;\&\; x_1 = y_1 \;\&\; x_2 = y_2,$$

$$\begin{aligned} x_0 + x_1\lambda a + x_2\lambda b &> y_0 + y_1\lambda a + y_2\lambda b \\ &\iff [x_0 > y_0 \;\&\; x_1 = y_1 \;\&\; x_2 = y_2] \\ &\qquad\qquad \vee [x_1 > y_1 \;\&\; x_2 = y_2] \vee [x_2 > y_2]. \end{aligned}$$

(2) *If $d^{2m+3} < a$ and $\lambda = 1 + d^{m+1}$, then there is an embedding*

$$\pi : \mathbf{N}_d \upharpoonright G_m(a, b) \rightarrowtail \mathbf{N}_d$$

such that $\pi a = \lambda a, \pi b = \lambda b$.

PROOF. (1) The identity $1 = a \cdot a - b$ witnesses that $a \perp\!\!\!\perp b$.

The second equivalence clearly implies the first, and it follows from the following proposition applied to $(x_0 - y_0)$, $(x_1 - y_1)$, $(x_2 - y_2)$.

If $|x|, |y|, |z| < \frac{a}{2}$ *and* $\lambda \geq 1$, *then* $x + y\lambda a + z\lambda b > 0$ *if and only if either* $x > 0$ *and* $y = z = 0$; *or* $y > 0$ *and* $z = 0$; *or* $z > 0$.

Proof. If $z = 0$, then the result follows from Lemma 5B.1, so assume $z \neq 0$ and compute:

$$x + y\lambda a + z\lambda b = x + y\lambda a + z\lambda(a^2 - 1) = (x - \lambda z) + y\lambda a + z\lambda a^2.$$

Now

$$\Big|(x - \lambda z) + y\lambda a\Big| = \lambda\Big|(\frac{x}{\lambda} - z) + ya\Big| < \lambda(a + \frac{a^2}{2}) < \lambda a^2 \leq \lambda |z| a^2,$$

and so $x + y\lambda a + z\lambda b$ and $\lambda z a^2$ have the same sign, which is the sign of z.

(2) Assume $d^{2m+3} < a$, set $\lambda = 1 + d^{m+1}$, and notice that $d^{2m} < \frac{1}{4}a$, so as in Lemma 5B.2, we can define the required embedding by

$$\pi\Big(\frac{x_0 + x_1 a + x_2 b}{d^m}\Big) = \frac{x_0 + x_1\lambda a + x_2\lambda b}{d^m}$$
$$\Big(|x_0|, |x_1|, |x_2| \leq d^{2m}, \frac{x_0 + x_1 a + x_2 b}{d^m} \in G_m(a, b)\Big),$$

using now (1) instead of Lemma 5B.1. ⊣

5D.2. **Theorem** (van den Dries and Moschovakis [2004]). *For all* $a > 2$,

$$\text{depth}_{\perp\!\!\!\perp}(\mathbf{N}_d, a, a^2 - 1) > \frac{1}{10 \log d} \log(a^2 - 1).$$

PROOF. Let $m = \text{depth}_{\perp\!\!\!\perp}(\mathbf{N}_0, a, a^2 - 1)$ for some $a > 2$. Since λa and $\lambda(a^2 - 1)$ are not coprime, part (2) of the preceding Lemma 5D.1 and the Embedding Test 4G.3 imply that

$$d^{2m+3} \geq a;$$

taking the logarithms of both sides and using the fact that $m \geq 1$ (by Problem x4E.4), we get

$$5m \log d \geq (2m + 3) \log d \geq \log a;$$

which with $\log(a^2 - 1) < 2 \log a$ gives the required

$$5m \log d > \frac{1}{2} \log(a^2 - 1).$$ ⊣

5D.3. **Corollary.** *Let E be the recursive program of* $\mathbf{N}_2$ *which decides* $a \perp\!\!\!\perp b$ *by adding to the Stein algorithm one step checking* $\gcd(a, b) = 1$. *For each* $d \geq 2$, *there is a constant* $K > 0$ *such that for all* $a > 2$,

$$l^s_E(a, a^2 - 1) \leq K \text{depth}_{\perp\!\!\!\perp}(\mathbf{N}_d, a, a^2 - 1).$$

In particular, the Stein algorithm is weakly optimal for coprimeness from Presburger primitives, for both the depth and calls complexity measures.

PROOF. Choose K_1 such that

$$l^s_E(a, b) \le K_1(\log a + \log b) \quad (a, b > 2),$$

and compute for $a > 2$:

$$l^s_E(a, a^2 - 1) < 2K_1 \log(a^2 - 1) < 20 \log dK_1 \text{depth}_{\perp\!\!\!\perp}(a, a^2 - 1). \qquad \dashv$$

Problems for Section 5D

Problem x5A.4 defines a recursive program from $\mathbf{N}_2$ which computes $\text{rem}(x, y)$ with $O(\log)$ complexity. The next problem claims that it is weakly depth-optimal from Presburger primitives—and a little more.

x5D.1. **Problem.** Prove that for each $d \ge 2$, there is a rational $r > 0$, such that

$$\text{for infinitely many pairs } (x, y), x \mid y \text{ and } \text{depth}_{|}(\mathbf{N}_d, x, y) \ge r \log y.$$

Infer that the recursive program for the remainder in Problem x5A.4 is weakly depth-optimal from $\boldsymbol{Lin}_d$.

HINT: Prove that when a, b, λ and μ satisfy suitable conditions, then the mapping

$$\frac{x_0 + x_1 a + x_2 b}{d^m} \mapsto \frac{x_0 + x_1 \lambda a + x_2 \mu b}{d^m}$$

is an embedding on $\mathbf{N}_d \upharpoonright G_m(a, b)$.

Busch [2007], [2009] has used "asymmetric" embeddings of this kind to derive lower bounds for several problems in number theory and algebra that are related to the Stein algorithm.

CHAPTER 6

LOWER BOUNDS FROM DIVISION WITH REMAINDER

We now add to the basic primitives of the Presburger structures *division with remainder*, i.e., the integer quotient and remainder operations. Set:

$$\boldsymbol{Lin}_0 = \{0, 1, =, <, +, \dot{-}\}, \quad \mathbf{N}_0 = (\mathbb{N}, \boldsymbol{Lin}_0),$$
$$\boldsymbol{Lin}_0[\div] = \boldsymbol{Lin}_0 \cup \{\mathrm{iq}, \mathrm{rem}\} = \{0, 1, =, <, +, \dot{-}, \mathrm{iq}, \mathrm{rem}\},$$
$$\mathbf{N}_0[\div] = (\mathbb{N}, \boldsymbol{Lin}_0[\div]) = (\mathbb{N}, 0, 1, =, <, +, \dot{-}, \mathrm{iq}, \mathrm{rem}).$$

Every expansion of a Presburger structure by division with remainder is obviously explicitly equivalent to $\mathbf{N}_0[\div]$, and so all the results of this chapter apply to these richer structures with only inessential changes in the constants.

The derivations of absolute lower bounds for unary relations from $\boldsymbol{Lin}_0[\div]$ is similar to those from $\boldsymbol{Lin}_d$ in Sections 5B and 5C and we will consider it first. For binary relations, however, like coprimeness, we need some new ideas as well as some elementary results from number theory.

6A. Unary relations from $\boldsymbol{Lin}_0[\div]$

We start with a representation of the numbers in $G_m(\mathbf{N}_0[\div], \vec{a})$ similar to that for $G_m(\mathbf{N}_d, \vec{a})$ in Lemma 5A.2, except that we cannot keep the denominators independent of $\vec{a}$.

For each tuple $\vec{a} = (a_1, \dots, a_n)$ of numbers and for each $h \geq 1$, we let

$$\text{(6A-1)} \quad C(\vec{a}; h) = \Big\{ \frac{x_0 + x_1 a_1 + \cdots + x_n a_n}{x_{n+1}} \in \mathbb{N} : x_0, \dots, x_{n+1} \in \mathbb{Z},$$
$$x_{n+1} > 0 \text{ and } |x_0|, \dots, |x_{n+1}| \leq h \Big\}.$$

The numbers in $C(\vec{a}; h)$ are said to have *height* (no more than) h with respect to $\vec{a}$, and, trivially,

$$x \leq h \Longrightarrow x \in C(a; h), \quad h \leq h' \Longrightarrow C(\vec{a}; h) \subseteq C(\vec{a}; h').$$

We need to estimate how much the height is increased when we perform various operations on numbers. The results are very simple for the primitives in $\boldsymbol{Lin}_0$:

6A.1. **Lemma.** *For all $\vec{a} = (a_1, \dots, a_n), h \geq 2$ and every k-ary operation in* ***Lin***$_0$,

$$X_1, \dots, X_k \in C(\vec{a}; h) \Longrightarrow f(X_1, \dots, X_k) \in C(\vec{a}; h^3).$$

PROOF is by direct computation. For example (taking $n = 2$ to keep the notation simple),

$$\frac{x_0 + x_1 a + x_2 b}{x_3} + \frac{y_0 + y_1 a + y_2 b}{y_3} = \frac{(y_3 x_0 + x_3 y_0) + (y_3 x_1 + x_3 y_1)a + (y_3 x_2 + x_3 y_2)b}{x_3 y_3},$$

and for the typical coefficient,

$$|y_3 x_0 + x_3 y_0| \leq h^2 + h^2 = 2h^2 \leq h^3. \qquad \dashv$$

There is no simple, general result of this type for division with remainder, and in this section we will consider only the simplest case $n = 1$, when $C(a; h)$ comprises the numbers of height h with respect to a single a. We start with the appropriate version of Lemma 5B.1.

6A.2. **Lemma** (***Lin***$_0[\div]$-Uniqueness). *If $|x_i|, |y_i| \leq h$ for $i \leq 2$, $\lambda \geq 1$ and $2h^2 < a$, then:*

$$\frac{x_0 + x_1 \lambda a}{x_2} = \frac{y_0 + y_1 \lambda a}{y_2} \iff y_2 x_0 = x_2 y_0 \;\&\; y_2 x_1 = x_2 y_1,$$

$$\frac{x_0 + x_1 \lambda a}{x_2} > \frac{y_0 + y_1 \lambda a}{y_2} \iff [y_2 x_1 > x_2 y_1] \vee [y_2 x_1 = x_2 y_1 \;\&\; y_2 x_0 > x_2 y_0].$$

In particular, if $x \in C(a; h)$ and $2h^2 < a$, then there are unique x_0, x_1, x_2 with no common factor other than 1 such that

$$\text{(6A-2)} \qquad x = \frac{x_0 + x_1 a}{x_2} \quad (|x_0|, |x_1|, |x_2| \leq h).$$

PROOF of the two equivalences is immediate from Lemma 5B.1, since

$$\frac{x_0 + x_1 a}{x_2} > \frac{y_0 + y_1 a}{y_2} \iff y_2 x_0 + y_2 x_1 a > x_2 y_0 + x_2 y_1 a.$$

The uniqueness of relatively prime x_0, x_1, x_2 which satisfy (6A-2) and these equivalences requires a simple divisibility argument, Problem x6A.1. $\dashv$

We will sometimes save a few words by referring to (6A-2) as the *canonical representation* of x in $C(a; h)$, when $2h^2 < a$.

6A.3. **Lemma.** *If $x, y \in C(a; h)$, $x \geq y > 0$, $h \geq 2$ and $h^9 < a$, then* $\mathrm{iq}(x, y), \mathrm{rem}(x, y) \in C(a; h^6)$.

PROOF. The hypothesis implies that $2h^2 < a$, and so we have canonical representations

$$x = \frac{x_0 + x_1 a}{x_2}, \quad y = \frac{y_0 + y_1 a}{y_2}$$

of x and y in $C(a; h)$. Suppose

$$x = yq + r \qquad (0 \leq r < y)$$

and consider two cases.

Case 1, $y_1 = 0$. Now $y = \dfrac{y_0}{y_2} \leq h$, and so

$$r = \mathrm{rem}(x, y) < h.$$

Solving for q (and keeping in mind that $r \in \mathbb{N}$), we have

$$q = \mathrm{iq}(x, y) = \frac{y_2}{y_0} \cdot \left[\frac{x_0 + x_1 a}{x_2} - r\right] = \frac{y_2}{y_0} \cdot \frac{(x_0 - x_2 r) + x_1 a}{x_2}$$

and the (potentially) highest coefficient in this expression is

$$|y_2 x_0 - y_2 x_2 r| \leq h^2 + h^2 h \leq 2h^3 \leq h^4 < h^6.$$

Case 2, $y_1 \neq 0$. We must now have $y_1 > 0$, otherwise $y < 0$ by Lemma 6A.2. Moreover,

$$y \cdot (2x_1 y_2) = \frac{2y_0 x_1 y_2 + 2y_1 x_1 y_2 a}{y_2} > \frac{x_0 + x_1 a}{x_2} \geq yq$$

by the same lemma, because $|y_2 x_1| < 2|x_2 y_1 x_1 y_2|$, and the lemma applies since (for the potentially largest coefficient),

$$2|2y_1 x_1 y_2|^2 \leq 2^3 h^6 \leq h^9 < a.$$

It follows that

$$q = \mathrm{iq}(x, y) \leq 2x_1 y_2 \leq h^3,$$

and then solving the canonical division equation for r, we get

$$r = \mathrm{rem}(x, y) = \frac{(y_2 x_0 - x_2 y_1 q) + (y_2 x_1 - x_2 y_1 q)a}{x_2 y_2}.$$

The (potentially) highest coefficient in this expression is

$$|y_2 x_0 - x_2 y_1 q| \leq h^2 + h^2 h^3 \leq h^6.$$ ⊣

6A.4. **Lemma.** *For every m, if*

$$2^{6^{m+2}} < a \text{ and } G_m(a) = G_m(\mathbf{N}_0[\div], a),$$

then $G_m(a) \subseteq C(a; 2^{6^m})$.

PROOF is by induction on m, the basis being trivial since $2^{6^0} = 2$ and $G_0(a) = \{a\} \subseteq C(a;2)$. For the induction step, set $h = 2^{6^m}$ to save typing exponents, and notice that $h \geq 2$, and

$$h^9 = \left(2^{6^m}\right)^9 = 2^{9 \cdot 6^m} < 2^{6^{m+2}} < a.$$

Thus by Lemmas 6A.1 and 6A.3, the value of any operation in $\mathbf{Lin}_0[\div]$ with arguments in $C(a;h)$ is in $C(a;h^6)$, and

$$h^6 = \left(2^{6^m}\right)^6 = 2^{6^{m+1}}. \qquad \dashv$$

6A.5. **Lemma** ($\mathbf{Lin}_0[\div]$-embedding). *If* $G_m(a) = G_m(\mathbf{N}_0[\div], a)$,

$$2^{6^{m+3}} < a \text{ and } a! \mid (\lambda - 1),$$

then there is an embedding

$$\pi : \mathbf{N}_0[\div] \restriction G_m(a) \rightarrowtail \mathbf{N}_0[\div]$$

such that $\pi(a) = \lambda a$.

PROOF. Set

$$h = 2^{6^{m+1}},$$

so that from the hypothesis (easily), $2h^2 < a$, and then by Lemma 6A.2, each $x \in C(a;h)$ can be expressed uniquely in the form

$$x = \frac{x_0 + x_1 a}{x_2}$$

with all the coefficients $\leq h$. We first set

$$\rho(x) = \rho\Big(\frac{x_0 + x_1 a}{x_2}\Big) = \frac{x_0 + x_1 \lambda a}{x_2} \qquad (x \in C(a;h)).$$

The values of $\rho(x)$ are all in $\mathbb{N}$, since

$$x_0 + x_1 \lambda a = x_0 + x_1 a + (\lambda - 1) x_1 a,$$

so that for any $x_2 \leq h \leq a$,

$$x_2 \mid x_0 + x_1 \lambda a \iff x_2 \mid x_0 + x_1 a,$$

by the hypothesis that $a! \mid (\lambda - 1)$. By another appeal to Lemma 6A.2, we verify that ρ is an injection, and it is order-preserving.

The required embedding is the restriction

$$\pi = \rho \restriction G_m(a),$$

and the verification that it respects all the operations in $\mathbf{Lin}_0$ follows along familiar lines. For addition, for example, set

$$h_1 = 2^{6^m},$$

and consider canonical representations

$$x = \frac{x_0 + x_1 a}{x_2}, \quad y = \frac{y_0 + y_1 a}{y_2}$$

of two numbers in $C(a; h_1)$. Adding the fractions, we get

$$x + y = \frac{x_0 + x_1 a}{x_2} + \frac{y_0 + y_1 a}{y_2} = \frac{(y_2 x_0 + x_2 y_0) + (y_1 x_1 + x_2 y_1) a}{x_2 y_2},$$

and we notice that the expression on the right is a canonical representation of $x + y$ in $C(a; h)$, since, with a typical coefficient,

$$|y_2 x_0 + x_2 y_0| \leq h_1^5 < h_1^6 = h.$$

This means that

$$\pi(x + y) = \frac{(y_2 x_0 + x_2 y_0) + (y_1 x_1 + x_2 y_1)\lambda a}{x_2 y_2} = \pi(x) + \pi(y),$$

as required.

The argument that π respects the integer quotient and remainder operations is a bit more subtle, primarily because these are defined together: we need to show that

$$\text{if } \mathrm{iq}(x, y) \in G_m(a), \text{ then } \rho(\mathrm{iq}(x, y)) = \mathrm{iq}(\rho(x), \rho(y)),$$

even if $\mathrm{rem}(x, y) \notin G_m(a)$, but we cannot define one without the other. This is why we introduced ρ, which is defined on the larger set $C(a; h) \supseteq G_{m+1}(a)$, and we proceed as follows.

Assume again canonical representations of x and y in $C(a; h_1)$, and also that $x \geq y \geq 1$, and consider the correct division equation

$$x = yq + r \quad (0 \leq r < y)$$

as in the proof of Lemma 6A.3. Recall the two cases in that proof.

Case 1, $y_2 = 0$. Now $r \leq h_1$, and

$$q = \mathrm{iq}(x, y) = \frac{(y_2 x_0 - y_2 x_2 r) + y_2 x_1 a}{x_2 y_0}$$

with all the coefficients $\leq h_1^5 < h_1^6 = h$, so that this is the canonical representation of q in $C(a; h)$. It follows that

$$\rho(r) = r, \quad \rho(q) = \frac{(y_2 x_0 - y_2 x_2 r) + y_2 x_1 \lambda a}{x_2 y_0},$$

so that, by direct computation,

(6A-3) $$\rho(x) = \rho(y)\rho(q) + \rho(r).$$

Moreover, ρ is order-preserving, so

$$0 \leq \rho(r) < \rho(y),$$

and (6A-3) is the correct division equation for $\rho(x), \rho(y)$. Thus

$$\rho(q) = \mathrm{iq}(\rho(x), \rho(y)), \quad \rho(r) = \mathrm{rem}(\rho(x), \rho(y))),$$

whether or not $\mathrm{iq}(x, y) \in G_m(a)$ or $\mathrm{rem}(x, y) \in G_m(a)$; but if it happens that $\mathrm{iq}(x, y) \in G_m(a)$, then

$$\pi(\mathrm{iq}(x, y)) = \rho(\mathrm{iq}(x, y)) = \mathrm{iq}(\rho(x), \rho(y)) = \mathrm{iq}(\pi(x), \pi(y)),$$

independently of whether $\mathrm{rem}(x, y) \in G_m(a)$ or not. The same argument works for $\pi(\mathrm{rem}(x, y))$ and completes the proof in this case.

Case 2 is handled in the same way, and we skip it. ⊣

Recall the definition of *good examples* in Section 5C.

6A.6. **Theorem.** *If $R(x)$ is a good example, then for all $a \geq 2$*

$$R(a) \Longrightarrow \mathrm{depth}_R(\mathbf{N}_0[\div], a) > \frac{1}{12} \log \log a.$$

PROOF. Suppose $R(a)$, let $m = \mathrm{depth}_R(\mathbf{N}_0[\div], a)$, and assume that

$$2^{6^{m+3}} < a.$$

If $\lambda(\mu)$ is the polynomial which witnesses the goodness of R and

$$\lambda = \lambda(a!),$$

then Lemma 6A.5 guarantees an embedding

$$\pi : \mathbf{N}_0[\div] \upharpoonright G_m(a) \rightarrowtail \mathbf{N}_0[\div],$$

with $\pi a = \lambda a$; and since $\neg R(\lambda a)$, the Homomorphism Test 4E.3 yields a contradiction, so that

$$2^{6^{m+3}} \geq a. \tag{6A-4}$$

Taking logarithms twice, we get from this

$$m + 3 \geq \frac{\log \log a}{\log 6};$$

and since $m \geq 1$ by Problem x4E.4, $4m \geq m + 3$, so that we get the required

$$m \geq \frac{\log \log a}{4 \log 6} > \frac{\log \log a}{12}.$$

⊣

Problems for Section 6A

x6A.1. **Problem.** Prove that if $x \in C(a;h)$ and $2h^2 < a$, then (6A-2) holds for uniquely determined, relatively prime x_0, x_1, x_2. HINT: By an (easy) extension of Bezout's Lemma, Problem x1C.13,

$$\gcd(x_0, x_1, x_2) = \alpha x_0 + \beta x_1 + \gamma x_2 \quad (\text{for some } \alpha, \beta, \gamma \in \mathbb{Z}).$$

Use this and the equivalences in Lemma 6A.2.

6B. Three results from number theory

We review in this section three elementary results from diophantine approximation, which give us just what we need to establish an analog of the $\boldsymbol{Lin}_0[\div]$-Uniqueness Lemma 6A.2 for canonical forms involving two numbers when these satisfy certain conditions. Those with some knowledge of number theory know these—in fact they probably know better proofs of them, which establish more; they should peruse the section quickly, just to get the terminology that we will be using—especially the definition of *difficult pairs* on page 181 which is not standard.

6B.1. **Theorem** (Pell pairs). *The pairs $(x_n, y_n) \in \mathbb{N}^2$ defined by the recursion*

$$\text{(6B-1)} \qquad (x_1, y_1) = (3, 2), \qquad (x_{n+1}, y_{n+1}) = (3x_n + 4y_n, 2x_n + 3y_n)$$

satisfy Pell's equation

$$\text{(6B-2)} \qquad x_n^2 - 2y_n^2 = 1,$$

and the inequalities

$$\text{(6B-3)} \qquad 2^n \leq 2 \cdot 5^{n-1} \leq y_n < x_n \leq 7^{n+1},$$

$$\text{(6B-4)} \qquad 0 < \frac{x_n}{y_n} - \sqrt{2} < \frac{1}{2y_n^2}.$$

PROOF. Equation (6B-2) is true for $n = 1$, and inductively:

$$x_{n+1}^2 - 2y_{n+1}^2 = (3x_n + 4y_n)^2 - 2(2x_n + 3y_n)^2 = x_n^2 - 2y_n^2 = 1.$$

For (6B-3), we first check that $2^n \leq 2 \cdot 5^{n-1}$ by a trivial induction on $n \geq 1$, and then, inductively again,

$$y_{n+1} = 2x_n + 3y_n \geq 5y_n \geq 5 \cdot 2 \cdot 5^{n-1} = 2 \cdot 5^n.$$

The last part of the triple inequality is proved similarly:

$$x_{n+1} = 3x_n + 4y_n \leq 7x_n \leq 7 \cdot 7^n = 7^{n+1}.$$

The crucial, last inequality (6B-4) holds for any pair of positive numbers which satisfies Pell's equation. To see this, suppose $x^2 - 2y^2 = 1$, and notice first that since

$$\frac{x^2}{y^2} = 2 + \frac{1}{y^2} > 2,$$

we have $\dfrac{x}{y} > \sqrt{2}$, and hence

$$\frac{x}{y} + \sqrt{2} > 2\sqrt{2} > 2;$$

now

$$(\frac{x}{y} - \sqrt{2})(\frac{x}{y} + \sqrt{2}) = \frac{1}{y^2}$$

yields the required

$$0 < \frac{x}{y} - \sqrt{2} = \frac{1}{(\frac{x}{y} + \sqrt{2})y^2} < \frac{1}{2y^2}.$$

⊣

In fact, the pairs (x_n, y_n) defined in (6B-1) comprise all positive solutions of Pell's equation, cf. Problem x6B.1.

Good approximations of irrationals. A pair of numbers (a, b) (or the proper fraction $\dfrac{a}{b}$) is a *good approximation of* an irrational number ξ, if $a \perp\!\!\!\perp b$ and

$$\left|\frac{a}{b} - \xi\right| < \frac{1}{b^2}. \tag{6B-5}$$

Theorem 6B.1 asserts in part that there are infinitely many good approximations of $\sqrt{2}$. This is true of all irrational numbers, and it is worth understanding it in the context of what we are doing, although we will never need it in its full generality.

6B.2. **Theorem** (Hardy and Wright [1938] Thm. 188). *For every irrational number $\xi > 0$, there are infinitely many pairs (x, y) of relatively prime natural numbers such that*

$$\left|\xi - \frac{x}{y}\right| < \frac{1}{y^2}.$$

Of the many proofs of this result, we outline one which (according to Hardy and Wright) is due to Dirichlet.

PROOF. For any real number ξ, let

$$\lfloor \xi \rfloor = \text{the largest natural number } \leq \xi$$

be the *house* of ξ and $\xi - \lfloor \xi \rfloor$ its *fractional part*, so that

$$0 \leq \xi - \lfloor \xi \rfloor < 1.$$

If we divide the half-open (real) unit interval into n disjoint, equal parts,

$$[0,1) = [0,\frac{1}{n}) \cup [\frac{1}{n},\frac{2}{n}) \cup \cdots \cup [\frac{n-1}{n},1),$$

then for every ξ, the fractional part $\xi - \lfloor \xi \rfloor$ will belong to exactly one of these subintervals. Now fix a number

$$n \geq 1,$$

and apply this observation to each of the $n+1$ numbers

$$0, \xi, 2\xi, \ldots, n\xi;$$

at least two of their fractional parts will be in the same subinterval of $[0,1)$, so that, no matter what the $n \geq 1$, we get

$$0 \leq j < k \leq n$$

such that

$$\Big|j\xi - \lfloor j\xi \rfloor - (k\xi - \lfloor k\xi \rfloor)\Big| < \frac{1}{n};$$

and setting $y = k - j$, $x = \lfloor k\xi \rfloor - \lfloor j\xi \rfloor$, we get

$$\Big|x - y\xi\Big| < \frac{1}{n}.$$

We may assume that x and y are relatively prime in this inequality, since if we divide both by $\gcd(x,y)$ the inequality persists. Moreover, since $0 < y < n$, we can divide the inequality by y to get

$$\Big|\frac{x}{y} - \xi\Big| < \frac{1}{ny} < \frac{1}{y^2}.$$

Notice that if $n = 1$, then this construction gives $y = 1$, $x = \lfloor \xi \rfloor$, and the rather trivial good approximation

$$\Big|\frac{\lfloor \xi \rfloor}{1} - \xi\Big| < \frac{1}{1^2}.$$

However, we have not yet used the fact that ξ is irrational, which implies that

$$0 < \Big|\frac{x}{y} - \xi\Big|,$$

so that there is a number

$$m > \frac{1}{|\frac{x}{y} - \xi|}.$$

We now repeat the construction with m instead of n, to get x_1, y_1 such that

$$\Big|\frac{x_1}{y_1} - \xi\Big| < \frac{1}{y_1^2} \text{ and } \Big|\frac{x_1}{y_1} - \xi\Big| < \frac{1}{my_1} \leq \frac{1}{m} < \Big|\frac{x}{y} - \xi\Big|,$$

so that $\dfrac{x_1}{y_1}$ is a better, good approximation of ξ; and repeating the construction indefinitely, we get infinitely many, distinct good approximations. ⊣

Next comes the most important result we need, which says, in effect, that algebraic irrational numbers cannot have "too good" approximations.

6B.3. **Theorem** (Liouville's Theorem). *Suppose ξ is an irrational root of an irreducible* (over $\mathbb{Q}$) *polynomial $f(x)$ with integer coefficients and of degree $n \geq 2$, and let*

$$c = \lceil \sup\{|f'(x)| \mid |x - \xi| \leq 1\} \rceil .$$

It follows that for all pairs (x, y) of relatively prime integers,

$$\left|\xi - \frac{x}{y}\right| > \frac{1}{cy^n}.$$

In particular, for all relatively prime (x, y),

$$\left|\sqrt{2} - \frac{x}{y}\right| > \frac{1}{5y^2}.$$

PROOF. We may assume that $|\xi - \frac{x}{y}| \leq 1$, since the desired inequality is trivial in the opposite case. Using the fact that $f(\xi) = 0$ and the Mean Value Theorem, we compute, for any $\frac{x}{y}$ within 1 of ξ,

$$|f(\frac{x}{y})| = |f(\xi) - f(\frac{x}{y})| \leq c|\xi - \frac{x}{y}|.$$

Moreover, $f(\frac{x}{y}) \neq 0$, since $f(x)$ does not have any rational roots, and $y^n f(\frac{x}{y})$ is an integer, since all the coefficients of $f(x)$ are integers and the degree of $f(x)$ is n; thus

$$1 \leq |y^n f(\frac{x}{y})| \leq y^n c|\xi - \frac{x}{y}|,$$

from which we get the desired inequality (noticing that it must be strict, since ξ is not rational).

For the special case $\xi = \sqrt{2}$, we have $f(x) = x^2 - 2$, so that $f'(x) = 2x$ and

$$c = \lceil \sup\{2x \mid |\sqrt{2} - x| \leq 1\} \rceil = \lceil 2(\sqrt{2} + 1) \rceil = 5. \quad ⊣$$

Liouville's Theorem implies that *good approximations of a non-rational algebraic number cannot be too well approximated by fractions with a much smaller denominator*. We formulate precisely the special case of this general fact that we need.

Difficult pairs. A pair of numbers (a, b) is *difficult* if $a \perp\!\!\!\perp b$,

$$2 \le b < a < 2b, \tag{6B-6}$$

and for all y, z,

$$0 < |z| < \frac{b}{\sqrt{10}} \implies \left|\frac{a}{b} - \frac{y}{z}\right| > \frac{1}{10z^2}. \tag{6B-7}$$

6B.4. **Lemma.** (1) *Every good approximation of $\sqrt{2}$ other than 1 is a difficult pair; in particular, every solution (a, b) of Pell's equation is a difficult pair.*

(2) *If (a, b) is a difficult pair, then for all y, z,*

$$0 < |z| < \frac{b}{\sqrt{10}} \implies |za + yb| > \frac{b}{10|z|}. \tag{6B-8}$$

PROOF. (1) Let (a, b) be a good approximation of $\sqrt{2}$ with $b \ge 2$. Then (6B-6) follows from

$$1 < \sqrt{2} - \frac{1}{4} \le \sqrt{2} - \frac{1}{b^2} < \frac{a}{b} < \sqrt{2} + \frac{1}{b^2} \le \sqrt{2} + \frac{1}{4} < 2.$$

To prove (6B-7), suppose $0 < |z| < \dfrac{b}{\sqrt{10}}$, and use Liouville's Theorem 6B.3:

$$\begin{aligned}\left|\frac{a}{b} - \frac{y}{z}\right| &\ge \left|\frac{y}{z} - \sqrt{2}\right| - \left|\frac{a}{b} - \sqrt{2}\right| \\ &> \frac{1}{5z^2} - \frac{1}{b^2} > \frac{1}{5z^2} - \frac{1}{10z^2} = \frac{1}{10z^2}.\end{aligned}$$

The result holds for all solutions of Pell's equation because the proof of (6B-4) was based only on the hypothesis $x^2 = 1 + 2y^2$.

(2) is very useful and easy: assuming the hypothesis of (6B-8),

$$|za + yb| = |z|b\left|\frac{a}{b} + \frac{y}{z}\right| > |z|b\frac{1}{10z^2} = \frac{b}{10|z|}. \qquad \dashv$$

We leave for the problems the similar proof that pairs (F_{k+1}, F_k) of successive Fibonacci numbers with $k \ge 3$ are also difficult.

Problems for Section 6B

x6B.1. **Problem.** Prove that the pairs of numbers (x_n, y_n) defined in the proof of Theorem 6B.1 comprise all the positive solutions of the Pell equation $a^2 = 2b^2 + 1$.

Recall from Problem x1C.8 that if

$$\varphi = \frac{1 + \sqrt{5}}{2}, \quad \hat{\varphi} = \frac{1 - \sqrt{5}}{2}$$

are the two solutions of the quadratic equation $x + 1 = x^2$, then

$$1 < \varphi < 2, \quad \hat{\varphi} < 0, \quad |\hat{\varphi}| < 1,$$

and the Fibonacci numbers are explicitly defined in terms of these,

$$F_k = \frac{\varphi^k - \hat{\varphi}^k}{\sqrt{5}}.$$

x6B.2. **Problem.** Prove that for $k \geq 1$,

$$\text{if } k \text{ is even, then } \varphi < \frac{F_{k+1}}{F_k}, \text{ and if } k \text{ is odd, then } \frac{F_{k+1}}{F_k} < \varphi.$$

HINT: Use the equation

$$\frac{F_{k+1}}{F_k} = \varphi R(k) \text{ where } R(k) = \frac{1 - \frac{\hat{\varphi}^{k+1}}{\varphi^{k+1}}}{1 - \frac{\hat{\varphi}^k}{\varphi^k}},$$

and compute the sign and size of $R(k)$ for odd and even k.

x6B.3. **Problem.** Prove that for all $n \geq 1$,

$$F_{n+1}F_{n-1} - F_n^2 = (-1)^n. \tag{6B-9}$$

Infer that

$$\left|\frac{F_{n+1}}{F_n} - \varphi\right| < \frac{1}{F_n^2} \quad (n \geq 1).$$

x6B.4. **Problem.** Prove that for every $n \geq 3$, the pair (F_{n+1}, F_n) is a difficult pair.

HINT: The golden mean φ is a root of the polynomial $f(x) = x^2 - x - 1$. Use Liouville's Theorem 6B.3 to show that for all coprime x, y,

$$\left|\frac{x}{y} - \varphi\right| > \frac{1}{5y^2},$$

and then imitate the proof of (1) of Lemma 6B.4 with φ in place of $\sqrt{2}$.

The definition of *difficult pair* is tailor made for the good approximations of $\sqrt{2}$, and it is only a coincidence that it also applies to pairs of successive Fibonacci numbers. It is, however, quite easy to fix the constants hard-wired in it so that it applies to the good approximations of any quadratic irrational, and then use it to extend the results in the next section to this more general case, cf. Problems x6B.5 and x6C.1*.

x6B.5. **Problem.** Suppose $\xi > 1$ is irrational, $C > 0$, $a \perp\!\!\!\perp b$, $2 \leq b$ and

$$\frac{1}{Cb^2} < \left|\xi - \frac{a}{b}\right| < \frac{1}{b^2}. \tag{$*$}$$

Let $\lfloor \xi \rfloor = M \geq 1$, so that

$$1 \leq M < \xi < M + 1.$$

Prove that:

(1) $a < (M + 2)b$.

(2) For all $z, y \in \mathbb{Z}$,

$$0 < |z| < \frac{b}{\sqrt{2C}} \Longrightarrow \left|\frac{a}{b} - \frac{y}{z}\right| > \frac{1}{2Cz^2} \Longrightarrow |za - yb| > \frac{b}{2C|z|}.$$

Prove also that for every quadratic irrational $\xi > 1$, $(*)$ holds for infinitely many coprime pairs a, b.

Beyond this, it is not clear whether the notion (or suitable adaptations of it to arbitrary algebraic real numbers) has any number-theoretic uses. But it is exactly what we need for the next, basic result.

6C. Coprimeness from $\boldsymbol{Lin}_0[\div]$

We can now combine the methods from Sections 5D and 6A, to derive a double-log lower bound for coprimeness from $\boldsymbol{Lin}_0[\div]$. The key is the following Uniqueness Lemma for linear combinations of a difficult pair.

6C.1. **Lemma.** *Suppose (a, b) is a difficult pair, $1 \le \lambda \in \mathbb{N}$, and*

$$|x_3 y_i|, |y_3 x_i| < \frac{\sqrt{b}}{2\sqrt{10}}$$

for $i = 0, 1, 2, 3$ with $x_3, y_3 > 0$. Then

$$\frac{x_0 + x_1\lambda a + x_2\lambda b}{x_3} = \frac{y_0 + y_1\lambda a + y_2\lambda b}{y_3}$$
$$\Longleftrightarrow [y_3x_0 = x_3y_0 \;\&\; y_3x_1 = x_3y_1 \;\&\; y_3x_2 = x_3y_2],$$

$$\frac{x_0 + x_1\lambda a + x_2\lambda b}{x_3} > \frac{y_0 + y_1\lambda a + y_2\lambda b}{y_3}$$
$$\Longleftrightarrow [y_3(x_1a + x_2b) > x_3(y_1a + y_2b)]$$
$$\text{or } \Big([y_3(x_1a + x_2b) = x_3(y_1a + y_2b)]$$
$$\& \; y_3x_0 > x_3y_0\Big).$$

PROOF. The claimed equivalences follow from the following two facts, applied to $(y_3x_0 - x_3y_0) + (y_3x_1 - x_3y_1)\lambda a + (y_3x_2 - x_3y_2)\lambda b$.

(1) *If $x + z\lambda a + y\lambda b = 0$ and $|x|, |y|, |z| < \dfrac{\sqrt{b}}{\sqrt{10}}$, then $x = y = z = 0$.*

Proof. Assume the hypothesis of (1). The case $y = z = 0$ is trivial, and if $z = 0$ and $y \neq 0$, then

$$b \leq \lambda|y|b = |x| < \frac{\sqrt{b}}{\sqrt{10}},$$

which is absurd. So we may assume that $z \neq 0$. Now the assumed bound on z and (6B-8) implies

$$|z\lambda a + y\lambda b| \geq |za + yb| > \frac{b}{10|z|} \geq |x|$$

the last because

$$|xz| \leq \frac{\sqrt{b}}{\sqrt{10}} \frac{\sqrt{b}}{\sqrt{10}} = \frac{b}{10};$$

and this contradicts the hypothesis $|z\lambda a + y\lambda b| = |-x|$.

(2) *If* $|x|, |y|, |z| < \dfrac{\sqrt{b}}{\sqrt{10}}$, *then*

$$x + z\lambda a + y\lambda b > 0 \iff [za + yb > 0] \vee [x > 0 \;\&\; z = y = 0].$$

Proof. If $z = 0$, then the equivalence follows from Lemma 5B.1; and if $z \neq 0$, then $|z\lambda a + y\lambda b| > |x|$ as above, and so adding x to $z\lambda a + y\lambda b$ cannot change its sign. ⊣

6C.2. **Lemma.** *Suppose* (a, b) *is a difficult pair,* $h \geq 2$, $X, Y \in C(a, b; h)$, *and* $h^{28} \leq b$. *Then* $\text{iq}(X, Y), \text{rem}(X, Y) \in C(a, b; h^{12})$.

PROOF. Let us notice immediately that (by a simple computation, using $2 \leq h$) the assumption $h^{28} \leq b$ implies that

$$h^2 < \frac{\sqrt{b}}{2\sqrt{10}}. \tag{6C-1}$$

This allows us to appeal to Lemma 6C.1 in several parts of the argument, and the more outrageous-looking $h^{28} \leq b$ will be needed for one more, specific application of the same Lemma. In effect, we just need to assume that h is sufficiently smaller than b to justify these appeals to Lemma 6C.1, and the 28th power is what makes this particular argument work.

It is enough to prove the result when $X \geq Y > 0$, since it is trivial when $X < Y$. Suppose

$$X = \frac{x_0 + x_1 a + x_2 b}{x_3}, \quad Y = \frac{y_0 + y_1 a + y_2 b}{y_3},$$

where all $|x_i|, |y_i| \leq h$ and $x_3, y_3 > 0$, and consider the correct division equation

$$\frac{x_0 + x_1 a + x_2 b}{x_3} = \frac{y_0 + y_1 a + y_2 b}{y_3} Q + R \quad (0 \leq R < Y). \tag{6C-2}$$

We must show that $Q, R \in C(a, b; h^{12})$.

Case 1, $y_1 a + y_2 b = 0$. Now (6C-2) takes the form

$$\frac{x_0 + x_1 a + x_2 b}{x_3} = \frac{y_0}{y_3} Q + R \quad (0 \le R < \frac{y_0}{y_3}),$$

so that $R < h$. Solving (6C-2) for Q, we get in this case

$$Q = \frac{y_3}{y_0} \frac{(x_0 - x_3 R) + x_1 a + x_2 b}{x_3} \in C(a, b; h^4). \tag{6C-3}$$

Case 2, $y_1 a + y_2 b \neq 0$. Then $y_1 a + y_2 b > 0$, by Lemma 6C.1, since $Y > 0$, using (6C-1). We are going to show that in this case

$$h^9 Y > X \tag{6C-4}$$

so that $Q \le h^9$. Assuming this, we can solve the division equation (6C-2) for R, to get

$$R = \frac{(x_0 y_3 - y_0 x_3 Q) + (x_1 y_3 - y_1 x_3 Q) a + (x_2 y_3 - y_2 x_3 Q) b}{x_3 y_3}; \tag{6C-5}$$

and from this, easily, $R \in C(a, b; h^{12})$.

We show (6C-4) by separately comparing the "infinite parts" (those involving a and b) of X and Y with b. Compute first:

$$y_3(x_1 a + x_2 b) \le |y_3 x_1| a + |y_3 x_2| b \le h^2 2b + h^2 b = 3h^2 b \le h^4 b, \tag{6C-6}$$

using $a < 2b$. On the other hand, if $y_2 = 0$, then $y_1 > 0$ and so

$$x_3(y_1 a + y_2 b) = x_3 y_1 a > b;$$

and if $y_2 \neq 0$, then by (6B-8),

$$(y_1 a + y_2 b) > \frac{b}{10|y_1|}, \quad \text{so that} \quad 10|y_1|(y_1 a + y_2 b) > b,$$

and hence (since $10 < 2^4$), in either case,

$$h^5(y_1 a + y_2 b) > b. \tag{6C-7}$$

Now (6C-6) and (6C-7) imply that

$$h^9 x_3(y_1 a + y_2 b) > h^4 h^5 (y_1 a + y_2 b) > h^4 b \ge y_3(x_1 a + x_2 b),$$

and we can finish the proof of (6C-4) with an appeal to Lemma 6C.1, provided that the coefficients of $h^9 Y$ and X in canonical form satisfy the hypotheses of this Lemma. For the worst case, the required inequality is

$$|x_3 h^9 y_i| \le \frac{\sqrt{b}}{2\sqrt{10}},$$

and it is implied by

$$h^{11} \le \frac{\sqrt{b}}{2\sqrt{10}};$$

if we square this and simplify (using that $40 < 2^6$), we see that it follows from the assumed $h^{28} \le b$. ⊣

6C.3. **Lemma** (Inclusion). *Suppose (a, b) is a difficult pair, and for any m, let $G_m(a, b) = G_m(\mathbf{N}_0[\div], a, b)$; it follows that*

$$\text{if } 2^{2^{4m+5}} \le a, \text{ then } G_m(a, b) \subseteq C(a, b; 2^{2^{4m}}).$$

PROOF is by induction on m, the case $m = 0$ being trivial. To apply Lemmas 6A.1 and 6C.2 at the induction step, we need to verify (under the hypothesis on a and m) the following two inequalities.

(1) $\left(2^{2^{4m}}\right)^{12} \le 2^{2^{4(m+1)}}$. This holds because

$$\left(2^{2^{4m}}\right)^{12} = 2^{12 \cdot 2^{4m}} < 2^{2^4 \cdot 2^{4m}} = 2^{2^{4(m+1)}}.$$

(2) $\left(2^{2^{4m}}\right)^{28} \le b$. So compute:

$$\left(2^{2^{4m}}\right)^{28} = 2^{28 \cdot 2^{4m}} < 2^{2^5 \cdot 2^{4m}} = 2^{2^{4m+5}} \le a. \qquad ⊣$$

6C.4. **Lemma.** *Suppose (a, b) is a difficult pair, $2^{2^{4m+6}} \le a$, and set $\lambda = 1 + a!$. Then there is an embedding*

$$\pi : \mathbf{N}_0[\div] \restriction C(a, b; 2^{2^{4m}}) \rightarrowtail \mathbf{N}_0[\div]$$

such that $\pi(a) = \lambda a$, $\pi(b) = \lambda b$.

PROOF. To simplify notation, let

$$h = 2^{2^{4m}}.$$

As in the proof of Lemma 6A.5, we will actually need to define the embedding on the larger substructure $\mathbf{N}_0[\div] \restriction C(a, b; h^{12})$, so let's first verify that the assumed bound on h is good enough to insure unique canonical forms in $C(a, b; h^{12})$. By Lemma 6C.1, we need to check that

$$\left(h^{12}\right)^2 < \frac{\sqrt{b}}{2\sqrt{10}},$$

which is equivalent to

$$4 \cdot 10 h^{48} < b; \tag{6C-8}$$

and this is true, because

$$4 \cdot 10 h^{49} < 2^2 \cdot 2^4 h^{49} \le h^{55} = \left(2^{2^{4m}}\right)^{55} < 2^{2^6 \cdot 2^{4m}} = 2^{2^{4m+6}} < a,$$

by the hypothesis, and it yields (6C-8) when we divide both sides by h.

Using Lemma 6C.1 now, we define

$$\rho : C(a, b; h^{12}) \to \mathbb{N},$$

in the expected way,

$$\rho\Big(\frac{x_0 + x_1 a + x_2 b}{x_3}\Big) = \frac{x_0 + x_1 \lambda a + x_2 \lambda b}{x_3},$$

and we verify as in the proof of Lemma 6A.5 that this is a well-defined, order-preserving injection, with values in $\mathbb{N}$ (since $h < a$, and so $x_3 \mid \lambda - 1$), and it respects all the operations in $\boldsymbol{Lin}_0$. We let

$$\pi = \rho \restriction G_m(a, b),$$

and all that remains is to show that π respects $\mathrm{iq}(x, y)$ and $\mathrm{rem}(x, y)$ when they are defined in $G_m(a, b)$. The key fact is that by Lemma 6C.2 and the bound on h^{12},

$$X, Y \in G_m(a, b) \Longrightarrow \mathrm{iq}(X, Y), \mathrm{rem}(X, Y) \in C(a, b; h^{12}).$$

Thus it is enough to show that if

$$X = YQ + R \quad (0 \le R < Y)$$

is the correct division equation for X, Y, then

(6C-9) $$\rho X = \rho Y \cdot \rho Q + \rho R \quad (0 \le \rho R < \rho Y)$$

is the correct division equation for $\rho X, \rho Y$. We distinguish two cases, following the proof of Lemma 6C.2.

Case 1, $y_1 a + y_2 b = 0$. Then $0 \le R < Y = \dfrac{y_0}{y_3} \le h$, so $\rho R = R$ and $\rho Y = Y$. Now $\rho R < \rho Y$ since ρ is order-preserving. The explicit formula (6C-3) for Q yields

$$\rho Q = \frac{y_3}{y_0} \frac{(x_0 - x_3 R) + x_1 \lambda a + x_2 \lambda b}{x_3},$$

and a direct computation with these expressions for ρR, ρY and ρQ yields (6C-9).

Case 2, $y_1 a + y_2 b > 0$. Now $Q \le h^9$, which implies $\rho Q = Q$. The explicit formula (6C-5) for R yields

$$\rho R = \frac{(x_0 y_3 - y_0 x_3 Q) + (x_1 y_3 - y_1 x_3 Q)\lambda a + (x_2 y_3 - y_2 x_3 Q)\lambda b}{x_3 y_3};$$

with these expressions for ρR and ρQ we get again (6C-9) by direct computation. ⊣

6C.5. **Theorem** (van den Dries and Moschovakis [2004]). *For all difficult pairs* (a, b),

(6C-10) $$\mathrm{depth}_{\perp\!\!\!\perp}(\mathbf{N}_0[\div], a, b) > \frac{1}{10} \log\log a.$$

PROOF. Let $m = \text{depth}_{\perp\!\!\!\perp}(\mathbf{N}_0[\div], a, b)$ for some difficult pair (a, b). If

$$2^{2^{4m+6}} \leq a,$$

then Lemma 6C.4 provides an embedding π which does not respect coprimeness at (a, b) since $\pi a = \lambda a$ and $\pi b = \lambda b$, with some λ. This contradicts the choice of m, and so

$$2^{2^{4m+6}} > a;$$

in other words

$$4m + 6 \geq \log\log a;$$

and since $m \geq 1$ by Problem x4E.4,

$$10m \geq 4m + 6 \geq \log\log a,$$

as required. ⊣

6C.6. **Corollary.** *For every difficult pair* (a, b),

$$\text{depth}_{\text{gcd}}(\mathbf{N}_0[\div], a, b) \geq \frac{1}{10}\log\log a.$$

PROOF. For any $\mathbf{U}$, every embedding $\pi : \mathbf{U} \rightarrowtail \mathbf{N}_0[\div]$ which respects $\gcd(a, b)$ also respects $a \perp\!\!\!\perp b$, so

$$\text{depth}_{\perp\!\!\!\perp}(\mathbf{N}_0[\div], a, b) \leq \text{depth}_{\text{gcd}}(\mathbf{N}_0[\div], a, b). \quad ⊣$$

6C.7. **Corollary.** *Pratt's algorithm is weakly* calls-*optimal for coprimeness in* $\mathbf{N}_0[\div]$.

PROOF is immediate from Problem x2E.19. ⊣

This corollary implies that Theorem 6C.5 is best possible (except, of course, for the specific constant 10), because the absolute lower bound it gives for all difficult pairs is matched by the Pratt algorithm on pairs of successive Fibonacci numbers. Note, however, that it does not rule out the possibility that the Main Conjecture in the Introduction holds for all uniform processes of $\mathbf{N}_\varepsilon$, even if we formulate it for coprimeness rather than the gcd—because it might hold with another, more restrictive or different notion of "difficult pair" which excludes pairs of successive Fibonacci umbers; in other words, we may have the wrong proof.

The most exciting possibility would be that the conjecture holds for deterministic uniform processes—or just deterministic programs—which would exhibit the distinction between determinism and nondeterminism in a novel context. I have no idea whether this holds or how to go about proving it.

Problems for Section 6C

x6C.1*. **Problem** (van den Dries and Moschovakis [2004], [2009]). For every quadratic irrational $\xi > 1$, there is a rational number $r > 0$ such that for all but finitely many good approximations (a, b) of ξ,

$$\text{depth}(\mathbf{N}[\div], \perp\!\!\!\perp, a, b) \geq r \log\log a. \tag{6C-11}$$

HINT: Use Problem x6B.5 to adjust the argument for difficult pairs in this section.

The $O(\log\log)$ bound in this problem is best possible, because of Pratt's algorithm.

CHAPTER 7

LOWER BOUNDS FROM DIVISION AND MULTIPLICATION

The arithmetic becomes substantially more complex—and a little algebra needs to be brought in—when we add both division with remainder and multiplication to the primitives of $\boldsymbol{Lin}_0$. We will only derive here lower bounds for *unary functions and relations* from

$$\boldsymbol{Lin}_0[\div, \cdot] = \boldsymbol{Lin}_0 \cup \{\mathrm{iq}, \mathrm{rem}, \cdot\} = \{0, 1, =, <, +, \dot{-}, \mathrm{iq}, \mathrm{rem}, \cdot\},$$

leaving the general problem open.

7A. Polynomials and their heights

We review briefly some elementary, basic results about the ring $K[T]$ of unary polynomials over a field K, and we also derive some equally simple facts about the ring $\mathbb{Z}[T]$ of polynomials over the integers.

To fix terminology, a *polynomial* in the *indeterminate* (variable) T over a ring K is a term

$$X = x_0 + x_1 T + x_1 T^2 + \cdots + x_n T^n,$$

where $x_i \in K$ and $x_n \neq 0$ together with the *zero polynomial* 0. It is sometimes useful to think of X as an infinite sum of monomials $x_i T^i$, in which $x_i \neq 0$ for only finitely many i; however we do this, the *degree* of a non-zero X is the largest power of T which appears in X with a non-zero coefficient, and it is 0 when $X = x_0$ is just an element of K. We do not assign a degree to the zero polynomial.[37]

Two polynomials are equal if they are literally identical as terms, i.e., if the coefficients of like powers are equal.

[37]The usual convention is to set $\deg(0) = -\infty$, which saves some considerations of cases in stating results.

The *sum*, *difference* and *product* of two polynomials are defined by the performing the obvious operations on the coefficients and collecting terms:

$$\begin{aligned} X + Y &= \textstyle\sum_i (x_i + y_i)T^i, & \deg(X + Y) &\le \max(\deg(X), \deg(Y)) \\ -X &= \textstyle\sum_i (-x_i)T^i, & \deg(-X) &= \deg(X) \\ XY &= \textstyle\sum_k \left(\sum_{i=0}^{i=k} x_i y_{k-i}\right) T^k & \deg(XY) &= \deg(X) + \deg(Y). \end{aligned}$$

The last formula illustrates the occasional usefulness of thinking of a polynomial as an infinite sum with just finitely many non-zero terms.

With these operations, the set $K[T]$ of polynomials over a ring K is a (commutative) ring over K. For the more interesting division operation, we need to assume that K is a field.

7A.1. **Theorem** (The Division Theorem for polynomials). *If K is a field, and $X, Y \in K[T]$ such that $\deg(X) \ge \deg(Y)$ and $Y \neq 0$, then there exist unique polynomials $Q, R \in K[T]$ such that*

(7A-1) $$X = YQ + R \quad \text{and } R = 0 \text{ or } \deg(R) < \deg(Y).$$

PROOF is by induction on the difference $d = n - m$ of the degrees of the given polynomials, $n = \deg(X)$, $m = \deg(Y)$.

At the basis, if $m = n$, then

$$X = Y\frac{x_n}{y_n} + R$$

with R defined by this equation, so that either it is 0 or its degree is less than n, since X and $\dfrac{x_n}{y_n}Y$ have the same highest term $x_n T^n$.

In the induction step, with $d = n - m > 0$, first we divide X by YT^d, the two having the same degree:

$$X = YT^d Q_1 + R_1 \quad (R_1 = 0 \text{ or } \deg(R_1) < n).$$

If $R_1 = 0$ or $\deg(R_1) < m$, we are done; otherwise $\deg(R_1) \ge \deg(Y)$ and we can apply the induction hypothesis to get

$$R_1 = YQ_2 + R_2 \quad (R_2 = 0 \text{ or } \deg(R_2) < \deg(Y)).$$

We now have

$$X = Y(T^d Q_1 + Q_2) + R_2 \quad (R_2 = 0 \text{ or } \deg(R_2) < \deg(Y)),$$

which is what we needed.

We skip the proof of uniqueness, which basically follows from the construction. ⊣

We call (7A-1) the *correct division equation* (**cde**) for X, Y.

This basic fact does not hold for polynomials in $\mathbb{Z}[T]$: for example, if $X = 3$ and $Y = 2$, then there are no Q, R which satisfy (7A-1), simply because 2 does not divide 3 in $\mathbb{Z}$. To get at the results we need, it is most convenient to work with the larger ring $\mathbb{Q}[T]$, but study a particular "presentation" of it, in which the concept if *height* is made explicit.

The *height* of a non-zero integer polynomial is the maximum of the absolute values of its coefficients,

$$\text{height}(x_0 + x_1 T + \cdots + x_n T^n) = \max\{|x_i| \mid i = 0, \ldots, n\} \quad (x_i \in \mathbb{Z}).$$

To extend the definition to $\mathbb{Q}[T]$, we let for each $n, h \in \mathbb{N}$,

$$\text{(7A-2)} \quad Q_n(T; h) = \Big\{ \frac{x_0 + x_1 T + x_2 T^2 + \cdots + x_n T^n}{x^*} \mid x_0, \ldots, x_n, x^* \in \mathbb{Z}, x^* > 0 \text{ and } |x_0|, \ldots, |x_n|, |x^*| \leq h \Big\}.$$

This is the set of polynomials in the indeterminate T over $\mathbb{Q}$, with degree n and *height* no more than h. When the degree is not relevant, we skip the subscript,

$$Q(T; h) = \bigcup_n Q_n(T; h);$$

and in computing heights, it is sometimes convenient to use the abbreviation

$$X : h \iff X \in Q(T; h).$$

The *canonical form* (7A-2) gives a unique height(X) if the coefficients x_i have no common factor with x^*, but this is not too important: most of the time we only care for an upper bound for height(X) which can be computed without necessarily bringing X to canonical form. Notice however, that (as a polynomial over $\mathbb{Q}$),

$$X = \frac{3 + 2T}{6} = \frac{1}{2} + \frac{1}{3}T,$$

but the height of X is neither 3 nor $\frac{1}{2}$; it is 6.

It is very easy to make "height estimates" for sums and products of polynomials:

7A.2. **Lemma.** *If X, Y are in $\mathbb{Q}[T]$ with respective degrees n and m and $X : H$, $Y : h$, then*

$$X + Y : 2Hh, \quad XY : (n + m)Hh.$$

PROOF. For addition,

$$X + Y = \frac{(y^* x_0 + x^* y_0) + (y^* x_1 + x^* y_1)T + \cdots}{x^* y^*},$$

and every term in the numerator clearly has absolute value $\leq 2Hh$. For multiplication,

$$XY = \frac{\sum_{k=0}^{n+m} \left(\sum_{i=0}^{i=k} x_i y_{k-i} \right) T^k}{x^* y^*}.$$

For $k < n + m$, the typical coefficient in the numerator can be estimated by

$$\left| \sum_{i=0}^{i=k} x_i y_{k-i} \right| \le \sum_{i=0}^{i=k} Hh \le (k+1)Hh \le (n+m)Hh;$$

and if $k = n + m$, then

$$\left| \sum_{i=0}^{i=n+m} x_i y_{k-i} \right| = |x_n y_m| \le Hh < (n+m)Hh$$

since $x_i = 0$ when $i > n$ and $y_j = 0$ when $j > m$. ⊣

The next result is a version of the Division Theorem 7A.1 for $\mathbb{Q}[T]$ which supplies additional information about the heights.

7A.3. **Lemma** (Lemma 2.3 of Mansour, Schieber, and Tiwari [1991b]). *Suppose X and Y are polynomials with integer coefficients,*

$$\deg(X) = n \ge m = \deg(Y), \ X : H, \ Y : h,$$
$$\textit{and } X = YQ + R \quad \text{with } R = 0 \text{ or } \deg(R) < \deg(Y).$$

Then

$$Q = \frac{Q_1}{y_m^{d+1}}, \qquad R = \frac{R_1}{y_m^{d+1}},$$

where $d = n - m$ and Q_1, R_1 are in $\mathbb{Z}[T]$ with height $\le H(2h)^{d+1}$. It follows that

$$Q, R : H(2h)^{d+1}.$$

PROOF is by induction on d.

BASIS, $\deg(X) = \deg(Y) = n$. In this case

$$X = Y\frac{x_n}{y_n} + R$$

with R defined by this equation, so that either it is 0 or it is of degree $< n$. Now $Q_1 = \dfrac{x_n}{y_n}$ has height $\le H$, and

$$R_1 = y_n X - x_n Y$$

so that the typical coefficient of R_1 is of the form $y_n x_i - x_n y_i$, and the absolute value of this is bounded by $2Hh = H(2h)^{0+1}$.

INDUCTION STEP, $d = \deg(X) - \deg(Y) = n - m > 0$. Consider the polynomial

(7A-3) $$Z = y_m X - x_n Y T^d$$

whose degree is $< n = m + d$ since the coefficient of T^n in it is $y_m x_n - x_n y_m$. If $Z = 0$ or $\deg(Z) < m$, then

$$X = Y\frac{x_n T^d}{y_m} + \frac{Z}{y_m} = Y\frac{x_n y_m^d T^d}{y_m^{d+1}} + \frac{y_m^d Z}{y_m^{d+1}}$$

is the **cde** for X, Y, and it follows easily that

$$Q_1 = x_n y_m^d T^d : Hh^d < H(2h)^{d+1},$$
$$R_1 = y_m^d Z = y_m^{d+1} X - x_n y_m^d Y T^d : H(2h)^{d+1}.$$

This proves the lemma for this case. If $\deg(Z) \geq m$, then the Induction Hypothesis applies to the pair Z and Y since, evidently,

$$\deg(Z) - \deg(Y) < n - m = d.$$

Now

$$\text{height}(Z) \leq hH + Hh = H(2h),$$

and so the Induction Hypothesis yields

$$Z = Y\frac{Q_2}{y_m^d} + \frac{R_2}{y_m^d}, \tag{7A-4}$$

with

$$Q_2, R_2 : H(2h)(2h)^d = H(2h)^{d+1}.$$

Solving (7A-3) for X, we get

$$\begin{aligned} X &= \frac{1}{y_m} Z + \frac{x_n}{y_m} T^d Y \\ &= \frac{1}{y_m}\Big(Y\frac{Q_2}{y_m^d} + \frac{R_2}{y_m^d}\Big) + \frac{x_n}{y_m} T^d Y \\ &= \frac{Q_2 + x_n y_m^d T^d}{y_m^{d+1}} Y + \frac{R_2}{y_m^{d+1}} \end{aligned}$$

which is the **cde** for X, Y. We have already computed that $R_2 : H(2h)^{d+1}$. To verify that $Q_2 + x_n y_m^d T^d : H(2h)^{d+1}$, notice that $\deg(Q_2) < d$, since the opposite assumption implies with (7A-4) that $\deg(Z) \geq m + d = n$, which contradicts the definition of Z; thus the coefficients of $Q_2 + x_n y_m^d T^d$ are the coefficients of Q_2 and $x_n y_m^d$, and they all have height $\leq H(2h)^{d+1}$, as required. ⊣

7A.4. **Theorem.** *If* $X, Y : h$, $\deg(X) \geq \deg(Y)$ *and* (7A-1) *holds, then*

$$Q, R : (2h)^{2n+5}.$$

PROOF. Theorem 7A.3 applied to $y^* X$ and $x^* Y$ (with height $\leq h^2$) yields

$$y^* X = x^* Y \frac{Q}{(x^* y_m)^{d+1}} + \frac{R}{(x^* y_m)^{d+1}}$$

with $Q, R : h^2(2h^2)^{d+1} < (2h)^{2d+4}$; and if we divide by y^*, we get that

$$X = Y\frac{x^* Q}{y^*(x^* y_m)^{d+1}} + \frac{R}{y^*(x^* y_m)^{d+1}},$$

which is the **cde** for X, Y, and such that (with $d \leq n$)

$$\frac{x^*Q}{y^*(x^*y_m)^{d+1}}, \frac{R}{y^*(x^*y_m)^{d+1}} : (2h)^{2n+5}. \qquad \dashv$$

7B. Unary relations from $\mathit{Lin}_0[\div, \cdot]$

We establish here suitable versions of Lemma 6A.5 and Theorem 6A.6 for the structure

$$\mathbf{N}_0[\div, \cdot] = (\mathbf{N}_0, \mathit{Lin}_0[\div, \cdot]) = (\mathbb{N}, 0, 1, =, <, +, \dot{-}, \mathrm{iq}, \mathrm{rem}, \cdot)$$

with a $\sqrt{\log\log}$ bound.

Set, for any $a, n, h \in \mathbb{N}$,

$$\text{(7B-1)} \quad Q_n(a; h) = \Big\{ \frac{x_0 + x_1 a + x_2 a^2 + \cdots + x_n a^n}{x^*} \in \mathbb{N}$$
$$\mid x_0, \ldots, x_n, x^* \in \mathbb{Z}, x^* > 0 \text{ and } |x_0|, \ldots, |x_n|, |x^*| \leq h \Big\}.$$

These are the values for $T := a$ of polynomials in $Q_n(T; h)$, but only those which are natural numbers; and they are the sort of numbers which occur (with various values of h) in $G_m(a) = G_m[(\mathbf{N}_0[\div, \cdot], a)$. To simplify dealing with them, we will be using the standard notations

$$\text{(7B-2)} \quad x = f(a) = \frac{x_0 + x_1 a + x_2 a^2 + \cdots + x_n a^n}{x^*},$$
$$y = g(a) = \frac{y_0 + y_1 a + y_2 a^2 + \cdots + y_m a^m}{y^*},$$

where it is assumed that $x_n, y_m \neq 0$ (unless, of course, $x = 0$, in which case, by convention, $n = 0$ and $x_0 = 0$). It is also convenient to set $x_i = 0$ for $i > n$, and similarly for y_j, and to use the same abbreviations we set up for $Q_n(T; h)$, especially

$$Q(a; h) = \bigcup_n Q_n(a; h), \quad x : h \iff x \in Q(a; h).$$

7B.1. **Lemma.** *With x and y as in* (7B-2), *if $h \geq 2$ and $x, y \in Q_n(a; h)$, then $x + y, x \dot{-} y \in Q_n(a; h^3)$, and $xy \in Q_{2n}(a; nh^3)$.*

PROOF. These are all immediate, using Lemma 7A.2. ⊣

The analogous estimates for $\mathrm{iq}(x, y)$ and $\mathrm{rem}(x, y)$ are substantially more complex, and we need to establish first the uniqueness of the representations (7B-2) when h is small relative to a.

7B.2. **Lemma.** (1) *With all $x_i \in \mathbb{Z}$ and $a > 2$, if $|x_i| < a$ for $i \leq n$, then*

$$x_0 + x_1 a + \cdots + x_n a^n = 0 \iff x_0 = x_1 = \cdots = x_n = 0;$$

and if, in addition, $x_n \neq 0$, *then*

$$x_0 + x_1 a + \cdots + x_n a^n > 0 \iff x_n > 0.$$

(2) *With* x *and* y *as in* (7B-2) *and assuming that* $2h^2 < a$:

$$x = y \iff m = n \;\&\; (\forall i \leq n)[y^* x_i = x^* y_i],$$
$$x > y \iff (\exists k \leq n)[(\forall i > k)[x_i = y_i] \;\&\; x_k > y_k].$$

In particular,

$$x > 0 \iff x_n > 0.$$

PROOF. (1) If $x_n \neq 0$, then

$$|x_0 + x_1 a + \cdots x_{n-1} a^{n-1}| \leq (a-1)(1 + a + a^2 + \cdots + a^{n-1})$$
$$= (a-1)\frac{a^n - 1}{a - 1} = a^n - 1 < a^n \leq |x_n| a^n,$$

and so adding $x_0 + x_1 a + \cdots x_{n-1} a^{n-1}$ to $x_n a^n$ cannot change its sign or yield 0.

(2) follows immediately from (1). ⊣

7B.3. **Lemma.** *Let* $c \geq 61$, $d \geq 33$, *and assume that* $h \geq 2$ *and* $h^{c(n+1)} < a$. *If* $x, y \in Q_n(a; h)$ *and* $x \geq y > 0$, *then* $\mathrm{iq}(x, y), \mathrm{rem}(x, y) \in Q_n(a; h^{d(n+1)})$.

PROOF. How large c and d must be will be determined by the proof, as we put successively stronger conditions on h to justify the computations. To begin with, assume

(H1) $\qquad c \geq 3$, so that $2h^2 \leq h^3 \leq h^{c(n+1)} < a$,

and Lemma 7B.2 guarantees that the canonical representations of x an y in $Q_n(a; h)$ are unique. By the same lemma,

$$n = \deg(f(T)) \geq \deg(g(T)) = m,$$

and so we can put down the correct division equation for these polynomials in $\mathbb{Q}[T]$,

(7B-3) $\quad f(T) = g(T)Q(T) + R(T)$
$$(R(T) = 0 \text{ or } \deg(R(T)) < \deg(g(T)).$$

We record for later use the heights from Lemma 7A.4,

(7B-4) $\qquad Q(T), R(T) : (2h)^{2n+5} \leq h^{4n+10}.$

From (7B-3) we get

(7B-5) $\qquad f(a) = g(a)Q(a) + R(a) \quad (R(a) = 0 \text{ or } R(a) < g(a)),$

where the condition on $R(a)$ is trivial if $R(a) \leq 0$, and follows from the corresponding condition about degrees in (7B-3) by Lemma 7B.2, provided that the height of $R(a)$ is sufficiently small, specifically

$$2\Big(h^{4n+10}\Big)^2 < a;$$

so here we assume

(H2) $\quad c \geq 21$, so that $2\Big(h^{4n+10}\Big)^2 \leq hh^{8n+20} = h^{8n+21} < h^{21(n+1)} < a.$

However, (7B-5) need not be the correct division equation for the numbers $f(a), g(a)$, because $Q(a)$ might not be integral or $R(a)$ might be negative. For an example where both of these occur, suppose

$$f(a) = a^2 - 1, \quad g(a) = 2a \text{ with } a \text{ odd},$$

in which case (7B-3) and (7B-5) take the form

$$T^2 - 1 = 2T(\frac{T}{2}) - 1, \quad a^2 - 1 = 2a(\frac{a}{2}) - 1.$$

To correct for this problem, we consider four cases.

Case 1, $Q(a) \in \mathbb{N}$ and $R(a) \geq 0$. In this case (7B-5) is the **cde** for $f(a)$ and $g(a)$, and from (7B-4),

$$Q(a), R(a) : h^{4n+10};$$

thus we assume at this point

(H3) $\quad d \geq 10$, so that $h^{4n+10} \leq h^{d(n+1)}$.

Case 2, $Q(a) \in \mathbb{N}$ but $R(a) < 0$. From (7B-5) we get

$$f(a) = g(a)[Q(a) - 1] + \underbrace{g(a) + R(a)},$$

and we contend that this is the **cde** for $f(a), g(a)$ in $\mathbb{N}$, if c is large enough. We must show that

$$(1) \quad 0 \leq g(a) + R(a) \text{ and } (2) \quad g(a) + R(a) < g(a),$$

and (2) is immediate, because $R(a) < 0$. For (1), notice that the leading coefficient of $g(T) + R(T)$ is the same as the leading coefficient of $g(T)$ (because $\deg(R(T) < \deg(g(T))$, and that it is positive, since $g(a) > 0$. To infer from this that $g(a) + R(a) > 0$ using Lemma 7B.2, we must know that

$$2\text{height}(g(a) + R(a))^2 < a,$$

and from Lemma 7B.1,

$$2\text{height}(g(a) + R(a))^2 \leq h\Big((h^{4n+10})^3\Big)^2 = h^{24n+61},$$

and so for this case we assume that

(H4) $$c \geq 61, \text{ so that } h^{24n+61} \leq h^{c(n+1)} < a.$$

Using Lemma 7B.1, easily (and overestimating again grossly)

$$Q(a) - 1, g(a) + R(a) : (h^{4n+10})^3 = h^{12n+30} \leq h^{30(n+1)};$$

and so we also assume

(H5) $$d \geq 30, \text{ so that } h^{30(n+1)} \leq h^{d(n+1)}.$$

CASE 3, $Q(a) = \dfrac{Q_1(a)}{z} \notin \mathbb{N}$, and $Q_1(a) \geq z > 1$. We note that

$$Q_1(a) : h^{4n+10}, \quad z \leq h^{4n+10},$$

and both $Q_1(a)$ and z are positive, and so we can put down the **cde** for them in $\mathbb{N}$:

$$Q_1(a) = zQ_2 + R_2 \quad (0 < R_2 < z),$$

where we know that $R_2 > 0$ by the case hypothesis. From this it follows that

$$R_2 < z \leq h^{4n+10}, \text{ and } Q_2 = \frac{Q_1(a) - R_2}{z} : \left(h^{4n+10}\right)^3 = h^{12n+30}$$

by Lemma 7B.1 again. Replacing these values in (7B-5), we get

(7B-6) $$f(a) = g(a)Q_2 + \underbrace{g(a)\frac{R_2}{z} + R(a)}$$

and the number above the underbrace is in $\mathbb{Z}$, as the difference between two numbers in $\mathbb{N}$. This number is the value for $T := a$ of the polynomial

$$g(T)\frac{R_2}{z} + R(T)$$

whose leading coefficient is that of $g(T)$—since $\deg(R(T)) < \deg(g(T))$—and hence positive. We would like to infer from this that

$$g(a)\frac{R_2}{z} + R(a) > 0,$$

using Lemma 7B.2, and so we make sure that its height is suitably small. From the two summands, the second has the larger height, h^{4n+10}, and so by Lemma 7B.1, the height of the sum is bounded by

$$\left(h^{4n+10}\right)^3 = h^{12n+30};$$

and to apply Lemma 7B.2, we must insure that

$$2(h^{12n+30})^2 = 2h^{24n+60} < a,$$

and so for this step we assume that

(H6) $c \geq 61$, so that $2h^{24n+60} \leq h^{24n+61} \leq h^{c(n+1)}$.

This number is also less than $g(a)$, again because its leading coefficient is that of $g(a)$ multiplied by $\frac{R_2}{z} < 1$. It follows that this is the correct division equation for $f(a), g(a)$, so it is enough to compute the height of the quotient (above the underbrace) since we have already computed that

$$Q_2 : h^{12n+30} \leq h^{30(n+1)};$$

using the known heights on $g(a)$, R_2, z and $R(a)$, it is easy to check that

$$g(a)\frac{R_2}{z} + R(a) : ((h)^{4n+11})^3 = h^{12n+33},$$

so at this point we assume that

(H7) $d \geq 33$, so that $h^{12n+30}, h^{12n+33} \leq h^{d(n+1)}$.

CASE 4, $Q(a) = \dfrac{Q_1(a)}{z} \notin \mathbb{N}$, and $Q_1(a) < z$. Since $Q_1(a) > 0$, this case hypothesis implies that $\deg(Q_1(T)) = 0$, so that $\deg(f(T)) = \deg(g(T))$. By now familiar arguments, this means that

$$f(a) \leq \frac{y^* x_n}{x^* y_n} g(a)$$

and so the quotient of these two numbers is some number

$$Q \leq \frac{y^* x_n}{x^* y_n} \leq h^2.$$

Thus (7B-5) takes the form

$$f(a) = g(a)Q + R \quad \text{with } 0 \leq Q \leq h^2,$$

from which it follows immediately that

$$R = f(a) - g(a)Q : (h^5)^3 = h^{15},$$

and the hypotheses we have already made on d insure $h^{15} \leq h^{d(n+1)}$, so that we need not make any more.

In fact then, the lemma holds with

$$c \geq 61, \quad d \geq 33. \qquad \dashv$$

7B.4. **Lemma** (***Lin***$_0[\div, \cdot]$**-embedding**). *Let $e = 61$ and for any m,*

$$G_m(a) = G_m(\mathbf{N}_0[\div, \cdot], a).$$

(1) *If $2^{2^{e(m+1)^2}} < a$, then $G_m(a) \subseteq Q_{2^m}(a; 2^{2^{em^2}})$.*

(2) *If $2^{2^{e(m+2)^2}} < a$ and $a! \mid (\lambda - 1)$, then there is an embedding*

$$\pi : \mathbf{N}_0[\div, \cdot] \restriction G_m(a) \rightarrowtail \mathbf{N}_0[\div, \cdot]$$

such that $\pi(a) = \lambda a$.

PROOF of (1) is by induction on m, the basis being trivial. To apply Lemmas 7B.1 and 7B.3 at the induction step, we need the inequalities

$$2^m\left(2^{2^{em^2}}\right)^3 \le 2^{2^{e(m+1)^2}}, \quad \left(2^{2^{em^2}}\right)^{61(2^m+1)} < a,$$

$$\left(2^{2^{em^2}}\right)^{33(2^m+1)} \le 2^{2^{e(m+1)^2}},$$

and these are easily verified by direct computation.

(2) is proved very much like Lemma 6C.4, the only subtlety being that we need to start with an injection

$$\rho : G_{m+1}(a) \rightarrowtail \mathbb{N}$$

on the larger set, which (by (1)) contains $\mathrm{iq}(x, y)$ and $\mathrm{rem}(x, y)$ for all $x, y \in G_m(a)$. The stronger hypothesis on m and a imply that the numbers in $G_{m+1}(a)$ have unique canonical forms in

$$Q_{2^{m+1}}(a; 2^{2^{e(m+1)^2}});$$

and so we can set (with $n = 2^m$)

$$\rho\left(\frac{x_0 + x_1 a + \cdots x_n a^n}{x^*}\right) = \frac{x_0 + x_1 \lambda a + \cdots x_n \lambda a^n}{x^*},$$

take $\pi = \rho \restriction G_m(a)$ and finish the proof as in Lemma 6C.4. ⊣

7B.5. **Theorem.** (Mansour, Schieber, and Tiwari [1991b], van den Dries and Moschovakis [2009]). *If $R(x)$ is a good example, then for all $a \ge 2$*

$$R(a) \Longrightarrow \mathrm{depth}_R(\mathbf{N}_0[\div, \cdot], a) > \frac{1}{24}\sqrt{\log\log a}.$$

PROOF. By the Homomorphism Test 4E.3 and the preceding Lemma, we know that if $m = \mathrm{depth}_R(\mathbf{N}_0[\div, \cdot], a)$, then

$$2^{2^{e(m+2)^2}} > a,$$

with $e = 61$. Taking logarithms twice, then the square root and finally using the fact that $3m \ge m + 2$ by Problem x4E.4, we get the required

$$3m \ge m + 2 \ge \sqrt{\frac{\log\log a}{e}} \ge \frac{1}{8}\sqrt{\log\log a},$$

the last step justified because $\sqrt{e} = \sqrt{61} < 8$. ⊣

Mansour, Schieber, and Tiwari [1991b] express their results using specific computation models and they establish $O(\sqrt{\log\log a})$ lower bounds for decision trees and RAMS, for a class of unary relations which includes square-free; Meidânis [1991] derives a similar lower bound for primality. The methods used in these papers are very different from the direct applications of the Embedding Test that we have been using, which makes it very difficult to compare their results with ours.

As we mentioned in Footnote 34 on page 146, Mansour, Schieber, and Tiwari [1991b] also derive (for their models) a triple-log lower bound for coprimeness from $\mathbf{\mathit{Lin}}_0[\div, \cdot]$ on infinitely many inputs, and van den Dries has proved (using output complexity) an $O(\sqrt{\log\log})$-lower bound for computing $\gcd(x, y)$ on infinitely many inputs. We list as an open problem what would be the best (intrinsic) relevant result.

Problems for Section 7B

x7B.1. **Open problem.** Prove that for some infinite set of pairs $A \subset \mathbb{N} \times \mathbb{N}$ and some $r > 0$,

$$(a, b) \in A \Longrightarrow \mathrm{depth}_{\perp\!\!\!\perp}(\mathbf{N}_0[\div, \cdot], a, b) > r\sqrt{\log\log\max(a, b)}.$$

CHAPTER 8

NON-UNIFORM COMPLEXITY IN $\mathbb{N}$

A computer has finite memory, and so it can only store and operate on a finite set of numbers. Because of this, complexity studies which aim to be closer to the applications are often restricted to the analysis of algorithms on structures with universe the finite set

$$[0, 2^N) = \{x \in \mathbb{N} : x < 2^N\}$$

of *N-bit numbers* for some fixed (large) N, typically restrictions to $[0, 2^N)$ of expansions of $\mathbf{N}_d$ or $\mathbf{N}_0$, e.g., $\mathbf{N}_0[\div], \mathbf{N}_0[\div, \cdot]$, etc. The aim now is to derive lower bounds for the *worst case behavior* of such algorithms as functions of N; and the field is sometimes called *non-uniform complexity theory*, since, in effect, we allow for each N the use of a different algorithm which solves a given problem in $\mathbf{A} \upharpoonright [0, 2^N)$.

For each structure $\mathbf{A} = (\mathbb{N}, \mathbf{\Phi})$ with universe $\mathbb{N}$, each relation $R \subseteq \mathbb{N}^n$ and each N, let

$$\text{(8-7)} \quad \text{depth}_R(\mathbf{A}, 2^N) = \max\left\{\text{depth}_R(\mathbf{A} \upharpoonright [0, 2^N), \vec{x}) : x_1, \dots, x_n < 2^N\right\}$$

and similarly for $\text{values}_R(\mathbf{A}, 2^N), \text{calls}_R(\mathbf{A}, 2^N)$. These are the *intrinsic* (worst case) *non-uniform bit complexities* of R from the primitives of $\mathbf{A}$, at least those of them for which we can derive lower bounds. As it turns out, the results and the proofs are essentially the same for $\mathbf{N}_d$, except for the specific constants which are now functions of N (and somewhat smaller). For $\mathbf{N}_0[\div]$ and $\mathbf{N}_0[\div, \cdot]$, however, we need a finer analysis and we can only derive lower bounds for the larger complexity $\text{values}_R(\mathbf{A}, 2^N)$, primarily because there is "less room" in $[0, 2^N)$ for embeddings which exploit the uniformity assumption in the definition of intrinsic complexities. It is this new wrinkle in the proofs that is most interesting in this brief chapter.

8A. Non-uniform lower bounds from $\textbf{\textit{Lin}}_d$

We will show here that the intrinsic lower bounds from $\textbf{\textit{Lin}}_d$ of Chapter 5 hold also in the non-uniform case, with somewhat smaller constants.

This means (roughly) that for these problems, the lookup algorithm in Problem x8A.1 is weakly optimal for depth intrinsic bit complexity in $\mathbf{N}_d$.

8A.1. **Theorem** (van den Dries and Moschovakis [2009]). *If $N \geq 3$, then*

$$\text{depth}_{\text{Prime}}(\mathbf{N}_d, 2^N) > \frac{N}{5\log d}.$$

PROOF. Suppose $p < 2^N$ is prime and let

$$m = \text{depth}_{\text{Prime}}(\mathbf{N}_d \restriction [0, 2^N), p), \quad \lambda = 1 + d^{m+1}.$$

If

$$(a) \quad d^{2m+2} < p \text{ and } (b) \quad \frac{x_0 + x_1\lambda p}{d^m} < 2^N \text{ for all } |x_0|, |x_1| \leq d^{2m},$$

then the proof of Lemma 5B.2 would produce an embedding

$$\pi : \mathbf{G}_m(\mathbf{N}_d \restriction [0, 2^n), p) \rightarrowtail \mathbf{N}_d \restriction [0, 2^N)$$

which does not respect the primality of p, yielding a contradiction. So for every prime $p < 2^N$, one of (a) or (b) must fail. To exploit this alternative, we need to apply it to primes not much smaller than 2^N, but small enough so that (b) holds, and to find them we appeal to Bertrand's Postulate, Hardy and Wright [1938, Theorem 418]; this guarantees primes between l and $2l$ when $l \geq 3$. So choose p such that

$$2^{N-1} < p < 2^N,$$

which exists because $2^{N-1} > 3$ when $N \geq 3$.

Case 1, $d^{2m+2} \geq p$, and so $d^{2m+2} > 2^{N-1}$. Using as always the fact that $m \geq 1$, this gives easily $m > \frac{N}{5\log d}$.

Case 2, for some x_0, x_1 with $|x_0|, |x_1| \leq d^{2m}$, we have

$$\frac{x_0 + x_1(1 + d^{m+1})p}{d^m} \geq 2^N.$$

Compute (with $m \geq 1$):

$$\frac{x_0 + x_1(1 + d^{m+1})p}{d^m} < \frac{d^{2m} + d^{2m}(1 + d^{m+1})2^N}{d^m}$$
$$< d^m + d^m d^{m+2} 2^N \leq d^{2m+3} 2^N \leq d^{5m} 2^N$$

and so the case hypothesis implies that $m > \frac{N}{5\log d}$ again, as required. ⊣

Similar mild elucidations of the proofs we have given extend all the lower bound results about $\mathbf{N}_d$ in Chapter 5 to intrinsic bit complexity, and they are simple enough to leave for the problems.

Problems for Section 8A

x8A.1. **Problem** (The lookup algorithm). If $n \geq 1$, then every n-ary relation R on $\mathbb{N}$ can be decided for inputs $x_1, \dots, x_n < 2^N$ by an explicit $\mathbf{N}_b$-term $E^N(\vec{x})$ of depth $\leq N$. It follows that

$$\text{depth}_R(\mathbf{N}_b, 2^N) \leq N.$$

HINT: Set $\text{Eq}_i(x) \iff x = i$, prove that this relation is defined by an $\mathbf{N}_b$-term of depth $\leq N$ when $i < 2^N$ and build (by induction on N) a "table" of data from which a decision procedure for any relation on $[0, 2^N)$ can read-off.

Recall the definition of good examples in Subsection 5C.

x8A.2. **Problem** (van den Dries and Moschovakis [2009]). Suppose R is a good example such that for some k and all sufficiently large m, there exists some x such that

$$R(x) \;\&\; 2^m \leq x < 2^{km}.$$

Prove that for all $d \geq 2$ there is some $r > 0$ such that for sufficiently large N,

$$\text{depth}_R(\mathbf{N}_d, 2^N) > rN,$$

and verify that all the good examples in Problem x5C.4 satisfy the hypothesis.

x8A.3. **Problem** (van den Dries and Moschovakis [2009]). Prove that for some $r > 0$ and all sufficiently large N,

$$\text{depth}_{\perp\!\!\!\perp}(\mathbf{N}_d, 2^N) > rN.$$

8B. Non-uniform lower bounds from $\boldsymbol{Lin}_0[\div]$

The methods of the preceding section do not extend immediately to $\mathbf{N}_0[\div]$, because the constant $\lambda = 1 + a!$ that we used in the proof of Lemma 6A.5 is too large: a direct adaptation of that proof to the non-uniform case leads to a $\log\log N$ lower bound for $\text{depth}_{\text{Prime}}(\mathbf{N}_0[\div], 2^N)$ which is way too low. In fact, it does not seem possible to get a decent lower bound for $\text{depth}_{\text{Prime}}(\mathbf{N}_0[\div], 2^N)$ with this method, but a small adjustment yields a lower bound for the values- and hence the calls- intrinsic bit complexities.

We start with the required modification of Lemma 6A.5.

8B.1. **Lemma.** *Suppose* $\mathbf{U} \subseteq_p \mathbf{N}_0[\div]$ *is generated by* a, $m = \text{depth}(\mathbf{U}, a)$, *and* $v = \text{values}(\mathbf{U}, a)$. *If* $2^{6^{m+3}} < a$, *then there is a number* $\lambda \leq 2^{v6^{m+2}}$ *and an embedding* $\pi : \mathbf{U} \rightarrowtail \mathbf{N}_0[\div]$ *such that* $\pi(a) = \lambda a$.

PROOF. As in the proof of Lemma 6A.5, the assumed inequality on m implies that each $x \in U \subseteq G_m(\mathbf{N}_0[\div], a)$ can be expressed uniquely as a proper fraction of the form

$$x = \frac{x_0 + x_1 a}{x_2} \quad (|x_i| \leq 2^{6^{m+1}}), \tag{8B-1}$$

and we can set

$$\text{denom}(x) = \text{the unique } x_2 \leq 2^{6^{m+1}} \text{ such that (8B-1) holds} \quad (x \in U).$$

We claim that the conclusion of the lemma holds with

$$\lambda = 1 + \textstyle\prod_{x \in U} \text{denom}(x) = 1 + \prod_{x \in U, x \neq a} \text{denom}(x) \leq 1 + \left(2^{6^{m+1}}\right)^{\nu} < 2^{\nu 6^{m+2}}, \tag{8B-2}$$

where the two products are equal because $\text{denom}(a) = 1$ and $\text{values}(\mathbf{U}, a) = |U| - 1$. To prove this we follow very closely the proof of Lemma 6A.5: we set

$$\rho(x) = \rho\left(\frac{x_0 + x_1 a}{x_2}\right) = \frac{x_0 + x_1 \lambda a}{x_2};$$

check that this is a well defined injection on U which takes values in $\mathbb{N}$, because

$$x \in U \Longrightarrow \text{denom}(x) \mid (\lambda - 1);$$

and finally verify that it is an embedding from $\mathbf{U}$ to $\mathbf{N}_0[\div]$ exactly as in the proof of Lemma 6A.5. ⊣

8B.2. **Theorem** (van den Dries and Moschovakis [2009]). *If $N \geq 8$, then*

$$\text{values}_{\text{Prime}}(\mathbf{N}_0[\div], 2^N) > \frac{1}{10} \log N. \tag{8B-3}$$

PROOF. Let $k = \lfloor \frac{N}{2} \rfloor - 1$ so that

$$k + 1 \leq \frac{N}{2} < k + 2 \text{ and so } k > \frac{N}{2} - 2.$$

The hypothesis on N yields $2^k > 4$, so Bertrand's Postulate insures that there exists some prime p such that $2^k < p < 2^{k+1}$. This is the prime we want. Let $\mathbf{A} = \mathbf{N}_0[\div] \restriction [0, 2^N)$ (to simplify notation) and choose $\mathbf{U} \subseteq_p \mathbf{A}$ so that

$$\mathbf{U} \Vdash_c^{\mathbf{A}} \text{Prime}(p), \quad \text{values}_{\text{Prime}}(\mathbf{A}, p) = \text{values}(\mathbf{U}, p),$$

and set $m = \text{depth}(\mathbf{U}, p)$. (It could be that $m > \text{depth}_{\text{Prime}}(\mathbf{A}, p)$.) Let

$$\lambda = 1 + \textstyle\prod_{x \in U} \text{denom}(x) = 1 + \prod_{x \in U, x \neq p} \text{denom}(x)$$

as above. If

$$2^{6^{m+3}} < p \text{ and } \frac{x_0 + x_1 \lambda p}{x_2} < 2^N \text{ whenever } |x_i| \leq 2^{6^{m+1}},$$

then $\lambda < 2^{\nu 6^{m+2}}$ and the argument in the proof of Lemma 8B.1 produces an embedding $\pi : \mathbf{U} \rightarrowtail \mathbf{A}$ which does not respect the primality of p, contrary to the choice of $\mathbf{U}$; so one of the two following cases must hold.

Case 1: $2^{6^{m+3}} \geq p > 2^k$. This gives $6^{m+3} > k > \frac{N}{2} - 2$, and this easily implies (with $m \geq 1$) that $\nu \geq m > \frac{\log N}{10}$ in this case, cf. Problem x8B.1.

Case 2: For some x_0, x_1, x_2 with $|x_i| \leq 2^{6^{m+1}}$,

$$\frac{x_0 + x_1 \lambda p}{x_2} \geq 2^N.$$

Compute:

$$\frac{x_0 + x_1 \lambda p}{x_2} \leq 2^{6^{m+1}} + 2^{6^{m+1}} \cdot 2^{\nu 6^{m+2}} \cdot 2^{\frac{N}{2}} \leq 2 \cdot 2^{6^{m+1} + \nu 6^{m+2}} \cdot 2^{\frac{N}{2}}$$
$$\leq 2^{2\nu 6^{m+2} + 1} \cdot 2^{\frac{N}{2}} \leq 2^{3\nu 6^{m+2}} \cdot 2^{\frac{N}{2}}.$$

So the case hypothesis gives $2^{3\nu 6^{m+2}} \cdot 2^{\frac{N}{2}} \geq 2^N$ which gives $3\nu 6^{m+2} \geq \frac{N}{2}$ and then $\nu 6^{m+3} \geq N$. This is the basic fact about the non-uniform, intrinsic bit complexity of primality from $\{=, <, +, \dot{-}, \text{iq}, \text{rem}\}$ and it can be used to derive a lower bound for the measure induced by the substructure norm

$$\mu(\mathbf{U}, a) = \text{values}(\mathbf{U}, a) \cdot 6^{\text{depth}(\mathbf{U}, a) + 3}.$$

To derive an easier to understand lower bound for the values complexity, we compute: $6^\nu 6^{\nu+3} \geq \nu 6^{m+3} \geq N$; so $6^{2\nu+3} \geq N$; so $(2\nu + 3)\log 6 \geq \log N$, and since $\nu > 0$, as usual, $5\nu \geq 2\nu + 3$ and so

$$\nu \geq \frac{1}{5 \log 6} \log N > \frac{1}{10} \log N. \qquad \dashv$$

It should be clear that the numerology in this proof was given in detail mostly for its amusement value, since from the first couple of lines in each of the cases one sees easily that $\nu > r \log N$ for some r. Moreover, one can certainly bring that $\frac{1}{10}$ up quite a bit, with cleverer numerology, a more judicious choice of p, or by weakening the result to show that (8B-3) holds only for very large N. The problems ask only for these more natural (if slightly weaker) results and leave it up to the solver whether they should indulge in manipulating numerical inequalities.

Problems for Section 8B

x8B.1. **Problem.** Prove that if $m \geq 1$, $N \geq 1$ and $6^{m+3} > \frac{N}{2} - 2$, then $m > \frac{\log N}{10}$. Hint: Check that $6^{6m} \geq 6^{5m+1} > 2 \cdot 6^{m+3} + 4 > N$.

x8B.2. **Problem.** Suppose R is a good example with the property that for some k and every $m \geq 1$, there is some x such that $R(x)$ and

$$2^{6^m} < x < 2^{6^{km}}.$$

Prove that for some $r > 0$ and sufficiently large N,

$$\text{values}_R(\mathbf{N}[\div], 2^N) > r \log N.$$

Verify also that the good examples in Problem x5C.4 satisfy the hypothesis.

x8B.3. **Problem** (van den Dries and Moschovakis [2009]). Prove that for some $r > 0$ and all sufficiently large N,

$$\text{values}_{\perp\!\!\perp}(\mathbf{N}[\div], 2^N) > r \log N.$$

HINT: Use the largest good approximation (a, b) of $\sqrt{2}$ with $a < 2^{\frac{N}{2}}$.

x8B.4. **Open problem.** Prove that for some $r > 0$ and all sufficiently large N,

$$\text{depth}_{\perp\!\!\perp}(\mathbf{N}[\div], 2^N) > r \log N.$$

x8B.5. **Problem.** Derive a lower bound for $\text{values}_R(\boldsymbol{Lin}_0[\div, \cdot], 2^N)$ when R is a good example or $\perp\!\!\perp$.

CHAPTER 9

POLYNOMIAL NULLITY (0-TESTING)

The complexity of polynomial evaluation

(9-4) $$V_F(x_0, \dots, x_n, y) = x_0 + x_1 y + \cdots + x_n y^n$$

in a field is perhaps the simplest problem in algebraic complexity and it has been much studied since its formulation by Ostrowski [1954]. Our (limited) aim in this chapter is to prove some intrinsic lower bound results about the (plausibly simpler) *nullity* (or 0-*testing*) relation

(9-5) $$N_F(z, x_1, \dots, x_n, y) \iff z + x_1 y + \cdots + x_n y^n = 0,$$

all of them establishing the intrinsic Calls-optimality of Horner's rule from various primitives on generic inputs; the upper bounds on these complexities are listed in Problem x4E.6*.[38]

Recall from the discussion on page 25, that "officially" a field F is a structure

$$\mathbf{F} = (F, 0, 1, +, -, \cdot, \div, =)$$

satisfying the standard axioms. It is a partial structure, because $x \div 0$ is not defined; and we will usually talk of a field F rather than $\mathbf{F}$, using the structure notation only when it is important, for example when we need to refer to partial substructures $\mathbf{U} \subseteq_p \mathbf{F}$ which are not fields.

9A. Preliminaries and notation

For any field F and indeterminates $\vec{u} = u_1, \dots, u_k$, $F[\vec{u}]$ is the ring of all polynomials with coefficients in F and $F(\vec{u})$ is the field of all rational functions in $\vec{u}$ with coefficients in F. The terms in a sequence $\vec{a} = (a_1, \dots, a_k)$ in some field $K \supset F$ are *algebraically independent* (or *generic*) over F, if for all $x_1, \dots, x_k \in F$,

$$x_1 a_1 + x_2 a_2 + \ldots + x_k a^k = 0 \Longrightarrow (x_1 = \ldots = x_k = 0);$$

---

[38] I am grateful to Tyler Arant for checking and correcting many errors in the first draft of this chapter; and, of course, I am fully responsible for the inevitable remaining errors.

and in that case, the extensions $F[\vec{a}], F(\vec{a}) \subseteq K$ are naturally isomorphic with $F[\vec{u}]$ and $F(\vec{u})$ respectively by the *relabelling isomorphism* determined by $a_i \mapsto u_i$, for $i = 1, \ldots, k$. Similarly, if $K_1, K_2 \supset F$ and $\vec{a} \in K_1^n, \vec{b} \in K_2^n$ are algebraically independent over F, then the relabelling $a_i \mapsto b_i$ induces an isomorphism

$$\lambda : F(\vec{a}) \rightarrowtail\!\!\!\!\twoheadrightarrow F(\vec{b}) \quad (\lambda(a_i) = b_i, i = 1, \ldots n)$$

which fixes F, and similarly for the polynomial rings $F[\vec{a}]$ and $F[\vec{b}]$. We will often establish some facts about one of these rings or fields and then quote them for the other, often without explicit mention, and we will also use the same terminology for these isomorphic structures: if, for example, a, b are complex numbers which are algebraically independent (over the prime field $\mathbb{Q}$), then the members of $\mathbb{Q}(a, b)$ are "the rational functions of a, b".

A *partial ring homomorphism*

$$\pi : \mathbf{F}_1 \rightharpoonup \mathbf{F}_2$$

on one field to another is a partial function whose domain of convergence Domain(π) is a subring R_1 of F_1 (with 1) and which respects as usual the ring operations, including $\pi(1) = 1$. Notice that for every $\mathbf{U} \subseteq_p \mathbf{F}_1$,

(9A-1) *if π is total on U and $\Big(0 \neq x \in U \Longrightarrow \pi(x) \neq 0\Big)$,*

then $\pi \restriction U : \mathbf{U} \to \mathbf{F}_2$ is a homomorphism,

i.e., $\pi \restriction U$ preserves not only the ring operations, but also all divisions in eqdiag($\mathbf{U}$). This is because if $(\div, u, v, w) \in$ eqdiag($\mathbf{U}$), then $\pi(u), \pi(v), \pi(w)$ are all defined and $v \neq 0$, so $\pi(v) \neq 0$ by the hypothesis of (9A-1); and since $vw = u$, we have $\pi(v)\pi(w) = \pi(u)$, which then gives $\pi(w) = \frac{\pi(u)}{\pi(v)}$.

If $F(v, \vec{u}) \subseteq K$ for some field K and $\alpha \in K$, then the substitution $v \mapsto \alpha$ induces a partial ring homomorphism

(9A-2) $$\rho_\alpha\Big(\frac{\chi_n(v, \vec{u})}{\chi_d(v, \vec{u})}\Big) = \frac{\chi_n(\alpha, \vec{u})}{\chi_d(\alpha, \vec{u})} \quad \Big(\frac{\chi_n(v, \vec{u})}{\chi_d(v, \vec{u})} \in F(v, \vec{u})\Big)$$

with

$$\text{Domain}(\rho_\alpha) = \Big\{\frac{\chi_n(v, \vec{u})}{\chi_d(v, \vec{u})} : \chi_d(\alpha, \vec{u}) \neq 0\Big\} \subseteq F(v, \vec{u}).$$

Notice that $F[v, \vec{u}] \subset$ Domain(ρ_α) and $\rho_\alpha(u_i) = u_i$. We will sometimes call ρ_α "the substitution" $(v \mapsto \alpha)$, and the only homomorphisms we will need are compositions of such substitutions.

9A.1. **The Substitution Lemma.** *Suppose F, K are fields, $v, \vec{u}$ are indeterminates, U is a finite subset of $F(v, \vec{u})$, $K \supseteq F(v, \vec{u})$ and $\{\alpha_t\}_{t \in I}$ is an infinite set of distinct elements of K. It follows that for all but finitely many $t \in I$, the* *partial ring homomorphism* $\rho_t : F \rightharpoonup K$ *induced by the substitution* $v \mapsto \alpha_t$ *is total and injective on* U.

PROOF. It is convenient to prove first a

Sublemma. If $U' \subset F[v, \vec{u}]$ *is any finite set of polynomials, then for all but finitely many* $t \in I$,

$$\Big(\chi(v, \vec{u}) \in U' \text{ and } \chi(v, \vec{u}) \neq 0\Big) \Longrightarrow \chi(\rho_t(v), \vec{u}) \neq 0.$$

Proof of the Sublemma. Write each $\chi(v, \vec{u}) \in U'$ as a polynomial in v,

$$\chi(v, \vec{u}) = \chi_0(\vec{u}) + \chi_1(\vec{u})v + \cdots + \chi_l(\vec{u})v^l,$$

so that if $a_1, \ldots, a_k$ are its roots in K, then

$$\chi(v, \vec{u}) = (v - a_1)\cdots(v - a_k)\phi(v, \vec{u}) \quad (\phi(v, \vec{u}) \neq 0, \text{ for all } v \in K).$$

Now $\rho_t(\chi(v, \vec{u})) = (\alpha_t - a_1)\cdots(\alpha_t - a_l)\phi(\alpha_t, \vec{u})$, and this can only be 0 if $\alpha_t = a_i$ for some i. The conclusion then holds for any t such that α_t is not a root a_i in K of a non-0 polynomial in U'. ⊣ (Sublemma)

We now fix a specific representation of the form

$$\text{(9A-3)} \qquad \chi(v, \vec{u}) = \frac{\chi_n(v, \vec{u})}{\chi_d(v, \vec{u})} \quad (\chi_d(v, \vec{u}) \neq 0)$$

for each $\chi(v, \vec{u}) \in U$, and we apply this sublemma to the finite set U' comprising all polynomials in one of the forms

$$(i)\ \chi_d(v, \vec{u}), \qquad (ii)\ \chi_n(v, \vec{u})\chi'_d(v, \vec{u}) - \chi'_n(v, \vec{u})\chi_d(v, \vec{u})$$

with $\chi(v, \vec{u}), \chi'(v, \vec{u}) \in U$. ⊣

9B. Generic $\{\cdot, \div\}$-optimality of Horner's rule

Several versions of the $\{\cdot, \div\}$-*optimality* of Horner's rule for polynomial evaluation were proved by Pan [1966], who established his result for algorithms expressed by *computation sequences* (page 91) and introduced the *method of substitution*, i.e., the use of partial ring homomorphisms induced by substitutions as above.

9B.1. **Theorem** (Bürgisser and Lickteig [1992]).[39] *If* F *is a field of characteristic* 0, $n \geq 1$, *and* $z, x_1, \ldots, x_n, y \in F$ *are algebraically independent* (over $\mathbb{Q}$), *then* (with $\vec{x} = x_1, \ldots, x_n$),

$$\text{(9B-1)} \qquad \text{calls}(\cdot, \div)(\mathbf{F}, N_F, z, \vec{x}, y) = n.$$

In particular, (9B-1) *holds for the reals* $\mathbb{R}$ *and the complexes* $\mathbb{C}$ *with algebraically independent* $z, \vec{x}, y$.

[39] The proof in Bürgisser and Lickteig [1992] is for *algebraic decision trees*, what we called primitive decision trees for fields on page 91.

This follows from the upper bounds in Problem x4E.6*, the Homomorphism Test 4E.3 and the following

9B.2. **Lemma.** ***If*** *F is a field of characteristic* 0, $n \geq 1$, $z, \vec{x}, y$ *are algebraically independent*, $\mathbf{U} \subseteq_p \mathbf{F}$ *is finite, generated by* $(U \cap \mathbb{Q}) \cup \{z, \vec{x}, y\}$ *and* $\text{calls}(\cdot, \div)(\mathbf{U}, z, \vec{x}, y) < n$,

then *there is a partial ring homomorphism* $\pi : F \rightharpoonup F$ *which is the identity on* $\mathbb{Q}(y)$, *it is total and injective on* U, *and it satisfies*

$$\pi(z) + \pi(x_1)y^1 + \cdots + \pi(x_n)y^n = 0.$$

We will derive this from a substantially stronger result which is shown by induction on n; the appropriate lemma is an elaboration of the construction in Winograd [1967], [1970], which extends and generalizes Pan's results.

For $\mathbf{U} \subseteq_p \mathbf{F}(z, \vec{x}, y)$, we will denote entries in eqdiag(**U**) using infix notation, $a + b = c$ for $(+, a, b, c)$, $a \cdot b = c$ for $(\cdot, a, b, c)$, etc. We write $\{\cdot, \div\}$ for *multiplications and divisions*, and we define the *trivial* $\{\cdot, \div\}$ (with respect to y) by the following rules:

$$a \cdot b = c \text{ is trivial if } a \in F \text{ or } b \in F \text{ or } a, b \in F(y);$$
$$a \div b = c \text{ is trivial if } b \in F \text{ or } a, b \in F(y).$$

All additions, subtractions and inequations in $\mathbf{F}(z, \vec{x}, y)$ are also trivial.

9B.3. **Lemma.** ***Suppose*** *F is an infinite field*, $n \geq 1$, $z, \vec{x} = x_1, \ldots, x_n, y$ *are indeterminates*, $\mathbf{U} \subseteq_p \mathbf{F}(z, \vec{x}, y)$ *is finite*, $\psi_1, \ldots, \psi_n \in F(y)$ *and the following conditions hold*:

(1) **U** *is generated by* $(F \cap U) \cup \{z, \vec{x}, y\}$.

(2) *For any* $f_1, \ldots, f_n \in F$,

$$f_1\psi_1 + \cdots + f_n\psi_n \in F \Longrightarrow f_1 = \cdots = f_n = 0.$$

(3) *There are no more than* $n - 1$ *non-trivial* $\{\cdot, \div\}$ *in* eqdiag(**U**).

Then *there is a partial ring homomorphism*

$$\pi : F(z, \vec{x}, y) \rightharpoonup F(\vec{x}, y)$$

which is the identity on $F(y)$, *it is total and injective on* U, *and it satisfies*

$$\pi(z) = \pi(x_1)\psi_1 + \cdots + \pi(x_n)\psi_n. \tag{9B-2}$$

It follows that $\pi \restriction U : \mathbf{U} \rightarrowtail \mathbf{F}(\vec{x}, y)$ *is an embedding which satisfies* (9B-2).

PROOF is by induction on n, but it is useful to consider first a case which covers the basis and also arises in the induction step.

Preliminary case: there are no non-trivial $\{\cdot, \div\}$ *in* **U**. It follows that every $X \in U$ is uniquely of the form

$$X = f_0 z + \textstyle\sum_{1 \leq i \leq n} f_i x_i + \phi(y) \tag{9B-3}$$

with $f_i \in F, \phi(y) \in F(y)$. If π is the partial ring homomorphism induced by the substitution

$$z \mapsto \sum_{1 \le i \le n} x_i \psi_i,$$

then π is the identity on $F(\vec{x}, y)$ and it is total on U, because the only divisions in eqdiag(**U**) are with both arguments in $F(y)$ or the denominator in F. So it is enough to check that it is injective on the set of all elements of the form (9B-3) and that it satisfies (9B-2).

To check injectivity, suppose that

$$\pi(X) = f_0\Big(\sum_{1\le i\le n} x_i\psi_i\Big) + \sum_{1\le i\le n} f_i x_i + \phi(y)$$
$$= f_0'\Big(\sum_{1\le i\le n} x_i\psi_i\Big) + \sum_{1\le i\le n} f_i' x_i + \phi'(y) = \pi(X')$$

so that

$$(f_0 - f_0')\sum_{1\le i\le n} x_i\psi_i + \sum_{1\le i\le n}(f_i - f_i')x_i + (\phi(y) - \phi'(y))$$
$$= \sum_{1\le i\le n}\Big((f_0 - f_0')\psi_i + (f_i - f_i')\Big)x_i + (\phi(y) - \phi'(y)) = 0.$$

This yields $\phi(y) = \phi'(y)$ and for each i, $(f_0 - f_0')\psi_i + (f_i - f_i') = 0$; and since no ψ_i is in F by (2) of the hypothesis, this implies that $f_0 = f_0'$, and finally that $f_i - f_i'$ for each i.

The identity (9B-2) is trivial because $\pi(z) = x_1\psi_1 + \cdots + x_n\psi_n$ and $\pi(x_i) = x_i$.

Basis, $n = 1$. This is covered by the preliminary case.

Induction Step, $n > 1$. If the preliminary case does not apply, then there is at least one non-trivial $\{\cdot, \div\}$ in eqdiag(**U**); so there is a least $m > 0$ such that some $\chi \in G_m(\mathbf{U}, z, \vec{x}, y)$ is a non-trivial product or quotient of elements of $\mathbf{G}_{m-1}(\mathbf{U}, z, \vec{x}, y)$ in which all $\{\cdot, \div\}$ are trivial; and so there is at least one non-trivial $\{\cdot, \div\}$ in eqdiag(**U**) of the form

(9B-4) $\quad (f_0'z + \sum_{1\le i\le n} f_i'x_i + \phi'(y)) \circ (f_0 z + \sum_{1\le i\le n} f_i x_i + \phi(y)) = \chi$

where $\circ$ is $\cdot$ or $\div$. We consider cases on how this can arise.

Case 1: There is some $i \ge 1$ such that $f_i \ne 0$, and the first factor in (9B-4) *is not in F*. We assume without loss of generality that $f_1 \ne 0$, and then dividing the equation by f_1 we put the second factor in the form

(9B-5) $$f_0 z + x_1 + \sum_{2\le i\le n} f_i x_i + \phi(y).$$

By the Substitution Lemma 9A.1, there is some $\overline{f} \in F$ such that the substitution

$$\rho_1(x_1) = \overline{f} - f_0 z - \sum_{2\le i\le n} f_i x_i - \phi(y)$$

induces an isomorphism

$$\rho_1 \restriction U : \mathbf{U} \rightarrowtail\!\!\!\!\twoheadrightarrow \rho_1[\mathbf{U}] = \mathbf{U}_1 \subseteq_p \mathbf{F}(z, x_2, \dots, x_n, y).$$

Notice that ρ_1 does not introduce any new non-trivial multiplication or division (because it is the identity on $F(y)$), and it turns the chosen operation in $\mathbf{U}$ into a trivial one since

$$\rho_1(f_0 z + x_1 + \textstyle\sum_{2 \le i \le n} f_i x_i + \phi(y)) = \overline{f}.$$

So there are fewer than $n - 1$ $\{\cdot, \div\}$ in eqdiag($\mathbf{U}_1$), and $\mathbf{U}_1$ is generated by $z, x_2, \dots, x_n, y$ and $\{\overline{f}\} \cup (F \cap U)$.

Applying Lemma 9A.1 again, fix some $\overline{g} \in F$ such that the substitution

$$\rho_2(z) = \frac{1}{1 + f_0 \psi_1}\Big((\overline{f} - \phi(y))\psi_1 + \overline{g} z\Big)$$

induces an isomorphism

$$\rho_2 \restriction U_1 : \mathbf{U}_1 \rightarrowtail\!\!\!\!\twoheadrightarrow \rho_2[\mathbf{U}_1] = \mathbf{U}_2 \subseteq_p \mathbf{F}(z, x_2, \dots, x_n, y).$$

This too does not introduce any non-trivial multiplications or divisions, and $\mathbf{U}_2$ is generated by $z, x_2, \dots, x_n, y$ and $F \cap U_2$. The required partial ring homomorphism is the composition

$$\pi = \sigma \circ \rho_2 \circ \rho_1 : F(z, \vec{x}, y) \rightharpoonup F(\vec{x}, y)$$

of the three substitutions, where σ is guaranteed by the induction hypothesis so that $\sigma \restriction U_2 : \mathbf{U}_2 \rightarrowtail \mathbf{F}(x_2, \dots, x_n, y)$ and

$$\overline{g}\sigma(z) = \textstyle\sum_{2 \le i \le n} (\psi_i - f_i \psi_1)\sigma(x_i).$$

This exists because the functions

$$\frac{1}{\overline{g}}(\psi_i - f_i \psi_1) \quad (i = 2, \dots, n)$$

satisfy (2) in the theorem.

To see that π has the required property, notice first that

$$\pi(z) = \sigma(\rho_2(z))$$

because $\rho_1(z) = z$. Using the corresponding properties of ρ_2 and σ, we get:

$$\begin{aligned}
&\pi(x_1)\psi_1 + \textstyle\sum_{2 \le i \le n} \pi(x_i)\psi_i \\
&\quad = \sigma\Big(\rho_2(\overline{f} - \phi(y) - \textstyle\sum_{2 \le i \le n} f_i x_i - f_0 z)\Big)\psi_1 + \textstyle\sum_{2 \le i \le n} \sigma(x_i)\psi_i \\
&\quad = \sigma\Big(\overline{f} - \phi(y) - \textstyle\sum_{2 \le i \le n} f_i x_i - f_0 \rho_2(z)\Big)\psi_1 + \textstyle\sum_{2 \le i \le n} \sigma(x_i)\psi_i \\
&\quad = (\overline{f} - \phi(y))\psi_1 - f_0 \psi_1 \sigma(\rho_2(z)) + \textstyle\sum_{2 \le i \le n} (\psi_i - f_i \psi_1)\sigma(x_i) \\
&\quad = (\overline{f} - \phi(y))\psi_1 - f_0 \psi_1 \sigma(\rho_2(z)) + \overline{g}\sigma(z).
\end{aligned}$$

So what we need to check is the equation

$$\begin{aligned} \sigma(\rho_2(z)) &= (\overline{f} - \phi(y))\psi_1 - f_0\psi_1\sigma(\rho_2(z)) + \overline{g}\sigma(z) \\ \text{equivalently } (1 + f_0\psi_1)\sigma(\rho_2(z)) &= (\overline{f} - \phi(y))\psi_1 + \overline{g}\sigma(z) \\ \text{equivalently } (1 + f_0\psi_1)\rho_2(z) &= (\overline{f} - \phi(y))\psi_1 + \overline{g}z, \end{aligned}$$

and the last is immediate from the definition of $\rho_2(z)$. (Note that we use repeatedly the fact that σ is injective on U_2 and the identity on $F(y)$.)

Case 2: $f_1 = \cdots = f_n = 0$, $f_0 \neq 0$, *and the first factor in* (9B-4) *is not in* F. We may assume without loss of generality that $f_0 = 1$, and so the second factor has the form

$$z + \phi(y).$$

By Lemma 9A.1, choose some $\overline{f} \in F$ such that the substitution

$$\rho_1(z) := \overline{f} - \phi(y)$$

induces an isomorphism

$$\rho_1 : \mathbf{U} \rightarrowtail\!\!\!\!\twoheadrightarrow \rho_1[\mathbf{U}] = \mathbf{U}_1 \subseteq_p \mathbf{F}(\vec{x}, y).$$

There is one fewer non-trivial operation in eqdiag($\mathbf{U}_1$), since ρ_1 does not introduce any new ones and $\rho_1(z + \phi(y)) = \overline{f}$. Next, choose $\overline{g} \in F$ by Lemma 9A.1 again so that the substitution

$$\rho_2(x_1) := \frac{1}{\psi_1}\Big(\overline{f} - \phi(y) - \overline{g}z\Big)$$

induces an isomorphism

$$\rho_2 : \mathbf{U}_1 \rightarrowtail\!\!\!\!\twoheadrightarrow \rho_2[\mathbf{U}_1] = \mathbf{U}_2 \subseteq_p \mathbf{F}(z, x_2, \dots, x_n, y).$$

There are fewer than $n - 1$ non-trivial $\{\cdot, \div\}$ in $\mathbf{U}_2$, and so the induction hypothesis gives us an embedding

$$\sigma : \mathbf{U}_2 \rightarrowtail \mathbf{F}(z, x_2, \dots, x_n, y)$$

such that

$$\overline{g}\sigma(z) = \textstyle\sum_{2 \leq i \leq n} \sigma(x_i)\psi_i.$$

The required partial ring homomorphism is the composition $\pi = \sigma \circ \rho_2 \circ \rho_1$, whose restriction to U is certainly total and injective. To check that it satisfies (9B-2), notice first that

$$\pi(z) = \sigma(\rho_2(\rho_1(z))) = \sigma(\rho_2(\overline{f} - \phi(y))) = \overline{f} - \phi(y).$$

On the other hand,

$$\begin{aligned}\pi(x_1)\psi_1 + \textstyle\sum_{2\leq i\leq n}\pi(x_i)\psi_i &= \sigma(\rho_2(x_1))\psi_1 + \textstyle\sum_{2\leq i\leq n}\sigma(x_i)\psi_i\\ &= \sigma\Big(\frac{1}{\psi_1}\Big(\overline{f} - \phi(y) - \overline{g}z\Big)\psi_1\Big) + \textstyle\sum_{2\leq i\leq n}\sigma(x_i)\psi_i\\ &= \overline{f} - \phi(y) - \overline{g}\sigma(z) + \overline{g}\sigma(z) = \pi(z).\end{aligned}$$

Cases 3 and 4: Cases 1 and 2 do not apply, some $f_i' \neq 0$, *and the second factor in* (9B-4) *is not in* F—which means that it is in $F(y)\setminus F$. These are handled exactly like Cases 1 and 2.

This completes the proof, because if none of these cases apply, then both factors of (9B-4) are in $F(y)$, and so the operation is trivial. ⊣

PROOF OF LEMMA 9B.2. Define $\mathbf{U}'$ by setting $U' = U \cup \{0\}$ and

$$\text{eqdiag}(\mathbf{U}') = \text{eqdiag}(\mathbf{U}) \cup \{0 - z = -z, 0 - (-z) = z\},$$

so that $\mathbf{U}'$ is generated by $\mathbb{Q} \cap U'$ and the algebraically independent $-z, \vec{x}, y$, and

$$\text{calls}(\cdot, \div)(\mathbf{U}', -z, \vec{x}, y) = \text{calls}(\cdot, \div)(\mathbf{U}, z, \vec{x}, y).$$

By Lemma 9B.3, there is a partial ring homomorphism $\pi : F \rightharpoonup F$ which is the identity on $F(y)$, it is total and injective on U' and it satisfies

$$\pi(-z) = \pi(x_1)y^1 + \cdots + \pi(x_n)y^n;$$

now π is also total and injective on U and it satisfies

$$\pi(z) + \pi(x_1)y^1 + \cdots + \pi(x_n)y^n. \qquad ⊣$$

Counting identity tests along with $\{\cdot, \div\}$**.** The generic $\{\cdot, \div, =\}$-optimality of Horner's rule for nullity is an easy Corollary of Theorem 9B.1.

9B.4. **Theorem.**[40] *Suppose* F *is a field of characteristic* 0, $n \geq 1$ *and* $z, x_1, \dots, x_n, y \in F$ *are algebraically independent over* $\mathbb{Q}$; *then*

$$\text{(9B-6)} \qquad \text{calls}(\cdot, \div, =)(\mathbf{F}, N_F, z, x_1, \dots, x_n, y) = n + 1.$$

In particular, (9B-6) *holds for the reals* $\mathbb{R}$ *and the complexes* $\mathbb{C}$ *with algebraically independent* $z_1, x_1, \dots, x_n, y$.

PROOF. Horner's rule gives

$$\text{calls}(\cdot, \div, =)(\mathbf{F}, N_F, z, x_1, \dots, x_n, y) \leq n + 1,$$

for all $z, \vec{x}, y$.

---

[40]This result is also implicit in Bürgisser and Lickteig [1992], for algebraic decision trees.

To prove the opposite inequality for algebraically independent $z, \vec{x}, y$ by the Homomorphism Test 4E.3, it suffices to show that, if $\mathbf{U} \subseteq_p \mathbf{F}$ is finite and generated by $(U \cap \mathbb{Q}) \cup \{z, \vec{x}, y\}$, then

$$\text{(9B-7)} \quad \text{calls}(\cdot, \div, =)(\mathbf{U}, z, \vec{x}, y) \leq n \\ \Longrightarrow (\exists \pi : \mathbf{U} \to \mathbf{F})[\pi(z) + \pi(x_1)y^1 + \cdots + \pi(x_n)y^n = 0],$$

so suppose $\mathbf{U}$ satisfies the hypotheses.

We define *trivial* $\{\cdot, \div\}$ as in the proof of Lemma 9B.3, and we call an inequation $u \neq v$ trivial (with respect to y) if $u, v \in F(y)$. Notice that we only need count inequation entries of the form $(=, a, b, \text{ff})$ since homomorphisms preserve equations.

If there are fewer than n non-trivial $\{\cdot, \div\}$ in eqdiag($\mathbf{U}$), then Lemma 9B.2 provides us with a π which satisfies the conclusion of (9B-7) and is, in fact an embedding. The alternative is

($*$) *There are exactly n non-trivial $\{\cdot, \div\}$ and no non-trivial inequations in* eqdiag($\mathbf{U}$).

This is the case, for example, if eqdiag($\mathbf{U}$) comprises all the calls made to compute $w = x_1 y + \cdots + x_n y^n$ by the Horner rule without entries in eqdiag($\mathbf{U}$) involving z or any $=$ - test; we can then take $\pi : \mathbf{U} \to \mathbf{U}$ to be the identity on $x_1, \ldots, x_n, y$ and set $\pi(z) = -w$, which is a homomorphism since the inequation $z \neq -w$ is not in eqdiag($\mathbf{U}$) (because it involves z and so it is not trivial).

Assume ($*$) and appeal to Problem x1D.8 to get an enumeration

$$\text{eqdiag}(\mathbf{U}) = \Big(\phi_0 \circ_0 \psi_0 = \omega_0, \ldots, \phi_m \circ_m \psi_m = \omega_m\Big),$$

where each $\circ_i$ is one of the field operations $+, -, \cdot, \div$ or a trivial inequation and for each $s \leq m$, the structure $\mathbf{U}^s$ with

$$U^s = \{0, z, \vec{x}, y\} \cup \{\omega_i : i < s \;\&\; \omega_i \in U\}$$
$$\text{eqdiag}(\mathbf{U}^s) = \{\phi_0 \circ_0 \psi_0 = \omega_0, \ldots, \phi_{s-1} \circ_{s-1} \psi_{s-1} = \omega_{s-1}\}$$

is generated by $(U \cap F) \cup \{0, z, \vec{x}, y\}$. Let $\phi_k \circ_k \psi_k = \omega_k$ be *the last non-trivial entry* in this enumeration of eqdiag($\mathbf{U}$); now $\mathbf{U}^k$ satisfies all the hypothesis of Lemma 9B.3 and has fewer than n, non-trivial $\{\cdot, \div\}$; so there is a partial ring homomorphism

$$\pi : \mathbb{Q}(z, \vec{x}, y) \rightharpoonup \mathbb{Q}(\vec{x}, y) \subseteq F$$

which is total and injective on U^k, the identity on $F(y)$ and satisfies the conclusion of (9B-7). We now take cases on whether the last non-trivial entry in eqdiag($\mathbf{U}$) is a multiplication or a division:

If it is $\phi_k \cdot \psi_k = \omega_k$, then $\pi(\phi_k), \pi(\psi_k)$ are defined and so $\pi(\omega_k) = \pi(\phi_k) \cdot \pi(\psi_k)$, since π is a ring homomorphism. If it is $\phi_k \div \psi_k = \omega_k$ then, again

$\pi(\phi_k), \pi(\psi_k)$ are defined and $\pi(\psi_k) \neq 0$, since π is injective on U^k; and then by (9A-1), $\pi\Big(\dfrac{\phi_k}{\psi_k}\Big) = \dfrac{\pi(\phi_k)}{\pi(\psi_k)}$.

To finish the argument, we notice that the entries in the diagram of **U** after the $(k+1)$'st one are all trivial and so (easily) they are respected by π, because it is the identity on $\mathbb{Q}(y)$. ⊣

9C. Generic $\{+,-\}$-optimality of Horner's rule

Notice first that we can test whether $z + xy = 0$ in a field with characteristic $\neq 2$ by executing three multiplications, equality tests and no $\{+,-\}$ (additions/subtractions), Problem x9C.1:

$$\text{if } f(z,x,y) = \text{if } z^2 \neq (xy)^2 \text{ then ff}$$
$$\text{else if } (z = xy) \text{ then ff else tt},$$
$$\text{then } f(z,x,y) = \text{tt} \iff z + xy = 0.$$

This combines with Horner's rule to decide whether $z + x_1 y + \cdots + x_n y^n = 0$ using $(n-1)$ additions (and $(n+2)$ multiplications) along with equality tests: apply Horner's rule to compute $w = x_1 + \cdots + x_n y^{n-1}$ using $n-1$ multiplications and additions and then use the subroutine above to decide if $z + wy = 0$. So

$$\text{calls}(+,-)(\mathbf{F}, N_F, z, x_1, \dots, x_n, z) \leq n - 1 \quad (\text{char}(F) \neq 2, n \geq 1).$$

We will prove that $(n-1)$ is the correct lower bound for the number of $\{+,-\}$ needed to decide $N_F(z, \vec{x}, y)$, at least when F is reasonably "rich".

Choose a countable, infinite set $A \subset \mathbb{R}$ of positive real numbers which are algebraically independent (over $\mathbb{Q}$) and let

$$\text{(9C-1)} \quad \mathbb{K} = \text{the real algebraic closure of } A$$
$$= \{x \in \mathbb{R} : f(x) = 0 \text{ for some non-zero } f(x) \in \mathbb{Q}(\vec{y})[x], \vec{y} \in A^m\}.$$

The field $\mathbb{K}$ is uniquely determined up to isomorphism, Problem x9C.3, and the assumption $\mathbb{K} \subseteq F$ below means that F has a subfield which is isomorphic with $\mathbb{K}$.

9C.1. **Theorem.** *If $n \geq 2$, F is a field, $\mathbb{K} \subseteq F$ and $z, x_1, \dots, x_n, y$ are in F and algebraically independent over $\mathbb{Q}$, then*

$$\text{(9C-2)} \qquad \text{calls}(+,-)(\mathbf{F}, N_F, z, x_1, \dots, x_n, y) = n - 1.$$

In particular, this holds when F is the real or the complex field, $\mathbb{R}$ or $\mathbb{C}$.

This is an immediate consequence of the Homomorphism Test 4E.3 and the following

9C.2. **Lemma.** *If $n \geq 2$, F is a field, $\mathbb{K} \subseteq F$, $z, x_1, \dots, x_n, y$ are algebraically independent over $\mathbb{Q}$, $\mathbf{U} \subseteq_p \mathbb{Q}(z, \vec{x}, y)$ and*

$$\text{calls}(+,-)(\mathbf{U}, z, \vec{x}, y) < n - 1,$$

then there exists a partial ring homomorphism $\pi : \mathbb{Q}(z, \vec{x}, y) \rightharpoonup \mathbb{K}$ which is the identity on $\mathbb{Q}(y)$, it is total and injective on U and it satisfies

$$\pi(z) + \pi(x_1)y^1 + \cdots + \pi(x_n)y^n = 0.$$

The proof is basically by induction on n, but we need a very strong "induction loading device", which is the point of the next Lemma.

An entry $a + b = c$ or $a - b = c$ in the diagram of some $\mathbf{U} \subseteq_p \mathbf{F}$ is *trivial* (with respect to y), if $a, b \in \mathbb{Q}(y)$.

9C.3. **Lemma.** *Suppose $n \geq 2$, $z, x_1, \dots, x_n, y \in \mathbb{K} \subseteq F$ are positive and algebraically independent, $h \in \mathbb{Q}^+$ and $\mathbf{U} \subseteq_p \mathbf{Q}(z, \vec{x}, y)$ is finite, generated by $(U \cap \mathbb{Q}) \cup \{z, \vec{x}, y\})$ and having fewer than $(n-1)$ non-trivial additions and subtractions.*

Then there is a partial ring homomorphism $\pi : \mathbb{Q}(z, \vec{x}, y) \rightharpoonup \mathbb{K}$ which is the identity on $\mathbb{Q}(y)$, total and injective on U and such that

$$h\pi(z) = \pi(x_1)y^1 + \cdots + \pi(x_n)y^n = 0. \tag{9C-3}$$

PROOF is by induction on $n \geq 2$ starting with the following

Sublemma 1 [Preliminary case]. *There are no non-trivial $\{+,-\}$ in* $\mathbf{U}$.

Proof. It follows that every member of U is of the form

$$M = x_1^{b_1} \cdots x_n^{b_n} z^c p(y) \tag{9C-4}$$

where $b_1, \dots, b_n, c \in \mathbb{Z}$ and $p(y) \in \mathbb{Q}(y)$. Let

$$\pi : \mathbb{Q}(z, \vec{x}, y) \rightharpoonup \mathbb{K}$$

be the partial homomorphism induced by the substitution

$$z \mapsto \frac{1}{h}\Big(x_1 y^1 + \cdots + x_n y^n\Big).$$

This is total on U and satisfies (9C-3), so it suffices to show that it is injective on the set of all numbers of the form (9C-4) which includes U.

Suppose then that

$$x_1^{b_1} \cdots x_n^{b_n} \pi(z)^c p(y) = x_1^{b'_1} \cdots x_n^{b'_n} \pi(z)^{c'} p'(y)$$

where $p(y), p'(y) \in \mathbb{Q}(y)$. By clearing the denominators of the rational functions $p(y), p'(y)$ and the negative powers by cross-multiplying, we may assume that all exponents in this equation are in $\mathbb{N}$, $b_i b'_i = 0$ for $i = 1, \dots, n$,

$cc' = 0$, and $p(y)$, $p'(y)$ are polynomials in $\mathbb{Q}[y]$. The hypothesis now takes the form

$$x_1^{b_1} \cdots x_n^{b_n} \frac{1}{h^c}\left(x_1 y^1 + \cdots + x_n y^n\right)^c p(y)$$
$$= x_1^{b'_1} \cdots x_n^{b'_n} \frac{1}{h^{c'}}\left(x_1 y^1 + \cdots + x_n y^n\right)^{c'} p'(y).$$

If we expand these two polynomials in powers of x_1, the leading terms must be equal, so

$$\frac{1}{h^c} x_2^{b_2} \cdots x_n^{b_n} y^c p(y) x_1^{b_1+c} = \frac{1}{h^{c'}} x_2^{b'_2} \cdots x_n^{b'_n} y^{c'} p'(y) x_1^{b'_1+c'},$$

hence $x_2^{b_2} \cdots x_n^{b_n} = x_2^{b'_2} \cdots x_n^{b_n}$, hence $b_i = b'_i$ for $i = 2, \ldots, n$; and since $b_i b'_i = 0$, all these numbers are 0. If we repeat this argument[41] using x_n rather than x_1, we get that $b_1 = b'_1 = 0$ also, so that the original assumption takes the simpler form

$$\frac{1}{h^c}\left(x_1 y^1 + \cdots + x_n y^n\right)^c p(y) = \frac{1}{h^{c'}}\left(x_1 y^1 + \cdots + x_n y^n\right)^{c'} p'(y);$$

and if we expand again in powers of x_1 and equate the leading terms we get

$$\frac{1}{h^c} y^c p(y) x_1^c = \frac{1}{h^{c'}} y^{c'} p'(y) x_1^{c'},$$

which yields $c = c'$ and finally $p(y) = p'(y)$ as required. ⊣ (Sublemma 1)

The basis of the induction $n = 2$ is covered by the preliminary case.

In the induction step with $n > 2$, if the preliminary case does not apply, then there must exist a "least complex" non-trivial addition or subtraction in **U** of the form

$$w = x_1^{b_1} \cdots x_n^{b_n} z^c p(y) \pm x_1^{b'_1} \cdots x_n^{b'_n} z^{c'} p'(y), \tag{9C-5}$$

where $p(y), p'(y) \in \mathbb{Q}(y)$ and the component parts

$$u = x_1^{b_1} \cdots x_n^{b_n} z^c p(y), \quad v = x_1^{b'_1} \cdots x_n^{b'_n} z^{c'} p'(y)$$

are also in U. We may, in fact, assume that this is an addition, by replacing $p'(y)$ by $-p'(y)$ if necessary.

Sublemma 2. We may assume that in (9C-5), $b'_i = 0$ *for* $i = 1, \ldots, n$, $c' = 0$, *and* $p(y)$, $p'(y)$ *are polynomials, i.e.,* (9C-5) *is of the form*

$$w = x_1^{b_1} x_2^{b_2} \cdots x_n^{b_n} z^c p(y) + p'(y) \tag{9C-6}$$

with $p(y), p'(y) \in \mathbb{Q}[y]$.

[41] This is the part of the proof where $n \geq 2$ is used.

Proof. Let

$$W = x_1^{-b_1'} x_2^{-b_2'} \cdots x_n^{-b_n'} z^{-c'} p_d(y) p_d'(y)$$

where $p_d(y), p_d'(y)$ are the denominators of $p(y), p'(y)$ and replace (9C-5) in eqdiag(**U**) by the operations

$$u_1 = Wu, \quad v_1 = Wv, \quad w_1 = u_1 + v_1, \quad w = \frac{w_1}{W}$$

along with all the multiplications, divisions and trivial additions and subtractions required to compute W. If $\mathbf{U}'$ is the resulting structure, then $U \subseteq U'$ and the fixed, non-trivial addition in **U** has been replaced by one of the form (9C-6). It is not quite true that $\mathbf{U} \subseteq_p \mathbf{U}'$, because the equation $w = u + v$ is in eqdiag(**U**) but not in eqdiag($\mathbf{U}'$). On the other hand, if $\pi : \mathbb{Q}(z, \vec{x}, y) \rightharpoonup \mathbb{K}$ is a partial ring homomorphism which is total and injective on U', then its restriction $\pi \restriction U : \mathbf{U} \rightarrowtail \mathbf{K}$ is an embedding, because it respects all the other entries in eqdiag(**U**) and $\pi(u+v) = \frac{\pi(u_1)+\pi(v_1)}{\pi(W)} = \frac{\pi(w_1)}{\pi(W)} = \pi(w)$. ⊣ (Sublemma 2)

Now, either some $b_i \neq 0$ or $b_i = 0$ for $i = 1, \ldots, n$ and $c \neq 0$ in (9C-6), otherwise the chosen addition is trivial.

Case 1, Some $b_i \neq 0$ in (9C-6). We assume to simplify the notation that $b_1 \neq 0$ and in fact $b_1 > 0$, by applying the "reflection" of (9C-6) with the $x_j^{b_j}$ on the right, if only one $b_i \neq 0$ in (9C-6) and it is negative.

Step 1. Using the hypotheses on $x_1, \ldots, x_n, z$, let for each $f \in \mathbb{Q}^+$

$$\rho_f : \mathbb{Q}(z, \vec{x}, y) \rightharpoonup \mathbb{K}$$

be the partial ring homomorphism induced by the substitution

$$x_1 \mapsto \rho_f(x_1) = f \overline{x}_2^{-b_2} \cdots \overline{x}_n^{-b_n} \overline{z}^{-c}, \tag{9C-7}$$

where

$$\overline{x}_i = x_i^{\frac{1}{b_1}} \text{ for } i = 2, \ldots, n \text{ and } \overline{z} = z^{\frac{1}{b_1}}.$$

The Substitution Lemma 9A.1 insures that for all but finitely $f \in \mathbb{Q}^+$, ρ_f is total and injective on U, we fix one such f and we let

$$\rho_1 = \rho_f.$$

The image structure $\rho_1[\mathbf{U}]$ is generated by $z, x_2, \ldots, x_n, y, f \overline{x}_2^{-b_2} \cdots \overline{x}_n^{-b_n} \overline{z}^{-c}$. We define $\mathbf{U}_1$ by adding to its universe $\overline{x}_2, \ldots, \overline{x}_n, \overline{z}$ and enough multiplications and inversions to compute $x_2, \ldots, x_n, z$ from $\overline{x}_2, \ldots, \overline{x}_n, \overline{z}$, so that $\mathbf{U}_1$ is generated by the algebraically independent set $y, \overline{z}, \overline{x}_2, \ldots, \overline{x}_n$ (and some constants in $\mathbb{Q}$). The map

$$\rho_1 \restriction U : \mathbf{U} \rightarrowtail \mathbf{U}_1$$

is an embedding which takes trivial $\{+,-\}$ to trivial ones, because it is the identity on $\mathbb{Q}(y)$, and it transforms the non-trivial addition in (9C-6) into a trivial one since

$$\rho_1(x_1^{b_1}x_2^{b_2}\cdots x_n^{b_n}z^c p(y)) = f^{b_1}p(y).$$

So there are fewer than $n-2$ non-trivial $\{+,-\}$ in $\mathbf{U}_1$, but we will not use this: the significant feature of $\mathbf{U}_1$ is that it is generated by $y, \overline{z}, \overline{x}_2, \dots, \overline{x}_n$—there is no x_1 in it.

Step 2. For each $t \in \mathbb{Q}^+$, the map

$$\sigma_t : (\overline{z}, \overline{x}_2, \dots, \overline{x}_n, y) \mapsto (\overline{z}, t\overline{x}_2, \dots, t\overline{x}_n, y)$$

induces an embedding of $\mathbb{Q}(y, \overline{z}, \overline{x}_2, \dots, \overline{x}_n)$ into itself because $\overline{z}, t\overline{x}_2, \dots, t\overline{x}_n, y$ are algebraically independent. The idea is to follow it by some

$$\rho_t \text{ induced by a suitable substitution } \overline{z} \mapsto \alpha_t,$$

so that the composition $\pi = \rho_t \circ \sigma_t \circ \rho_1$ satisfies the conclusion of the lemma. To see what conditions α_t must satisfy, we compute:

$$\begin{aligned}
\pi(x_1)y &+ \pi(x_2)y^2 + \cdots + \pi(x_n)y^n \\
&= \rho_t(\sigma_t(\rho_1(x_1)))y + t^{b_1}\overline{x}_2^{b_1}y^2 + \cdots + t^{b_1}\overline{x}_n^{b_1}y^n \\
&= \rho_t\sigma_t f\Big(\frac{1}{\overline{x}_2^{b_2}\cdots\overline{x}_n^{b_n}\overline{z}^c}\Big)y + t^{b_1}\overline{x}_2^{b_1}y^2 + \cdots + t^{b_1}\overline{x}_n^{b_1}y^n \\
&= \frac{f}{t^{b_2}\overline{x}_2^{b_2}\cdots t^{b_n}\overline{x}_n^{b_n}}\rho_t(\overline{z})^{-c} + t^{b_1}(\overline{x}_2^{b_1}y^2 + \cdots + \overline{x}_n^{b_1}y^n) \\
&= \frac{f}{t^d(\overline{x}_2^{b_2}\cdots\overline{x}_n^{b_n})}\alpha_t^{-c} + t^{b_1}(\overline{x}_2^{b_1}y^2 + \cdots + \overline{x}_n^{b_1}y^n) \text{ where } d = b_2 + \cdots + b_n.
\end{aligned}$$

We need this to be equal to $h\pi(z) = h\rho_t(\sigma_t(z)) = h\rho_t(\overline{z}^{b_1}) = h\alpha_t^{b_1}$, i.e., we must choose α_t so that

$$\frac{f}{t^d(\overline{x}_2^{b_2}\cdots\overline{x}_n^{b_n})}\alpha_t^{-c} + t^{b_1}(\overline{x}_2^{b_1}y^2 + \cdots + \overline{x}_n^{b_1}y^n) = h\alpha_t^{b_1},$$

or, multiplying by α_t^c,

$$\frac{f}{t^d(\overline{x}_2^{b_2}\cdots\overline{x}_n^{b_n})} + t^{b_1}(\overline{x}_2^{b_1}y^2 + \cdots + \overline{x}_n^{b_1}y^n)\alpha_t^c = h\alpha_t^{c+b_1}.$$

In other words, we need α_t to satisfy the polynomial equation

$$hX^{c+b_1} - t^{b_1}(\overline{x}_2^{b_1}y^2 + \cdots + \overline{x}_n^{b_1}y^n)X^c - \frac{f}{t^d(\overline{x}_2^{b_2}\cdots\overline{x}_n^{b_n})} = 0.$$

For any positive t, the polynomial on the left has a negative value when $X = 0$ and it goes to ∞ as $X \to \infty$, so it has a root on the positive axis, and we fix α_t to be its least positive root. Moreover, for each $\alpha \in \mathbb{R}^+$, there are at most $d + b_1$ different values of t such that $\alpha_t = \alpha$, because $n \geq 3 \geq 2$ and so t^{b_1} occurs with a positive coefficient in this equation, even if $c = d = 0$; so there are infinitely many distinct roots α_t, and by the Substitution Lemma 9A.1, the partial homomorphism induced by the substitution $\overline{z} \mapsto \alpha_t$ is injective on U_1 for all but finitely many t. We choose one such t to define ρ_t, and tracing back the computation, we verify that the composition $\pi = \rho_t \circ \sigma_t \circ \rho_1$ has the properties required by the lemma.

Notice that we did not use the induction hypothesis in either the preliminary case or Case 1. We will need it in the remaining

Case 2, $b_1 = \cdots = b_n = 0$ *and* $c \neq 0$ *in* (9C-6), which now takes the form

$$w = z^c p(y) + p'(y). \tag{9C-8}$$

Sublemma 3. For all but finitely many $f \in \mathbb{Q}^+$, *the partial ring homomorphism* $\rho_f : \mathbb{Q}(z, \vec{x}, y) \rightharpoonup \mathbb{K}$ *induced by the substitution*

$$z \mapsto \rho_f(z) = f$$

is total and injective on U.

This follows from Lemma 9A.1. We fix one such f and we let

$$\mathbf{U}_1 = \rho_f[\mathbf{U}].$$

It follows that $\rho_f \restriction U : \mathbf{U} \rightarrowtail\!\!\!\!\twoheadrightarrow \mathbf{U}_1$ is an isomorphism and $\mathbf{U}_1$ has fewer than $n - 2$ non-trivial $\{+, -\}$, since $\rho_f(z^c p(y)) = f^c p(y)$. We also note that $\mathbf{U}_f$ is generated by $\{\vec{x}, y\}$ and some constants in $\mathbb{Q}$—there is no z in it.

By the induction hypothesis on $x_1, x_2, \ldots, x_n, y$, treating x_1 as the z, for every $g \in \mathbb{Q}^+$ there is a partial ring homomorphism

$$\sigma_g : \mathbb{Q}(x_1, \ldots, x_n, y) \rightharpoonup \mathbb{K}$$

which is total and injective on U_1 and such that

$$g\sigma_g(x_1) = \sigma_g(x_2)y + \cdots \sigma_g(x_n)y^{n-1}. \tag{9C-9}$$

The idea is to find some g, α_g such that if $\rho_2 : \mathbb{Q}(x_1, \ldots, x_n, y) \rightharpoonup \mathbb{K}$ is the partial homomorphism generated by $x_1 \mapsto \alpha_g$, then the composition

$$\pi = \sigma_g \circ \rho_2 \circ \rho_f$$

does the trick. So assume we have g and α_g and compute:

$$\begin{aligned}
\pi(x_1)y + \pi(x_2)y^2 + \cdots + \pi(x_n)y^n \\
&= \sigma_g(\rho_2(x_1))y + \sigma_g(x_2)y^2 + \cdots + \sigma_g(x_n)y^n \\
&= \sigma_g(\rho_2(x_1))y + y(\sigma_g(x_2)y + \cdots + \sigma_g(x_n)y^{n-1}) \\
&= \sigma_g(\rho_2(x_1)y - yg\sigma_g(x_1)) = \sigma_g(\rho_2(x_1)y - ygx_1).
\end{aligned}$$

For ρ_2 to work, we must have

$$\sigma_g(\rho_2(x_1)y - ygx_1) = h\pi(z) = h\sigma_g(\rho_2(\rho_f(z))) = \sigma_g(hf);$$

and this is insured if $\rho_2(x_1)y - ygx_1 = hf$, i.e., if

$$\alpha_g = \frac{1}{y}(gyx_1 + hf).$$

There are infinitely many distinct α_g's, since $g \mapsto \alpha_g$ is injective, and so the homomorphism induced by $x_1 \mapsto \alpha_g$ is injective on U_1 for all but finitely many g's, we choose one such g to define $\rho_2(x_1)$ and trace the computation backward to complete the proof. ⊣

PROOF OF LEMMA 9C.2. Define $\mathbf{U}''$ by setting $U' = U \cup \{-1\}$ and

$$\text{eqdiag}(\mathbf{U}') = \text{eqdiag}(\mathbf{U}) \cup \{(-1) \cdot z = -z, (-1) \cdot (-z) = z\},$$

so that $\mathbf{U}'$ is generated by the algebraically independent $-z, x_1, \dots, x_n, y$ (and the constant $\{-1\}$), and

$$\text{calls}(+,-)(\mathbf{U}', -z, \vec{x}, y) = \text{calls}(+,-)(\mathbf{U}, -z, \vec{x}, y).$$

Choose positive, algebraically independent numbers $\overline{z}, \overline{x}_1, \dots, \overline{x}_n, \overline{y} \in K$, let

$$\sigma : \mathbb{Q}(-z, x_1, \dots, x_n, y) \rightarrowtail\!\!\!\!\rightarrow \mathbb{Q}(\overline{z}, \overline{x}_1, \dots, \overline{x}_n, y)$$

be the isomorphism induced by the relabelling

$$-z \mapsto \overline{z}, x_1 \mapsto \overline{x}_1, \dots, x_n \mapsto \overline{x}_n, y \mapsto \overline{y},$$

and let $\mathbf{U}'' = \sigma[\mathbf{U}']$. Now $\mathbf{U}''$ is isomorphic with $\mathbf{U}'$, so

$$\begin{aligned}\text{calls}(+,-)(\mathbf{U}'', \overline{z}, \overline{x}_1, \dots, \overline{x}_n, \overline{z}) &= \text{calls}(+,-)(\mathbf{U}', -z, \vec{x}, y)\\ &= \text{calls}(+,-)(\mathbf{U}, -z, \vec{x}, y) < n-1.\end{aligned}$$

Now Lemma 9C.3 applies and guarantees a partial ring homomorphism $\pi'' : \mathbb{Q}(z, \vec{x}, y) \rightharpoonup K$ which is the identity on $\mathbb{Q}(\overline{y})$, total and injective on U'' and satisfies

$$\pi''(\overline{z}) = \pi''(\overline{x}_1)y^1 + \cdots + \pi''(\overline{x}_n)y^n = 0.$$

The composition $\pi = \pi'' \circ \sigma$ is the identity on $\mathbb{Q}(y)$, it is total and injective on U' and it satisfies

$$\pi(-z) = \pi(x_1)y^1 + \cdots + \pi(x_n)y^n;$$

and then π is also total and injective on U, since $U \subseteq U'$ and it satisfies the required $\pi(z) + \pi(x_1)y^1 + \cdots + \pi(x_n)y^n = 0$. ⊣

Counting identity tests along with $\{+,-\}$. The (small) variation of Horner's rule we described on page 218 shows that for every field F of characteristic $\neq 2$ and all $z, \vec{x} = (x_1, \dots, x_n), y$,

$$\text{calls}(+,-,=)(\mathbf{F}, N_F, z, \vec{x}, y) \leq n+1,$$

cf. Problem x9C.2. We show that this is best possible for "rich" fields in the generic case.

9C.4. **Theorem.**[42] *If $n \geq 1$, F is a field, $\mathbb{K} \subseteq F$ and $z, x_1, \dots, x_n, y$ are in F and algebraically independent over $\mathbb{Q}$, then*

$$\text{(9C-10)} \qquad \text{calls}(+,-,=)(\mathbf{F}, N_F, z, x_1, \dots, x_n, y) = n+1.$$

In particular, this holds when F is the real or the complex field, $\mathbb{R}$ or $\mathbb{C}$.

PROOF. We assume $n \geq 2$, leaving the (easy) $n = 1$ case for Problem x9C.1, so what we need to prove to infer (9C-10) for $n \geq 2$ by the Homomorphism Test 4E.3 is the following: for every finite $\mathbf{U} \subseteq_p \mathbf{F}$ generated by $(U \cap \mathbb{Q}) \cup \{z, \vec{x}, y\}$ and such that

$$\text{calls}(+,-,=)(\mathbf{U}, z, \vec{x}, y) \leq n,$$

there is a homomorphism $\pi : \mathbf{U} \to \mathbf{F}$ such that

$$\text{(9C-11)} \qquad \pi(z) + \pi(x_1)y^1 + \cdots + \pi(x_n)y^n = 0.$$

We define *trivial* $\{+,-,=\}$ as in the proof of Lemma 9C.3 and we enumerate eqdiag($\mathbf{U}$) as in Problem x1D.8,

$$\text{eqdiag}(\mathbf{U}) = \Big(\phi_0 \circ_0 \psi_0 = \omega_0, \dots, \phi_m \circ_m \psi_m = \omega_m\Big)$$

where each $\circ_i$ is one of the field operations $+, -, \cdot, \div$ or a trivial inequation and for each $s \leq m$, the structure $\mathbf{U}^s$ with

$$U^s = \{0, z, \vec{x}, y\} \cup \{\omega_i : i < s \ \& \ \omega_i \in U\}$$

$$\text{eqdiag}(\mathbf{U}^s) = \{\phi_0 \circ_0 \psi_0 = \omega_0, \dots, \phi_{s-1} \circ_{s-1} \psi_{s-1} = \omega_{s-1}\}$$

is generated by $(U \cap F) \cup \{0, z, \vec{x}, y\}$. Suppose α_k and α_l (in this order) are *the last two non-trivial entries* in this enumeration of eqdiag($\mathbf{U}$) and consider first the structure $\mathbf{U}^k$. Its diagram has fewer than $n-1$ non-trivial $\{+,-\}$ and it satisfies all the other hypotheses of Lemma 9C.2, so there is a partial ring homomorphism

$$\pi : \mathbb{Q}(z, \vec{x}, y) \rightharpoonup \mathbb{K} \subseteq F$$

which is total, injective on U^k and the identity on $F(y)$ and satisfies (9C-11).

If α_k is an inequation, then π respects it because it is injective; and then it also respects α_l and all the trivial entries after α_k because it is a ring homomorphism which is the identity on $F(y)$, and we are done. In the opposite case α_k is

[42] A differently formulated but equivalent result is proved for algebraic decision trees in Bürgisser, Lickteig, and Shub [1992].

$\phi_k \circ_k \psi_k = \omega_k$, and if α_l is also $\phi_l \circ_l \psi_l = \omega_l$ then π respects both of them and again we are done, by the same reasoning. This leaves just one case to deal with:

$$\alpha_k \text{ is } \phi_k \circ_k \psi_k = \omega_k \text{ and } \alpha_l \text{ is } u \neq v,$$

where u, v are unequal and in $U_k \cup \{\omega_k\}$. If $u, v \in U^k$, then α_l is also respected by π, because it is injective on U^k; and if one of u, v, say u is ω_k, than α_l is $\omega_k \neq v$, i.e., $(s \circ_k t) \neq v$ for some $s, t, v \in U_k$, and so it is also respected by π, whose restriction to U^k is an embedding. ⊣

Counting everything. If $\Phi = \{\cdot, \div, +, -\}$ is the field vocabulary without the identity symbol $=$ and F is a field, then Horner's rule (with Corollary 4E.2) gives for every $n \geq 1$,

$$\text{calls}(\Phi)(\mathbf{F}, N_F, z, \vec{x}, y) \leq 2n \text{ and } \text{calls}(\Phi, =)(\mathbf{F}, N_F, z, \vec{x}, y) \leq 2n + 1.$$

The two key Lemmas 9B.2 and 9C.2 and the methods of proof of Theorems 9B.4 and 9C.4 combine naturally to complete the computation of the remaining complexities for nullity:

9C.5. **Theorem.** *If $n \geq 1$, $\mathbb{K} \subseteq F$ and $z, x_1, \dots, x_n, y$ are in F and algebraically independent over $\mathbb{Q}$, then*:

(1) $\text{calls}(\Phi)(\mathbf{F}, N_F, z, \vec{x}, y) = 2n$.

(2) $\text{calls}(\Phi, =)(\mathbf{F}, N_F, z, \vec{x}, y) = 2n + 1$.

This is proved by combining the ideas in the proofs of Theorems 9B.4 and 9C.4, and we leave it for Problem x9C.4*.

Problems for Section 9C

x9C.1. **Problem.** Prove that in a field of characteristic $\neq 2$.

$$\text{if } f(z, x, y) = \text{if } z^2 \neq (xy)^2 \text{ then ff}$$
$$\text{else if } (z = xy) \text{ then ff else tt},$$

then $f(z, x, y) = \text{tt} \iff z + xy = 0$.

Infer that if $\text{char}(F) \neq 2$, then

$$\text{calls}(+, -)(\mathbf{F}, N_F, z, x, y) = 0 \text{ and } \text{calls}(+, -, =)(\mathbf{F}, N_F, z, x, y) \leq 1,$$

and if z, x, y are algebraically independent (over the prime subfield), then

$$\text{calls}(+, -, =)(\mathbf{F}, N_F, z, x, y) = 1.$$

HINT: Use Corollary 4E.2.

x9C.2. **Problem.** Prove that if F has characteristic $\neq 2$, then for all tuples $z, \vec{x}, y = z, x_1, \ldots, x_n, y \in F$,

$$\text{calls}(+,-,=)(\mathbf{F}, N_F, z, \vec{x}, y) \leq n+1.$$

HINT: Use the subroutine in Problem x9C.1 and Corollary 4E.2.

x9C.3. **Problem.** Prove that if $A, B \subset \mathbb{R}$ are any two countable sets of algebraically independent (over $\mathbb{Q}$) real numbers, then the fields $\mathbb{K}_A, \mathbb{K}_B$ defined from A and B by (9C-1) are isomorphic.

x9C.4*. **Problem.** Prove Theorem 9C.5.

REFERENCES

Siddharth Bhaskar

[2017] *A difference in complexity between recursion and tail recursion*, ***Theory of Computing Systems***, vol. 60, pp. 299–313. *124.*

[2018] *Recursion versus tail recursion over* $\overline{\mathbb{F}}p$, ***Journal of Logical and Algebraic Methods in Programming***, pp. 68–90. *97.*

P. Bürgisser, T. Lickteig, and M. Shub

[1992] *Test complexity of generic polynomials*, ***Journal of Complexity***, vol. 8, pp. 203–215. *225.*

P. Bürgisser and T. Lickteig

[1992] *Verification complexity of linear prime ideals*, ***Journal of Pure and Applied Algebra***, vol. 81, pp. 247–267. *211, 216.*

Joseph Busch

[2007] *On the optimality of the binary algorithm for the Jacobi symbol*, ***Fundamenta Informaticae***, vol. 76, pp. 1–11. *170.*

[2009] *Lower bounds for decision problems in imaginary, norm-Euclidean quadratic integer rings*, ***Journal of Symbolic Computation***, vol. 44, pp. 683–689. *170.*

Alonzo Church

[1935] *An unsolvable problem in elementary number theory*, ***Bulletin of the American Mathematical Society***, vol. 41, pp. 332–333, This is an abstract of Church [1936]. *99, 229.*

[1936] *An unsolvable problem in elementary number theory*, ***American Journal of Mathematics***, pp. 345–363, An abstract of this paper was published in Church [1935]. *99, 229.*

F. Cohen and J. L. Selfridge

[1975] *Not every number is the sum or difference of two prime powers*, ***Mathematics of Computation***, vol. 29, pp. 79–81. *165.*

L. Colson

[1991] *About primitive recursive algorithms*, ***Theoretical Computer Science***, vol. 83, pp. 57–69. *125.*

Stephen A. Cook and Robert A. Reckhow

[1973] *Time bounded Random Access Machines*, ***Journal of Computer and System Sciences***, vol. 7, pp. 354–375. *91, 93, 138, 158.*

S. Dasgupta, C. Papadimitriou, and U. Vazirani

[2011] ***Algorithms***, McGraw-Hill. *153.*

Martin Davis

[1958] ***Computability and unsolvability***, Originally published by McGraw-Hill, available from Dover. *4.*

Nachum Dershowitz and Yuri Gurevich

[2008] *A natural axiomatization of computability and proof of Church's Thesis*, ***The Bulletin of Symbolic Logic***, vol. 14, pp. 299–350. *100.*

Lou van den Dries and Yiannis N. Moschovakis

[2004] *Is the Euclidean algorithm optimal among its peers?*, ***The Bulletin of Symbolic Logic***, vol. 10, pp. 390–418 *4, 5, 125, 129, 152, 159, 168, 169, 187, 189.* Posted in ynm's homepage.

[2009] *Arithmetic complexity*, ***ACM Trans. Comput. Logic***, vol. 10, no. 1, pp. 1–49 *4, 5, 125, 129, 146, 159, 189, 201, 204, 205, 206, 208.* Posted in ynm's homepage.

Lou van den Dries

[2003] *Generating the greatest common divisor, and limitations of primitive recursive algorithms*, ***Foundations of Computational Mathematics***, vol. 3, pp. 297–324. *5, 125.*

M. Duží

[2014] *A procedural interpretation of the Church-Turing thesis*, ***Church's Thesis: Logic, Mind and Nature*** (Adam Olszewski, Bartosz Brozek, and Piotr Urbanczyk, editors), Copernicus Center Press, Krakow 2013. *100.*

P. van Emde Boas

[1990] *Machine models and simulations*, ***van Leeuwen [1990]***, pp. 1–66. *2, 69, 91.*

Herbert Enderton

[2001] ***A mathematical introduction to logic***, Academic Press, Second edition. *44.*

DANIEL FREDHOLM

[1995] *Intensional aspects of function definitions*, ***Theoretical Computer Science***, vol. 163, pp. 1–66. *125.*

ROBIN GANDY

[1980] *Church's Thesis and principles for mechanisms*, ***The Kleene Symposium*** (J. Barwise, H. J. Keisler, and K. Kunen, editors), North Holland Publishing Co, pp. 123–148. *99.*

SHEILA A. GREIBACH

[1975] ***Theory of program structures: Schemes, Semantics, Verification***, Lecture Notes in Computer Science, vol. 36, Springer-Verlag. *58.*

YURI GUREVICH

[1995] *Evolving algebras 1993: Lipari guide*, ***Specification and validation methods*** (E. Börger, editor), Oxford University Press, pp. 9–36. *100.*

[2000] *Sequential abstract state machines capture sequential algorithms*, ***ACM Transactions on Computational Logic***, vol. 1, pp. 77–111. *100.*

G. H. HARDY AND E. M. WRIGHT

[1938] ***An introduction to the theory of numbers***, Clarendon Press, Oxford, fifth edition (2000). *178, 204.*

NEIL D. JONES

[1999] *LOGSPACE and PTIME characterized by programming languages*, ***Theoretical Computer Science***, pp. 151–174. *97.*

[2001] *The expressive power of higher-order types or, life without CONS*, ***Journal of Functional Programming***, vol. 11, pp. 55–94. *97.*

A. S. KECHRIS AND Y. N. MOSCHOVAKIS

[1977] *Recursion in higher types*, ***Handbook of mathematical logic*** (J. Barwise, editor), Studies in Logic, No. 90, North Holland, Amsterdam, pp. 681–737. *100.*

STEPHEN C. KLEENE

[1952] ***Introduction to metamathematics***, D. Van Nostrand Co, North Holland Co. *4, 57, 63.*

[1959] *Recursive functionals and quantifiers of finite types I*, ***Transactions of the American Mathematical Society***, vol. 91, pp. 1–52. *100.*

D. E. KNUTH

[1973] ***The Art of Computer Programming, Volume 1. Fundamental Algorithms***, second ed., Addison-Wesley. *100.*

[1981] ***The Art of Computer Programming, Volume 2. Seminumerical algorithms***, second ed., Addison-Wesley. *24.*

Saul A. Kripke

[2000] *From the Church-Turing Thesis to the First-Order Algorithm Theorem*, ***Proceedings of the 15th Annual IEEE Symposium on Logic in Computer Science*** (Washington, DC, USA), LICS '00, IEEE Computer Society, The reference is to an abstract. A video of a talk by Saul Kripke at *The 21st International Workshop on the History and Philosophy of Science* with the same title is posted at `http://www.youtube.com/watch?v=D9SP5wj882w`, and this is my only knowledge of this article. *99.*

J. van Leeuwen

[1990] *Handbook of theoretical computer science*, vol. A, Algorithms and Complexity, Elsevier and the MIT Press. *230.*

John Longley and Dag Normann

[2015] ***Higher-order computability***, Springer. *100.*

Nancy A. Lynch and Edward K. Blum

[1979] *A difference in expressive power between flowcharts and recursion schemes*, ***Mathematical Systems Theory***, pp. 205–211. *95, 97.*

Zohar Manna

[1974] ***Mathematical theory of computation***, Originally published by McGraw-Hill, available from Dover. *20, 49.*

Yishay Mansour, Baruch Schieber, and Prasoon Tiwari

[1991a] *A lower bound for integer greatest common divisor computations*, ***Journal of the Association for Computing Machinery***, vol. 38, pp. 453–471. *146.*

[1991b] *Lower bounds for computations with the floor operation*, ***SIAM Journal on Computing***, vol. 20, pp. 315–327. *194, 201, 202.*

John McCarthy

[1960] *Recursive functions of symbolic expressions and their computation by machine, Part I*, ***Communications Of the ACM***, vol. 3, pp. 184–195. *49.*

[1963] *A basis for a mathematical theory of computation*, ***Computer programming and formal systems*** (P. Braffort and D Herschberg, editors), North-Holland, pp. 33–70. *49, 52.*

Gregory L. McColm

[1989] *Some restrictions on simple fixed points of the integers*, ***The Journal of Symbolic Logic***, vol. 54, pp. 1324–1345. *60.*

João Meidânis

[1991] *Lower bounds for arithmetic problems*, ***Information Processing Letters***, vol. 38, pp. 83–87. *201.*

YIANNIS N. MOSCHOVAKIS

[1984] *Abstract recursion as a foundation of the theory of algorithms*, ***Computation and proof theory*** (M. M. Richter et al., editors), vol. 1104, Springer-Verlag, Berlin, Lecture Notes in Mathematics, pp. 289–364. *3, 59.*

[1989a] *The formal language of recursion*, ***The Journal of Symbolic Logic***, vol. 54, pp. 1216–1252 *3, 49, 99, 100.* Posted in ynm's homepage.

[1989b] *A mathematical modeling of pure, recursive algorithms*, ***Logic at Botik '89*** (A. R. Meyer and M. A. Taitslin, editors), vol. 363, Springer-Verlag, Berlin, Lecture Notes in Computer Science, pp. 208–229 *100.* Posted in ynm's homepage.

[1998] *On founding the theory of algorithms*, ***Truth in mathematics*** (H. G. Dales and G. Oliveri, editors), Clarendon Press, Oxford, pp. 71–104 *3, 100, 101.* Posted in ynm's homepage.

[2001] *What is an algorithm?*, ***Mathematics unlimited – 2001 and beyond*** (B. Engquist and W. Schmid, editors), Springer, pp. 929–936 *3.* Posted in ynm's homepage.

[2003] *On primitive recursive algorithms and the greatest common divisor function*, ***Theoretical Computer Science***, vol. 301, pp. 1–30. *125.*

[2006] ***Notes on set theory, second edition***, Undergraduate texts in mathematics, Springer. *19.*

[2014] *On the Church-Turing Thesis and relative recursion*, ***Logic and Science Facing the New Technologies*** (Peter Schroeder-Heister, Gerhard Heinzmann, Wilfrid Hodges, and Pierre Edouard Bour, editors), College Publications, Logic, Methodology and Philosophy of Science, Proceedings of the 14th International Congress (Nancy), pp. 179–200 *99.* Posted in ynm's homepage.

YIANNIS N. MOSCHOVAKIS AND VASILIS PASCHALIS

[2008] *Elementary algorithms and their implementations*, ***New computational paradigms*** (S. B. Cooper, Benedikt Lowe, and Andrea Sorbi, editors), Springer, pp. 81—118 *101.* Posted in ynm's homepage.

A. M. OSTROWSKI

[1954] *On two problems in abstract algebra connected with Horner's rule*, ***Studies presented to R. von Mises***, Academic Press, New York, pp. 40–48. *209.*

V. YA. PAN

[1966] *Methods for computing values of polynomials*, ***Russian Mathematical Surveys***, vol. 21, pp. 105–136. *211.*

CHRISTOS H. PAPADIMITRIOU

[1994] ***Computational complexity***, Addison-Wesley. *4.*

MICHAEL S. PATTERSON AND CARL E. HEWITT

[1970] *Comparative schematology*, MIT AI Lab publication posted at http://hdl.handle.net/1721.1/6291 with the date 1978. *95, 97.*

RÓZSA PÉTER

[1951] ***Rekursive funktionen***, Akadémia Kiadó, Budapest. *58.*

GORDON PLOTKIN

[1977] *LCF considered as a programming language*, ***Theoretical Computer Science***, vol. 5, pp. 223–255. *3.*

[1983] *Domains*, Posted on Plotkin's homepage. *3.*

VAUGHAN PRATT

[1975] *Every prime has a succint certificate*, ***SIAM Journal of computing***, vol. 4, pp. 214–220. *143.*

[2008] *Euclidean gcd is exponentially suboptimal: why gcd is hard to analyse*, unpublished manuscript. *85.*

HARTLEY ROGERS

[1967] ***Theory of recursive functions and effective computability***, McGraw-Hill. *4.*

GERALD E. SACKS

[1990] ***Higher recursion theory***, Perspectives in Mathematical Logic, Springer. *100.*

D. S. SCOTT AND C. STRACHEY

[1971] *Towards a mathematical semantics for computer languages*, ***Proceedings of the symposium on computers and automata*** (New York) (J. Fox, editor), Polytechnic Institute of Brooklyn Press, pp. 19–46. *3.*

J. STEIN

[1967] *Computational problems associated with Racah Algebra*, ***Journal of Computational Physics***, vol. 1, pp. 397Ű–405. *24.*

A. P. STOLBOUSHKIN AND M. A. TAITSLIN

[1983] *Deterministic dynamic logic is strictly weaker than dynamic logic*, ***Information and Control***, vol. 57, pp. 48–55. *97.*

Z. W. SUN

[2000] *On integers not of the form* $\pm p^a \pm q^b$, ***Proceedings of the American Mathematical Society***, vol. 208, pp. 997–1002. *165.*

TERENCE TAO

[2011] *A remark on primality testing and decimal expansions*, ***Journal of the Australian Mathematical Society***, vol. 91, pp. 405–413. *165.*

ALFRED TARSKI

[1986] *What are logical notions?*, ***History and Philosophy of Logic***, vol. 7, pp. 143–154, edited by John Corcoran. *152.*

JERZY TIURYN

[1989] *A simplified proof of DDL < DL*, ***Information and Computation***, vol. 82, pp. 1–12. *97, 98.*

ANUSH TSERUNYAN

[2013] (1) *Finite generators for countable group actions;* (2) *Finite index pairs of equivalence relations;* (3) *Complexity measures for recursive programs*, ***Ph.D. Thesis***, University of California, Los Angeles, Kechris, A. and Neeman, I., supervisors. *118, 121, 123.*

J.V. TUCKER AND J.I. ZUCKER

[2000] *Computable functions and semicomputable sets on many-sorted algebras*, ***Handbook of Logic in Computer Science*** (S. Abramsky, D.M. Gabbay, and T.S.E. Maibaum, editors), vol. 5, Oxford University Press, pp. 317–523. *100.*

ALAN M. TURING

[1936] *On computable numbers with an application to the Entscheidungsproblem*, ***Proceedings of the London Mathematical Society***, vol. 42, pp. 230–265, *A correction*, ibid. volume 43 (1937), pp. 544–546. *99.*

S. A. WALKER AND H. R. STRONG

[1973] *Characterizations of flowchartable recursions*, ***Journal of Computer and System Science***, vol. 7, pp. 404–447. *97.*

SHMUEL WINOGRAD

[1967] *On the number of multiplications required to compute certain functions*, ***Proceedings of the National Academy of Sciences, USA***, vol. 58, pp. 1840–1842. *212.*

[1970] *On the number of multiplications required to compute certain functions*, ***Communications on pure and applied mathematics***, vol. 23, pp. 165–179. *212.*

Symbol index

General index